Advances in Analytics for Learning and Teaching

This book series highlights the latest developments of analytics for learning and teaching as well as providing an arena for the further development of this rapidly developing field.

It provides insight into the emerging paradigms, frameworks, methods, and processes of managing change to better facilitate organizational transformation toward implementation of educational data mining and learning analytics. The series accepts monographs and edited volumes focusing on the above-mentioned scope, and covers a number of subjects. Titles in the series *Advances in Analytics for Learning and Teaching* look at education from K-12 through higher education, as well as vocational, business, and health education. The series also is interested in teaching, learning, and instructional design and organization as well as data analytics and technology adoption.

Yuan 'Elle' Wang • Srećko Joksimović
Maria Ofelia Z. San Pedro • Jason D. Way
John Whitmer
Editors

Social and Emotional Learning and Complex Skills Assessment

An Inclusive Learning Analytics Perspective

Springer

Editors
Yuan 'Elle' Wang
Arizona State University
Scottsdale, AZ, USA

Srećko Joksimović
University of South Australia
Adelaide, SA, Australia

Maria Ofelia Z. San Pedro
ACT, Inc.
Iowa City, IA, USA

Jason D. Way
ACT, Inc.
Iowa City, IA, USA

John Whitmer
Davis, CA, USA

ISSN 2662-2122 ISSN 2662-2130 (electronic)
Advances in Analytics for Learning and Teaching
ISBN 978-3-031-06335-0 ISBN 978-3-031-06333-6 (eBook)
https://doi.org/10.1007/978-3-031-06333-6

This Springer imprint is published by the registered company Springer Nature Switzerland AG
The registered company address is: Gewerbestrasse 11, 6330 Cham, Switzerland

Contents

Contributors

Abayomi Akanji, MA, is an EdTech researcher and software engineer. He has broad experience leading and building offline and online tech solutions for eLearning, collaborative learning, and exam-prep. These platforms have supported over 500K African students to learn effectively and consequently pass their examinations.

Roger Azevedo is a professor in the Departments of Learning Sciences & Educational Research, Computer Science, and Internal Medicine at the University of Central Florida. He is a world-renowned expert in Self-Regulated Learning and has won multiple national and international grants, including a National Science Foundation grant on using real-time multichannel self-regulated learning data to enhance student learning and teachers' decision-making ($1.5M). His expertise will be invaluable to develop and analyze the items on self-regulated learning.

Ryan S. Baker is an associate professor at the University of Pennsylvania, and director of the Penn Center for Learning Analytics. Baker has developed models that can automatically detect student engagement in more than a dozen online learning environments, and has led the development of an observational protocol and app for field observation of student engagement that has been used by more than 150 researchers in 7 countries. He was the founding president of the International Educational Data Mining Society, is currently serving as editor of the journal *Computer-Based Learning in Context*, is associate editor of the *Journal of Educational Data Mining*, was the first technical director of the Pittsburgh Science of Learning Center DataShop, and currently serves as co-director of the MOOC Replication Framework (MORF). Baker has co-authored published papers with more than 400 colleagues.

Abhinava Barthakur is a researcher and a PhD candidate in the field of learning analytics. His research focuses on using large and diverse learning trace and assessment datasets about learners, with the goal of assessing and improving learning and learning experiences. More specifically, his PhD will contribute towards developing novel learning assessment models that is rooted in learning analytics and measurement theories. His research will utilize various data mining and machine learning algorithms to assess the acquisition of complex skills in online learning platforms. Also, Mr. Barthakur is affiliated with the Centre for Change and Complexity in Learning, UniSA and have contributed to various research learning analytics projects.

Brent Benson is Enterprise Architect for Harvard Business School Online, HBS's online learning initiative, where he works on innovative learning technologies, including the Course Platform asynchronous learning platform and the HBS Online data science practice, which uses data-driven methods to understand and improve learning experiences and to expand the scope and reach of Harvard Business School's case-learning approach.

Alex J. Bowers is Professor of Education Leadership at Teachers College, Columbia University. His research interests include organizational level data analytics, education data science, machine learning, organizational behavior, school and district leadership, data-driven decision making, high school dropouts and completion, educational assessment and accountability, education technology and school facilities financing. ORCID: 0000-0002-5140-6428

Ida Camacho is a principal developer and researcher on DiLab's Jill Watson project. As a member of the Jill Watson team, she works on developing a Virtual Teaching Assistant that can answer questions about a course. More recently, she has led the development of an Introduction Agent and researches NLP techniques to answer student introductions.

Ida holds an MS in computer science from Georgia Tech. Her research interests include neuroscience, cognitive science, natural language processing, conversational agents, and knowledge engineering. Outside of DiLab, Ida loves the NBA, gardening, traveling, and playing with her two dogs.

Christy Weng-Lam Cheong is head of the Academic Affairs Department at the Macao Polytechnic Institute. Her research areas include learning analytics, social and emotional learning, and higher education policies and administration.

Rhonda Christensen, PhD, is research professor in the Learning Technologies Department in the College of Information at the University of North Texas. She is a co-director of the Institute for the Integration of Technology into Teaching and Learning at UNT. She is lead PI on an NSF Research on Emerging Technologies for Teaching and Learning (RETTL) exploring issues related to equity in teacher education. She serves as the Chair of the Consultative Council for the Society for Information Technology in Teacher Education (SITE). Her research interests are the impact of technology integration in education, enhancing STEM education in middle schools, dispositions toward learning, simulated learning in preservice education, and mobile learning in education.

Elizabeth B. Cloude is a PhD candidate within the Department of Learning Sciences & Educational Research at the University of Central Florida. Her research is focused on collecting, processing, and analyzing multimodal sensor data (e.g., eye movements, human-system interactions, physiology, screen recordings, speech, facial expressions of emotions) using a dynamical systems perspective to study the complex nature of learning across learning activities that hold implications for AI in education and building intelligent adaptive systems. Her professional productivity includes 2 book chapters, 8 peer-reviewed journal articles (with an additional 8 under review or in preparation), 11 refereed conference proceedings, and 24 conference paper presentations, and 22 poster presentations at local, national, and international conferences, including the International conference on Affective Computing and Intelligent Interaction, the International Learning Analytics and Knowledge conference, the American Educational Research Association, the European Association for Research on Learning and Instruction, the International Society of Artificial Intelligence in Education, the Cognitive Science Society, as well as the Association for Educational Communications and Technology.

David H. Cropley is the Professor of Engineering Innovation at the University of South Australia. He specializes in helping people and organizations become better, more effective, problem solvers.Dr. Cropley joined the School of Engineering at the South Australian Institute of Technology (SAIT) in 1990, after serving for 4 years in the United Kingdom's Royal Navy, including deployments to the Caribbean and Middle East. Following the establishment of the University in 1991, he completed a PhD in measurement systems engineering in 1997 and obtained a graduate certificate in higher education in 2002.Dr. Cropley has taught a variety of courses on engineering, including digital electronics, linear electronics, systems engineering, and engineering design. He teaches creativity and innovation across programs, from gifted education through to management. Dr. Cropley's research interests range across creativity in education, the assessment of creativity and innovation in organizations, the role of creativity and innovation in terrorism and crime, and the nexus of creative problem-solving and engineering.Dr. Cropley is author/co-author of nine books including *Creativity and Crime: A Psychological Analysis* (Cambridge University Press, 2013); *Creativity in Engineering: Novel Solutions to Complex Problems* (Academic Press, 2015); *The Psychology of Innovation in Organizations* (Cambridge University Press, 2015), and *Femina Problematis Solvendis – Problem-Solving Woman: A History of the Creativity of Women* (Springer, 2020).Now a recognized expert in creative problem solving and innovation, Dr. David Cropley was a scientific consultant and on-screen expert for the Australian ABC TV Documentaries Redesign My Brain (2013), Life at 9 (2014) and Redesign My Brain, Series 2 (2015). For more than 20 years, he has helped schools and organizations in Australia, the US, and Europe develop more effective, creative problem solvers.

Mutlu Cukurova, PhD, is Associate Professor of Learning Technologies at the University College London, UK. His research focuses on the design, development and implementation of Artificial Intelligence and Analytics solutions to support human learning. Dr. Cukurova is engaged with UNESCO's unit for Technology and AI in Education, UCL's Grand Challenges on Transformative Technologies, is a chief research advisor at EDUCATE Ventures, an editor of the *British Journal of Educational Technology* and an AE of the *International Journal of Child-Computer Interaction.*

Samantha G. Daley is Assistant Professor of Education and Human Development at the University of Rochester. Her interdisciplinary work informs the design of inclusive learning environments in which students thrive both academically and emotionally. In both formal and informal learning contexts, she primarily focuses on students identified with disabilities and those with a history of academic difficulty or marginalization.

Shane Dawson is the executive dean of Education Futures, co-director of the Centre for Change and Complexity in Learning (C3L), and Professor of Learning Analytics at the University of South Australia. Shane has published widely on topics from creative capacity to social network analysis and the application of learner ICT interaction data to inform and benchmark teaching and learning quality. His current research interests relate to complex systems and education leadership to aid adoption and application of learning analytics at scale. With the support of many talented colleagues, Shane has been involved in the development of numerous open source software including the Online Video Annotations for Learning (OVAL), OnTask (a personalized learner feedback tool), and SNAPP, a social network visualization tool designed for teaching staff to better understand, identify, and evaluate student learning, engagement, academic performance, and creative capacity.

Yannis Dimitriadis is Full Professor of Telematics Engineering at Universidad de Valladolid (Spain). He received his PhD in 1992, and has made multiple research stays at UC Berkeley, EPFL, Boston University, etc., over the last 30 years. His research interests focus on the technological and conceptual support to teachers regarding learning design and orchestration, as well as computer-supported collaborative learning (CSCL), and the human perspectives in the design and use of information and communication technologies (ICT) in teaching and learning.

Florence Gabriel is a Lecturer in STEM education and a Fellow at the Centre for Change and Complexity in Learning at the University of South Australia (UniSA). She completed her PhD in educational neuroscience at the Université Libre de Bruxelles and the University of Cambridge. Before joining UniSA, she was a policy analyst with the OECD Directorate for Education and Skills. She studies the cognitive, metacognitive, motivational, and affective factors that influence students' learning and academic achievement, with a focus on mathematics education.

David Gibson is UNESCO Chair of Data Science in Higher Education Learning and Teaching and the director of Learning Futures at Curtin University. Professor Gibson has extensive experience in large-scale projects. He created and led a multi-million-dollar effort funded by NSF and he led a major university funded initiative (more than $3M) called Challenge, a scalable cloud-based application for gamified individual and team learning that captures high resolution data of user behaviors, communications and production artefacts and supports novel learning analytics. Professor Gibson will contribute to the item development and data analysis.

Ashok K. Goel is a professor in the School of Interactive Computing at Georgia Institute of Technology and the Chief Scientist with Georgia Tech's Center for 21st Century Universities. For more than 35 years, he has conducted research into cognitive systems at the intersection of AI and cognitive science with a focus on computational design and creativity. Over the last decade much of his research has focused on AI in education and education in AI. Ashok was a Co-Chair of the 41st Annual Conference of the Cognitive Science Society held in 2019. He is the founding editor of AAAI's Interactive AI Magazine established in 2020 and Editor Emeritus of AI Magazine. He is a Fellow of AAAI and a recipient of AAAI's Outstanding AI Educator Award. Ashok is the executive director of the new NSF National AI Research Institute on Adult Learning and Online Education.

Boris Goldowsky is the director of technology at CAST, where he works on a variety of educational research and development projects focused on Universal Design for Learning. He has worked to make educational software that supports students by allowing them to use multiple modalities to learn and to show their understanding.

Xingmin Guan is a graduate student in the Master of Science in Information Technology program in the Faculty of Education, University of Hong Kong. Her research interests are learning design as well as teaching and learning with information technology.

Neil T. Heffernan is a professor in the Computer Science Department at Worcester Polytechnic Institute, and directs the PhD and master's programs in learning sciences and technologies graduate program. Prof. Heffernan is well known for ASSISTments, an online tutoring program that he created. He enjoys doing educational data mining and running the ASSISTments system. ASSISTments has been used as a shared scientific instrument to conduct randomized controlled trials comparing different ideas on how to improve student learning. Prof. Heffernan enjoys supervising WPI students in creating ASSISTments content and features. He has six dozen papers in educational data mining, and 20+ papers in comparing different ways to optimize student learning.

Garron Hillaire is a post-doctoral researcher at the Teaching Systems Lab at the Massachusetts Institute of Technology where he investigates how to broaden participation in K-12 computer science education through innovative teacher education using simulations. His work focuses on innovative emotional measures and how they can be integrated into AI architecture to provide dynamic supports within simulation based learning.

Mohamed Houtti is a Computer Science PhD student in GroupLens, at the University of Minnesota. He does research in human-computer interaction, with a particular focus on 1) incorporating community-level values into the design of recommendation tools for online platforms like Wikipedia and 2) mitigating participation bias in professional video-conferencing meetings.

Xiao Hu is an associate professor in the Academic Unit of Human Communication, Development and Information Sciences, Faculty of Education, University of Hong Kong. Dr. Hu's research aims to improve people's learning and well-being by designing and developing intelligent systems. Her current projects focus on learning analytics, social and emotional learning, and cultural informatics.

Srećko Joksimović is a Senior Lecturer in Data Science at the Education Futures, University of South Australia. His research is centered around augmenting abilities of individuals to solve complex problems in collaborative settings. Srećko is particularly interested in evaluating the influence of contextual, social, cognitive, and affective factors on groups and individuals as they solve complex real-world problems. In so doing, he utilizes a wide range of methods from machine learning, artificial intelligence, and natural language processing, as well as data science and social computing in general.

Carmel Kent, PhD, Dr. Kent has over 20 years of industry and academic experience, having worked for IBM research, EdTech and healthcare provider companies and start-ups, as a software engineer, data scientist, entrepreneur, teacher, and a researcher. Carmel's research focus is on Artificial Intelligence in Education, online learning communities and learning analytics.

Gerald Knezek is Regents Professor of Learning Technologies at the University of North Texas, in Denton, Texas, USA. He is co-director of the Institute for the Integration of Technology into Teaching and Learning at UNT.

Marianne E. Krasny is a professor in the Department of Natural Resources and director of the Civic Ecology Lab at Cornell University. Her books include *Civic Ecology: Adaptation and Transformation from the Ground Up* (with K Tidball), *Urban Environmental Education Review* (with A Russ), *Communicating Climate Change: A Guide for Educators* (with A Armstrong and J Schuldt), *Grassroots to Global: Broader Impacts of Civic Ecology* and *Advancing Environmental Education Practice*. She leads efforts to teach environmental and climate education Massive Open Online Courses (MOOCs) for international audiences, and to facilitate online learning and resource sharing communities for more than 10,000 environmental activists and educators globally. Dr. Krasny was director of EPA's

National Environmental Education Training Program, and of the Garden Mosaics community gardening education program, and is a Public Voices Fellow and an International Fellow of the Royal Swedish Academy of Agriculture and Forestry.

Yue Li is a research associate in the Department of Natural Resources and the Environment at Cornell University. Her research focuses on professional development for formal and non-formal educators, Massive Open Online Course participants' learning and actions, and university students' community-engaged learning to address environmental and sustainability challenges. She bridges quantitative and qualitative methods to conduct interdisciplinary research in both online and face-to-face settings, and in different cultural contexts. Yue has taught or co-taught environmental and sustainability courses for high school students in China, undergraduate and graduate students at Cornell, and for a global audience. Her online courses have brought high quality and equitable learning opportunities to teachers and other professionals from all over the world. http://yueyuelee.me/

Fernando Marmolejo-Ramos is a research fellow in human and artificial cognition at the Centre for Change and Complexity in Learning (C3L) at the University of South Australia. Fernando has a Master of Applied Science (MAppSc, by research) in experimental psychology from the University of Ballarat (2005–2007) and a PhD (by research) in cognitive psychology from the University of Adelaide (2007–2011). From November 2014 to December 2016, he was a postdoctoral research fellow in the Department of Psychology at Stockholm University. His research interests include embodied cognition (e.g., embodiment of language and emotions) and applied statistics/methodology. More recently Fernando's interests include experimental education, statistical cognition, and intersections between concepts in human cognition and artificial intelligence.

Rebecca L. Marrone is a Research Fellow for the Centre for Change and Complexity in Learning (C3L) at the University of South Australia. Her research focuses on initiating and directing projects in creativity, educational psychology, and human and artificial cognition across varying educational contexts.Dr. Marrone completed her PhD in the School of STEM at the University of South Australia. Her PhD research investigated the role of creativity in encouraging female students to study Mathematics.Dr. Marrone has taught various courses on education and psychology and has given keynote addresses about creativity and education at institutions around the globe. In 2019, Dr. Marrone recorded a TedX talk titled 'We can all be STEM superheroes'.

Dana Murano is a Research Scientist in ACT's Center for Social, Emotional, and Academic Learning. She completed her PhD in educational psychology with a specialization in learning, development, and instruction. Her research focuses primarily on the development and assessment of social and emotional skills in students.

Paula Odriozola-González received her PhD in psychology from the University of Oviedo (Spain) in 2011. She is an assistant professor in the Department of Education at the University of Cantabria. Her main research interest focuses on the processes underlying mental health problems and effective therapeutic strategies to tackle them. Currently, she is investigating such mental disorders (and potential solutions) within academic research careers.

Oleksandra Poquet is a researcher at the Centre for Change and Complexity in Learning, Education Futures, University of South Australia. Dr. Poquet focuses on learning analytics, digital learning, and complexity science lens on learning and change. She uses large datasets of digital social data from massive open online courses (MOOCs), university online courses, reading annotation software in university and K12 settings, and phone call data between adult learners. Prior to her career in research, she was an adult educator in non-formal and formal education, with a graduate degree in educational sociology. Dr. Poquet has been actively involved with learning analytics, community, serving as vice president for the Society for Learning Analytics Research (2019–2021).

Luis P. Prieto is a senior research fellow in the School of Educational Sciences at Tallinn University (Estonia). He received his PhD from the University of Valladolid (Spain) in 2012, and was a Marie Curie fellow at the École Polytechnique Fédérale de Lausanne (Switzerland). His research interests include learning analytics, especially multimodal learning and teaching analytics, the study of teacher orchestration and computer-supported collaborative learning. He is also deeply interested in supporting doctoral students and supervisors in facing the productivity and mental health challenges of doing a PhD (a topic he blogs about at ahappyphd.org), and the role that technologies can play in such support.

Mladen Raković is a Research Fellow in the Centre for Learning Analytics at Monash University. Mladen's research interests span the fields of natural language processing, predictive modelling, computational linguistics, and educational psychology. He focuses on the development and evaluation of computer-based writing systems that monitor undergraduates' writing activity and generate learning analytics to promote self-regulated learning and deep engagement with disciplinary content.

Bart Rienties is Professor of Learning Analytics and program lead of the learning analytics and learning design research program at the Institute of Educational Technology at the Open University UK. As educational psychologist, he conducts multi-disciplinary research on work-based and collaborative learning environments and focuses on the role of social interaction in learning, which is published in leading academic journals and books.

María Jesús Rodríguez-Triana received her PhD in information and communication technologies from the University of Valladolid (Spain) in 2014, joining the École Polytechnique Fédérale de Lausanne (Switzerland) as a postdoctoral fellow.

She is an associate professor in the School of Digital Technologies at Tallinn University (Estonia). Currently, she is investigating the challenges and strategies towards the adoption of learning analytics in authentic scenarios such as classrooms, institutions, and communities of practice.

Maria Ofelia Z. San Pedro is a senior research scientist at ACT, Inc. in the Applied Research group, specializing in learning sciences and learning analytics. She researches on learning experiences, assessment, and instruction within online learning environments. She has more than 10 years of experience in quantitative and qualitative research and educational data science that use learning technologies and big data (i.e., in education) to drive insights and develop solutions.

George Siemens is director of Centre for Change and Complexity in Learning at the University of South Australia, and Professor at the University of Texas Arlington. His research focuses on how knowledge is created and how that process impacts individuals and organizations. His work is centered on learning analytics, networked learning (connectivism), and human and artificial cognition.

Kate E. Walton obtained her PhD in personality psychology from the University of Illinois, Urbana-Champaign. After a postdoctoral fellowship at the University of Minnesota, she joined the faculty of the Department of Psychology at St. John's University where she spent 11 years. In 2017, Kate joined the Center for Social, Emotional, and Academic Learning at ACT as a Principal Researcher. She recently took the position of director of research at Mindprint Learning where she continues her research on social and emotional skill assessment and development.

Qiaosi Wang is a PhD student in the School of Interactive Computing at Georgia Institute of Technology. Qiaosi is interested in understanding the design and impact of human-AI interaction and communication in community context. Recently, she has been conducting research into the human-centered design and impact of community-facing AI agents that can mediate online learners' social interaction process. Her work has been published and awarded at prestigious venues such as ACM CHI, ACM CSCW, ACM DIS, and ACM Learning@Scale. Qiaosi holds Bachelor of Science degrees in informatics and psychology from the University of Washington, Seattle.

Yuan 'Elle' Wang is a Research Scientist at ASU EdPlus Action Lab. Her current projects center on assessing social and emotional leaning skillsets and predicting learner longitudinal career development in large-scale online learning environments. Specifically, her projects take a comprehensive approach by linking three sources of learner data: pre-course learner motivation, within-course learner engagement, as well as post-course development. Measurement of post-course development reflects both individual learner career development as well as advancement of communities of practice. Her recent publications can be seen in the *Journal of Online Learning and Teaching* and the *Journal of Learning Analytics*.

Jason D. Way is a Senior Research Psychologist in ACT's Center for Social, Emotional, and Academic Learning. He earned his PhD in industrial/organizational psychology from the University of South Florida. His research interests include the assessment and development of social and emotional skills that lead to academic and workplace success.

John Whitmer EdD, is an educational data science leader with over a decade of experience conducting research and development using advanced algorithmic approaches. He is currently a Senior Fellow at the Institute for Education Statistics (IES) through the Federation for American Scientists. Prior to this role, John led teams of data scientists, research scientists, and machine learning engineers in large educational assessment companies (ACTNext), edTech providers (Blackboard), and educational institutions (California State University & California Community Colleges).

Philip H. Winne is Fraser University Distinguished Professor of Education, a Fellow of the Royal Society of Canada and previously held a Tier I Canada Research Chair for its maximum term. He designed the nStudy software suite to research and support self-regulating learners as fledgling learning scientists who pursue a personal research program to improve their learning skills.

Chapter 1
Re-contextualizing Inclusiveness & SEL in Learning Analytics

Yuan 'Elle' Wang and Srećko Joksimović

Abstract In this chapter, we review the brief history of SEL-focused research within the learning analytics field, followed by an introduction of the current state of inclusiveness and SEL. We then discuss the inclusiveness-driven development opportunities via (1) developing methods to enable SEL assessment at scale; (2) integrating more domain-based knowledge into diversifying the design of success metrics; and (3) incorporating diverse contextual learner features such as those associated with the COVID-19 pandemic. Finally, the four parts including 15 chapters in this book are introduced.

Keywords Learning analytics · Social and emotional learning · Inclusive learning · DEI

1.1 Introduction

1.1.1 SEL

Various past research studies reveal that success within school, workplace, and other aspects of life in general depend on skills beyond academic achievements (D'Mello, 2017; Duckworth & Yeager, 2015; Heckman et al., 2006). Non-cognitive skills (Heckman & Rubinstein, 2001), as a term, has been widely used to refer to social and emotional learning capabilities beyond academic knowledge. Non-cognitive skills consist of a diverse range of aspects including conscientiousness

Y. E. Wang (✉)
EdPlus Action Lab, Arizona State University, Scottsdale, AZ, USA

S. Joksimović
Education Futures, University of South Australia, Adelaide, SA, Australia
e-mail: srecko.joksimovic@unisa.edu.au

Y. E. Wang et al. (eds.), *Social and Emotional Learning and Complex Skills Assessment*, Advances in Analytics for Learning and Teaching,
https://doi.org/10.1007/978-3-031-06333-6_1

(Camara et al., 2015), growth mindset (Dweck, 2006), time management, and self-regulation (Gutman & Schoon, 2013), to name a few that have been well studied in recent years. Social and emotional learning (SEL) are often used interchangeably with the phrase "non-cognitive" attributes or 21st century skills. Scholars have noted that non-cognitive and cognitive aspects are closely related and far from being mutually exclusive; therefore, in this book, we will use the term "SEL" to refer to the attributes that have been traditionally called "non-cognitive" or "soft" skills in workplaces.

1.1.2 Recent Development of SEL & LA

As a burgeoning field, researchers and practitioners in learning analytics (LA) also started to analyze SEL-related questions by utilizing the fast-growing amount of learning data. SEL attributes have been measured either by using survey measures or derived from online behavior data such as those generated from learning management systems.

Survey Measures Researchers (e.g., Kizilcec & Schneider, 2015) investigated the relationship between SEL factors, such as learner motivation and self-efficacy, and learning outcomes using self-reported survey items. They found that different motivational types lead to varied types of engagement patterns. While these studies provide critical insights toward theory development, applying self-reported surveys alone limits what we can learn about online learners' non-cognitive capabilities. For example, self-reported surveys often have limited response rates, lower than 10 percent in MOOCs (e.g., Wang & Baker, 2015). Moreover, it is also arguable to what extent self-reported data provide unbiased data. Consequently, not reliably assessing online learners' SEL skills *at scale* undermines the extent to which learning analytics can be applied to assess and improve student learning.

Learning Analytics (LA) and Educational Data Mining (EDM) In addition to traditional survey measures, recent advances in the LA and EDM research fields provide innovative venues for investigation of students' non-cognitive skills (e.g., emotion, metacognition, motivation). Multimodal data, such as eye gazes, facial expressions of emotions, heart rate and electro-dermal activities (D'Mello et al., 2017) were investigated. In addition, toward scaling up, EDM researchers have developed "automated detectors" out of computer log data to infer student's emotions and engagement in real-time. These detectors, developed from a combination of expert field observation (e.g. Ocumpaugh et al., 2015) and data mining on log files, can accurately predict affect and engagement (Baker et al., 2012).

LA and EDM methods offer promising paths toward assessing online learners' non-cognitive skills, and ultimately toward deriving actionable applications to improve learner non-cognitive skills and subsequently improve cognitive outcomes as well. The direction toward developing tools and metrics to assess online learners' non-cognitive skills reliably and at scale remains an ongoing focus. This topic has been further discussed in Chap. 3 of this book.

1.2 SEL and Inclusiveness

Inclusive education focuses on addressing a wide spectrum of individual educational needs and promoting diversity to enrich learning experience for all learners and particularly focusing on those at risk of marginalization and social exclusion. Contemporary literature underscores the importance of inclusive education for supporting social and academic outcomes (e.g., Heckman & Rubinstein, 2001). In that sense, social and emotional learning has been utilized as a framework for enabling learners to build relationships, respect others, and develop healthy, collaborative, and supportive learning environments (Weissberg et al., 2015).

Although the line of inquiry with regard to the relationship between SEL skills *and* academic or professional success have been extensively investigated, the connection between various lines of SEL research and their influence on promoting diversity and inclusive excellence remains largely at the secondary level. This is particularly challenging in the context of online and higher education more generally. In other words, there is a lack of research investigating (1) how inclusion and diversity should be promoted in the context of online and higher education, (2) how inclusive excellence can be promoted via the lens of SEL research, and (3) what is the role of learning analytics in enhancing SEL skills and promoting inclusion and diversity?

Many have noted that the collateral benefits of these lines of inquiries are to expand our understanding on the diverse needs and challenges learners face, where raising awareness of this issue has been the primary goal. However, in order to develop and derive actionable insights such as informing intervention inclusive learning design and policy change, more empirical research is needed to directly target how SEL research can benefit toward promoting inclusive excellence and diversity as a primary target.

Before one can argue how SEL and LA research can benefit inclusive education, it is imperative to understand the complexity and diversity of learners and the various appropriate success metrics or learning outcomes for these learners. As such, in the following section, we discuss some of the common types of success metrics and learners in SEL-focused LA research studies.

1.2.1 Diversity of Success Metrics

SEL skills have been studied in relation to a variety of success metrics including academic outcomes, engagement and persistence, and professional development. One of the most common success metrics that have been used is academic success

(Farkas, 2003; Lleras, 2008; Duckworth & Seligman, 2005). It has been shown that the predictive power of non-cognitive skills on academic achievement across a wide range of educational settings is at least equal to or better than the predictive power of cognitive skills.

In addition to academic success, popular success metrics also encompass persistence in postsecondary settings with attendance and retention (e.g., Credé et al., 2010) and engagement. Moreover, SEL skills have also been linked to career advancement (Bowles & Gintls, 2002; Heckman et al., 2006), well-being (Cohen, 2006) and key 21st century competencies, such as critical thinking and problem solving (Buckingham Shum & Crick, 2016). In these settings, SEL attributes may serve both as predictors and proxies of success.

1.2.2 Diversity of Learners

Similar to the development of success metrics, the concept of learner diversity has been further enriched and acknowledged in recent years. A simple binary partition of K-12 and adult learners is far from sufficient. An emerging group of learners in postsecondary and online higher education consists of adult learners predominantly defined as those with adult responsibilities, such as working full-time, being financially dependent, have dependents or being a single parent, or a nontraditional trajectory in their educational experience (e.g., delayed enrollment into higher ed, or did not complete high school) (Horn, 1996). They comprise an important segment of student diversity that is usually overlooked – the nontraditional adult learners (NAL). In the US, NALs represent around 40% of the postsecondary population and are projected to grow much faster compared to traditional late adolescent students who enter postsecondary education (National Center for Education Statistics, 2009). Over 40% of higher education (first tier) institutions indicated that there was no consideration for older adult students in their "outreach, programs and services or financial aid" (Lakin et al., 2008). But adult learners seek higher education for a variety of reasons – related to career change, career mobility, retraining, or retirement (DiSilverstro, 2013). And in a predominantly work-based society and growth of NALs, the demand for postsecondary education will increasingly attract NALs that are qualitatively, developmentally, and socially very different (Chen, 2017) from the traditional-age, late adolescent student. Hence, such educational settings would require embracing this aspect of diversity in the type of learners they attract, admit, and provide support.

1.3 Opportunities for Inclusiveness in SEL-Focused LA Research

1.3.1 SEL Assessments at Scale *Online*

The availability of clickstream data from large-scale online learning platforms, such as Massive Open Online Courses (MOOCs), are expected to enable assessments at scale to advance the science of learning and learner success (Stokes, 2013). Although assessment of learner success should include both cognitive and non-cognitive aspects, more attention has been given toward measuring learners' cognitive abilities, oftentimes operationalized by test scores or the number of clicks in the environment (Reich, 2015), than on the non-cognitive abilities.

Much of the research studies on non-cognitive assessments are focused on theory development rather than actionable applications (Duckworth & Yeager, 2015) derived from practical measurements incorporating short term feedback on progress toward improvement. The lack of feedback in assessment limits applicability. A recent review of MOOC literature (Joksimovic et al., 2018) shows that existing research focuses primarily on understanding factors that explain academic learning outcomes, thus in most cases failing to include the assessment of non-cognitive skills.

1.3.2 Integrating Domain-Based Education Research

Domain-based education research can be better integrated into the SEL-focused LA research. Baker and Boser (2021) argued that domain-based knowledge including both the learning sequence and structure are critical toward building effective learning analytics models. Much existing SEL-focused LA research was carried out in the mathematics learning setting at the K-12 level. Insights obtained from these studies can be helpful as a guide for other domain areas and other age groups. However, relying heavily on knowledge generated from one or a few domains may discourage researchers and practitioners from less-studied domains and fields such as the various science subjects and non-STEM subjects. The lack of domain diversity may also hinder the long-term development and limit the scope of generalizability of SEL-focused LA research. Therefore, it is recommended that domain-specific knowledge be integrated early in the research ideation process. More granular domain-based structure and learning sequences should be considered. For example, research into students' SEL perspectives of a college biology course may benefit by considering special features unique to learning biology for college students in addition to using any general principles that can apply to all STEM college-level courses.

1.3.3 Incorporate Diverse Learner Features

Diverse features of learners beyond basic demographic features need to be considered. One the one hand, an exciting trend is seen that SEL-focused LA research includes different age groups. On the other hand, there is a tendency of oversimplification by only looking at basic demographic features such as learners' age and gender. For example, contextual factors such as the learner's home-based learning environment and adult learners' career and family responsibilities can provide critical insights into explaining the learning challenges that are not solely due to reasons associated with design features of learning tools. By incorporating variables that reflect learners' contextual factors may also help us better design learning environments for learners.

To conclude, LA research has enabled SEL-focused studies to be conducted at scale across a wide spectrum of age groups and domain areas. Along with the fast development, this sub-field has encountered numerous inclusiveness-driven development opportunities via (1) developing methods to enable SEL assessment at scale; (2) integrating more domain-based knowledge into diversifying the design of success metrics; and (3) incorporating diverse contextual learner features including those associated with unexpected large-scale influences from the COVID-19 pandemic.

In this book, Part I includes three chapters that provide an overview of SEL, LA, as well as the SEL-focused research development, followed by Part II which includes six chapters that zero in on six distinctive SEL attributes including empathy, creativity, motivation, self-regulated learning, and collaboration. Part III consists of three studies that looked at students in the K-12 settings with subjects including reading, mathematics, and biology. Finally, Part IV comprises four chapters that focus on the adult and professional learning settings including those designed for undergraduate students, graduate students, professional degree-seekers that pursue online MBA degrees, as well as a course designed for the general public that encourages actions toward protecting our environment.

References

Baker, R. S., & Boser, U. (2021). *High-leverage opportunities for learning engineering*. University of Pennsylvania Center for Learning Analytics.

Baker, R. S. J. D., Gowda, S. M., Wixon, M., Kalka, J., Wagner, A. Z., Salvi, A., Aleven, V., Kusbit, G., Ocumpaugh, J., & Rossi, L. (2012). Towards sensor-free affect detection in cognitive tutor algebra. In *Proceedings of the 5th international conference on educational data mining*, pp. 126–133.

Bowles, S., & Gintis, H. (2002). Schooling in capitalist America revisited. *Sociology of Education, 75*(1), 1–18.

Buckingham Shum, S., & Crick, R. D. (2016). Learning analytics for 21st century competencies. *Journal of Learning Analytics, 3*(2), 6–21.

Camara, W., O'Connor, R, Mattern, K, Hanson, M. A. (2015). *Beyond academics: A holistic framework for enhancing education and workplace success* (ACT research report 2015-4). Retrieved from http://www.act.org/content/dam/act/unsecured/documents/ACT_RR2015-4.pdf

Chen, J. C. (2017). Nontraditional adult learners: the neglected diversity in postsecondary education. *SAGE Open, 7*(1), 2158244017697161.

Cohen, J. (2006). Social, emotional, ethical, and academic education: Creating a climate for learning, participation in democracy, and well-being. *Harvard Educational Review, 76*(2), 201–237.

Credé, M., Roch, S. G., & Kieszczynka, U. (2010). Class attendance in college: A meta-analytic review of the relationship of class attendance with grades and student characteristics. *Review of Educational Research, 80*, 272–295.

D'Mello, S. (2017). Emotional learning analytics. In *Handbook of learning analytics* (p. 115).

D'Mello, S., Dieterle, E., & Duckworth, A. (2017). Advanced, analytic, automated (AAA) measurement of engagement during learning. *Educational Psychologist, 52*(2), 104–123.

DiSilvestro, F. R. (2013). Continuing higher education and older adults: A growing challenge and golden opportunity. *New Directions for Adult and Continuing Education, 140*, 79–87.

Duckworth, A. L., & Seligman, M. E. (2005). Self-discipline outdoes IQ in predicting academic performance of adolescents. *Psychological science, 16*(12), 939–944.

Duckworth, A. L., & Yeager, D. S. (2015). Measurement matters: Assessing personal qualities other than cognitive ability for educational purposes. *Educational Researcher, 44*(4), 237–251.

Dweck, C. S. (2006). *Mindset: The new psychology of success*. Random House. Chicago.

Farkas, G. (2003). Cognitive skills and noncognitive traits and behaviors in stratification processes. *Annual Review of Sociology, 29*, 541–562. https://doi.org/10.1146/annurev.soc.29.010202.100023

Gutman, L. M., & Schoon, I. (2013). *The impact of non-cognitive skills on outcomes for young people*. Education Empowerment Foundations. Retrieved from http://hdl.voced.edu.au/10707/287500

Heckman, J. J., & Rubinstein, Y. (2001). The importance of noncognitive skills: Lessons from the GED testing program. *The American Economic Review, 91*(2), 145–149.

Heckman, J. J., Stixrud, J., & Urzua, S. (2006). The effects of cognitive and noncognitive abilities on labor market outcomes and social behavior. *Journal of Labor Economics, 24*(3), 411–482.

Horn, L. (1996). *Nontraditional undergraduates, trends in enrollment from 1986 to 1992 and persistence and attainment among 1989–90 beginning postsecondary students* (NCES 97–578). U.S. Government Printing Office, U.S. Department of Education, National Center for Education Statistics.

Joksimović, S., Poquet, O., Kovanović, V., Dowell, N., Mills, C., Gašević, D., … Brooks, C. (2018). How do we model learning at scale? A systematic review of research on MOOCs. *Review of Educational Research, 88*(1), 43–86.

Kizilcec, R. F., & Schneider, E. (2015). Motivation as a lens to understand online learners: Toward data-driven design with the OLEI scale. *ACM Transactions on Computer-Human Interaction (TOCHI), 22*(2), 6.

Lakin, M. B., Mullane, L., & Robinson, S. (2008). *Mapping new directions: Higher education for older adults. Reinvesting in the third age: Older adults and higher education – Second report*. American Council on Education.

Lleras, C. (2008). Do skills and behaviors in high school matter? The contribution of noncognitive factors in explaining differences in educational attainment and earnings. *Social Science Research, 37*(3), 888–902.

National Center for Education Statistics. (2009). *Projections of educational statistics to 2018. Table 11. Actual and middle alternative projects numbers for total enrollment in all degree granting postsecondary institutions, by sex, age, and attendance status: Fall 1993 to Fall 2018*. Retrieved from https://nces.ed.gov/pubs2009/2009062.pdf

Ocumpaugh, J., Baker, R. S., Rodrigo, M. M., Salvi, A., Van Velsen, M., Aghababyan, A., & Martin, T. (2015). HART: The human affect recording tool. In *Proceedings of the 33rd annual international conference on the design of communication*, pp. 1–6.

Reich, J. (2015). Rebooting MOOC research. *Science, 347*(6217), 34–35.
Stokes, P. (2013). The particle accelerator of learning. *Inside Higher Ed, 22*.
Wang, Y., & Baker, R. (2015). Content or platform: Why do students complete MOOCs? *Journal of Online Learning and Teaching, 11*(1), 17.
Weissberg, R. P., Durlak, J. A., Domitrovich, C. E., & Gullotta, T. P. (2015). Social and emotional learning: Past, present, and future. In J. A. Durlak, C. E. Domitrovich, R. P. Weissberg, & T. P. Gullotta (Eds.), *Handbook of social and emotional learning: Research and practice* (pp. 3–19). Guilford.

Chapter 2
State of the Science on Social and Emotional Learning: Frameworks, Assessment, and Developing Skills

Jason D. Way, Dana Murano, and Kate E. Walton

Abstract This chapter provides an overview of the current research and thinking on social and emotional learning. First, it presents several frameworks for organizing the social and emotional learning space. Second, it summarizes research supporting the idea that social and emotional skills can change over time. Finally, it discusses traditional approaches to social and emotional learning assessment.

Keywords Social and emotional learning · Frameworks · Assessment · Development

In recent years, education has shifted from focusing solely on cognitive skills to an approach that considers the whole child (e.g., Aspen Institute, 2018; Lipnevich et al., 2016; Weissberg et al., 2015). This shift is supported by an increasingly large evidence base for the importance of social and emotional (SE) skills in fostering students' cognitive and non-cognitive development. SE skills can be defined as "individual characteristics that originate from biological predispositions and environmental factors, manifested as consistent patterns of thoughts, feelings, and behaviors, developed through formal and informal learning experiences, and that influence different outcomes throughout the individual's life" (John & DeFruyt,

J. D. Way (✉)
Center for Social, Emotional, and Academic Learning, ACT, Inc., Iowa City, IA, USA
e-mail: jason.way@act.org

D. Murano
ACT, Inc., Iowa City, IA, USA
e-mail: dana.murano@act.org

K. E. Walton
MindPrint Learning, Princeton, NJ, USA
e-mail: kate@mindprintlearning.com

Y. E. Wang et al. (eds.), *Social and Emotional Learning and Complex Skills Assessment*, Advances in Analytics for Learning and Teaching,
https://doi.org/10.1007/978-3-031-06333-6_2

2015, p. 4). Meta-analytic evidence shows that SE skills contribute to students' positive behavior, attitudes toward school, and academic performance in K-12 contexts (Durlak et al., 2011). SE skills remain important constituents of student success throughout the lifespan, with SE skills being strong predictors of student retention and academic performance in higher education (Robbins et al., 2004) and predictors of job performance (Barrick et al., 2001). In this chapter, we provide an overview of how SE skills are conceptualized and organized by various frameworks, discuss approaches to assessing SE skills, and summarize research describing how SE skills change over time. The goal of this chapter is to provide a solid foundation of the SE skill landscape upon which to build throughout the rest of this book.

2.1 Organizing Frameworks for Social and Emotional Skills

Although initiatives to help students develop SE skills in educational contexts are becoming increasingly popular, little consensus exists on an operational definition and organizing framework through which to conceptualize these skills (e.g., Jones et al., 2016). We chose the definition above as one that encompasses what we see as the important factors surrounding SE skills. An important consideration in understanding the SE skill landscape is that it is rife with that we know as the jangle fallacy (using different terms to describe the same thing). This fallacy first emerges in how we describe this group of skills we are discussing. Terms such as *psychosocial skills, non-cognitive skills, 21st century skills, social and emotional skills, personality traits,* and *behavioral skills* are all frequently used in the literature to refer to this same group of skills, and these terms are often used interchangeably between frameworks. We will use *social and* emotional skills throughout this chapter for consistency but will refer to bodies of other literature using different terminology regularly. Another instance of the jangle fallacy occurs when we consider how we define and label individual social and emotional skills. Multiple frameworks which organize and label skills exist, and in many instances, different terms are used interchangeably to label the same underlying constructs. For example, behaviors associated with the personality trait conscientiousness may be labeled "grit" in one framework, "organization" in another, and "sustaining effort" in a third. The following section reviews several prominent conceptualizations and organizing frameworks in the field. While this review is not exhaustive of all frameworks used in practice, we aim to cover several prominent frameworks in order to acquaint readers with the landscape.

2.1.1 *The Collaborative for Academic, Social, and Emotional Learning*

The Collaborative for Academic, Social, and Emotional Learning (CASEL) defines social and emotional learning (SEL) as "the process through which children and adults acquire and effectively apply the knowledge, attitudes, and skills necessary to develop healthy identities, manage emotions and achieve personal and collective goals, feel and show empathy for others, establish and maintain supportive relationships, and make responsible and caring decisions (CASEL, 2020). This framework is fairly prominent in the United States and referenced frequently by educators and researchers alike (see Fig. 2.1). Within this framework, SE skills are categorized into five competency areas: self-awareness, self-management, responsible decision making, relationship skills, and social awareness. These competencies can be developed in multiple contexts, including in classrooms via curriculum and instruction, in schools through schoolwide practices and policies, and in homes and communities through family and community partnerships. This integrated framework promotes intrapersonal, interpersonal, and cognitive competencies.

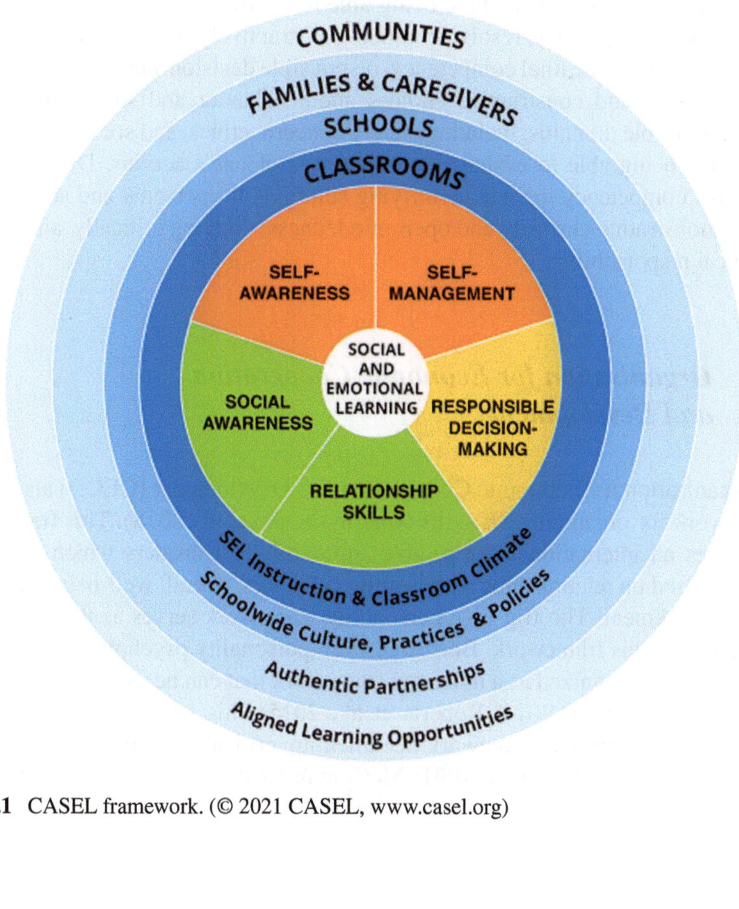

Fig. 2.1 CASEL framework. (© 2021 CASEL, www.casel.org)

Each of the framework's five competency areas covers one key aspect of social and emotional development. Both the self-awareness and self-management competencies target intrapersonal skills. The self-awareness competency is the ability to recognize one's emotions, thoughts, and values, and to identify how these factors influence behavior. This competency area also includes the individual's ability to assess his or her own strengths and limitations, social identities, sense of self-efficacy, and sense of purpose. The second competency related to intrapersonal skills, self-management, is the ability to successfully regulate one's emotions, thoughts, and behaviors. This is key in situations which require individuals to manage stress and control behavioral impulses. This competency also includes stress management, self-discipline, self-motivation, goal setting, and organizational skills. The framework also includes two competencies related to interpersonal skills: social awareness and relationship skills. Social awareness entails the abilities to understand the perspectives of and empathize with others, particularly those from diverse backgrounds. Social awareness also includes showing concern for others' feelings and recognizing situational demands and opportunities. Social awareness skills serve as the foundation for relationship skills, the fourth competency area. Relationship skills are the abilities to establish and maintain healthy and supportive relationships and to effectively navigate settings with diverse individuals and groups. Relationship skills include being able to communicate effectively, demonstrate cultural competency, resolve conflicts constructively, and seek and offer support when needed. The final competency, responsible decision-making, is the ability to make caring and constructive choices about behavior and social interactions based on multiple domains, including safety concern, ethics, and social norms. This also entails being able to evaluate consequences of one's actions. Discrete skills within this competency include identifying solutions for personal and social problems, demonstrating curiosity and open-mindedness, thinking critically, and assuming ethical responsibility.

2.1.2 Organisation for Economic Cooperation and Development

The Organisation for Economic Cooperation and Development (OECD) also offers a framework for organizing SE skills (Chernyshenko et al., 2018). This framework emphasizes an international perspective and carefully considers which skills are included based on relationships with healthy behaviors, overall well-being, and academic achievement. The Big Five personality framework serves as the theoretical foundation for this framework. Borrowed from personality psychology, this framework has been recognized as a universal framework that can be used to organize SE skills (Kyllonen et al., 2014; Roberts et al., 2015).This stems from 50 years of empirical support for the framework documenting critical educational and life outcomes (e.g., Barrick & Mount, 1991; McCrae & Costa, 1996; Poropat, 2009) and cross-cultural relevance (McCrae & Terracciano, 2005; Schmitt et al., 2007). The five factors are conscientiousness, agreeableness, emotional stability, openness to experience, and extraversion. Conscientiousness describes a person's tendency to be organized, dependable, diligent, hard-working, and achievement oriented. Agreeableness is a trait most prominent when considering interactions with others. It describes an individual's tendency to be friendly, helpful, empathetic, and trusting of others. Emotional stability, which is also referred to by its negative pole, neuroticism, describes a person's capability to cope with stressful situations and emotions, remain composed in times of change and uncertainty, and manage emotions. Openness to experience describes a person's curiosity, creativity, and interest in and acceptance of different cultures, ideas, values, and art. Extraversion describes a person's preference for social interactions with others, gregariousness, assertiveness, positive affect, and sensation-seeking (John & Srivastava, 1999).

In the OECD framework (see Fig. 2.2), SE skills are aligned with each of the factors of the Big Five, and each contains facet-level skills. SE skills were selected

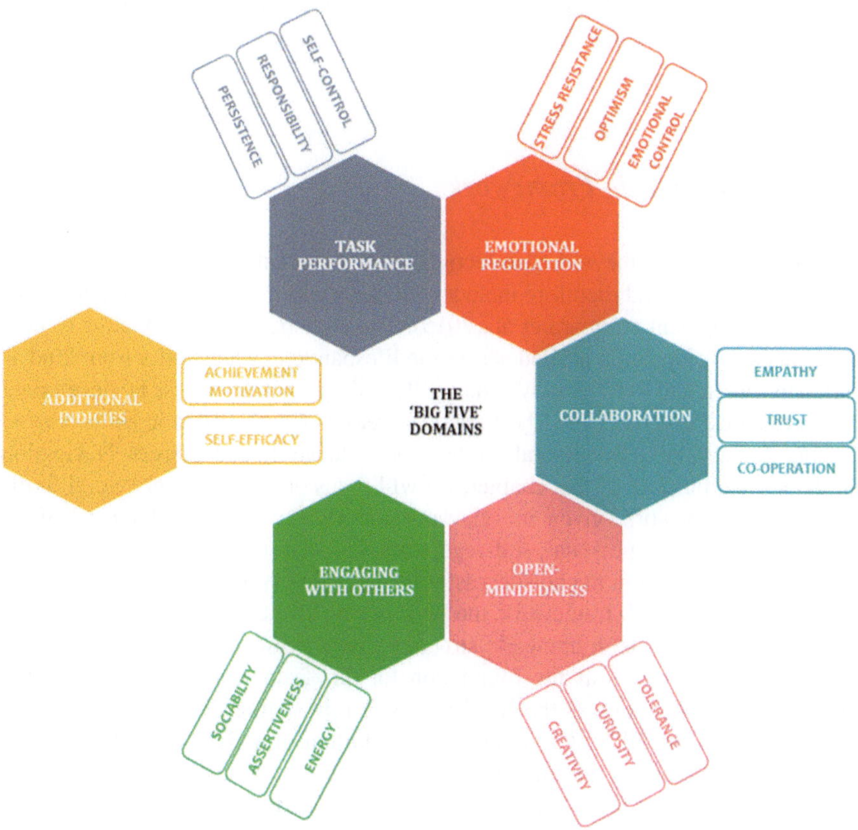

Fig. 2.2 OECD social and emotional skills. (OECD (2021), (About the OECD's Study on Social and Emotional Skills), https://www.oecd.org/education/ceri/social-emotional-skills-study/about/)

for inclusion in the framework based on demonstrated malleability, appropriateness for 10- and 15-year-old students, cross-cultural comparability, relevance, and predictive validity. The framework includes facets representing each of the Big Five factors, as well as facets from compound skills, which are more broad skill areas that do not align one-to-one with the Big Five factors. The five clusters within this framework are task performance, collaboration, emotional regulation, open-mindedness, and engaging with others. Task performance aligns with conscientiousness and includes achievement motivation, responsibility, persistence, and self-control. Collaboration aligns with agreeableness and includes empathy, trust, and cooperation. Emotional regulation aligns with emotional stability and includes stress resistance, optimism, and emotional control. Open-mindedness aligns with openness to experience and includes tolerance, curiosity, and creativity. Engaging with others aligns with extraversion and includes sociability, assertiveness, and energy. The compound skills selected for the framework and inclusion in the study are critical thinking, meta-cognition, and self-efficacy (Chernyshenko et al., 2018). Given their alignment with the Big Five, there is confidence about the five domains' generalizability and replication, and defining skills at the facet-level within the framework advances our understanding of SE skills beyond the factor-level structure.

2.1.3 ACT Holistic Framework

The ACT Holistic Framework also recognizes that student success is comprised of many components, and therefore includes SE skills within a broader framework for student development (Camara et al., 2015). The Holistic Framework aims to help individuals develop skills needed across the lifespan to educate shifts from kindergarten to career, and to ultimately achieve life and workplace success by developing a well-rounded area of skills. The four skill areas as Core Academic skills, Cross-Cutting Capabilities, Behavioral Skills, and Education and Career Navigation Skills. For the purpose of this chapter, we will focus only on the Behavioral Skills area of the framework, given its relevance for SE skills. The Behavioral Skills domain includes interpersonal, self-regulatory, and task-related behaviors that are important for adaptation to and successful performance in workplace settings.

Similar to the OECD framework, the Behavioral Skills component of the Holistic Framework derives its framework structure from personality psychology. The HEXACO model serves as the foundation for this framework, which shares the same five broad domains as the Big Five (with minor differences). An important differentiator of the HEXACO model is the addition of a sixth domain: honesty-humility. This describes an individual's tendency to act honestly, fairly, and ethically. The HEXACO model includes this additional domain to account for the emergence of a sixth factor in cross cultural research (Ashton & Lee, 2007). Both the HEXACO and Big Five models show empirical support linking personality structure to behaviors, and hence are used to categorize SE skills.

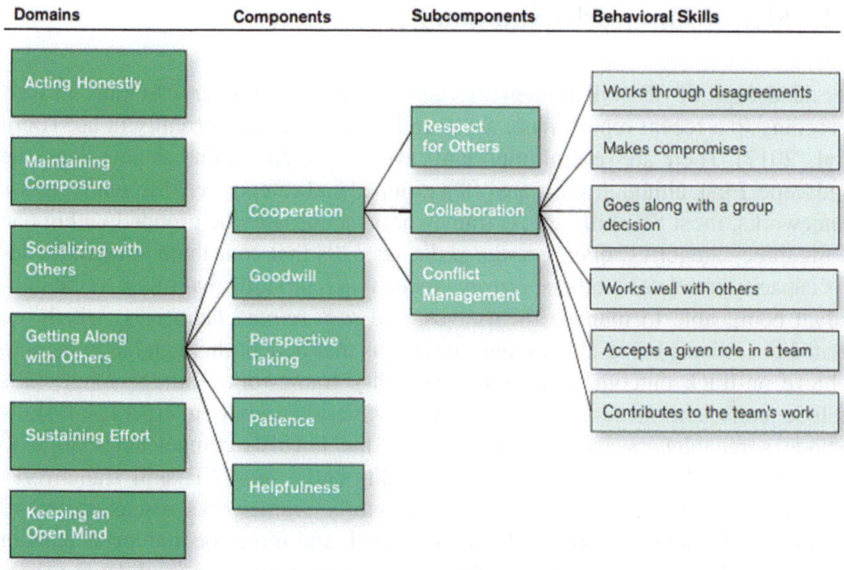

Fig. 2.3 ACT holistic framework behavioral skills. (© 2021 ACT, Inc. https://www.act.org/content/act/en/k12-educators-and-administrators/college-and-career-readiness/holistic-framework.html)

Another similarity between the OECD framework and the Behavioral Skills component of the Holistic Framework is that each SE skill maps to a HEXACO domain and contains subcomponents. There are six broad domains in the ACT Behavioral Skills Framework (see Fig. 2.3): Acting Honestly (honesty/humility), Maintaining Composure (emotional stability), Socializing with Others (extraversion), Getting Along with Others (agreeableness), Sustaining Effort (conscientiousness), and Keeping an Open Mind (openness to experience). Domains are arranged hierarchically, so that each higher-order domain has a set of components, subcomponents, and behavioral skills (see Casillas et al., 2015). For example, the Getting Along with Others domain includes the components of Cooperation, Goodwill, Perspective Taking, Patience, and Helpfulness. The subcomponent level of Cooperation includes Respect for Others, Collaboration, and Conflict Management. Each subcomponent then contains specific behavioral skills. Behaviors within the Collaboration subcomponent include working through disagreements, making compromises, working well with others, and contributing to a team's work. The framework includes 23 components and 50 subcomponents in total, each of which is mapped to specific behaviors, resulting in a framework that provides a rich level of detail at the skill level relative to many others of its kind.

2.2 SEL Frameworks Summary

The landscape of SE skill frameworks can certainly seem overwhelming. In fact, according to a recent report, there are over 136 frameworks currently in use (Berg et al., 2017). There are several important caveats to keep in mind as you digest the landscape. First, although skills are often grouped and categorized differently across frameworks, most SE skills across frameworks can broadly be grouped within one of the three categories: *interpersonal skills*, or skills that contribute to an individual's capacity to relate to others, *intrapersonal skills*, or skills that relate to the individual being able to manage themselves, and *task-related behaviors*, which are related to students being able to make decisions that benefit them and help them to achieve on-track outcomes. In both personality frameworks, intrapersonal skills include conscientiousness and emotional regulation, while interpersonal skills include agreeableness and extraversion, and the CASEL competencies align to these higher-order components as well. Additionally, most SE skill frameworks make use of a hierarchical structure. In all three frameworks discussed, broad domain-level categories exist at the highest level, and more specific levels contain narrower skills and competencies. The jangle fallacy emerges at both levels; there are many commonalities in skills included across frameworks, but they are often labeled and categorized differently. This can create confusion when attempting to compare skills from different frameworks, as it is not always clear how similar skills from different frameworks might be. Thus, care must be taken when choosing which framework and skills to measure, to ensure that the intended skill is truly the construct being assessed.

2.3 Social and Emotional Skill Development

Having provided an overview of what SE skills are, it is important to discuss that these skills are not static. SE skills are known to develop naturally over time and, as we will review later, in response to intervention. Although there are multiple ways to evaluate skill continuity and change (for a thorough review, see Walton & Billera, 2016), here we will review research relying on two approaches: rank-order stability and mean-level change. Rank-order stability, which is typically estimated with test-retest correlations, refers to the degree to which students maintain their relative positioning within their cohort. That is, if a child is the most rambunctious student in the class at one time point, is that child still among the most rambunctious at a later timepoint? If so, it would suggest that there is high rank-order stability for this characteristic. One study including first- and second-grade students (Hampson & Goldberg, 2006) provided estimates of five characteristics over a four-year period, and the test-retest correlations ranged from .36 to .55, which were on the same order of magnitude as those reported by Prinzie and Deković (2008) in their study of a similarly aged sample of students. Studies tracking the development in older students (i.e., 12- to 16-years-old) indicate that test-retest correlations increase in magnitude as students age (Klimstra et al., 2009; Pullman et al., 2006), reaching as high as .75. In sum, research demonstrates a fairly high degree of rank-order stability in school-aged children. The student who, relative to his/her peers, is rambunctious at an early age is likely to be among the more highly rambunctious students at an older age.

Significant rank-order stability does not preclude the possibility of mean-level change (i.e., the average amount a cohort develops) over time. Throughout development, a student may continue to be among the more rambunctious in his/her cohort, but that student may still become less rambunctious over time. That is, on average, the cohort may become calmer and more orderly as a whole. Indeed, research shows this to be true; personality traits, which some erroneously perceive as being stable or immutable over time, show significant growth over the life course, even into old age (Roberts et al., 2006). To cite some examples in school-aged children, gains of approximately a quarter of a standard deviation in extraversion and openness to experience during adolescence have been reported (Pullman et al., 2006).

The research cited above pertains to naturally occurring growth. We see that people tend to maintain their relative standing in their cohort while their skills and traits develop and change. Can anything be done to hasten the developmental process? There is an abundance of literature saying yes. In fact, several meta-analyses have been carried out to summarize the efficacy of interventions. In their meta-analysis, Roberts and colleagues showed that clinical interventions are effective in altering personality traits to a significant degree, even when the interventions last a brief eight weeks. School-based interventions targeting SE skills show similar promise. In the first such meta-analysis (Durlak et al., 2011), which included more than 270,000 students from 213 studies, students in SEL programs significantly improved on SE skills (compared with those in control conditions). Effects across SE skills for the total sample averaged .57. Since the release of Durlak and colleagues' meta-analysis, three additional meta-analyses (Sklad et al., 2012; Taylor et al., 2017; Wiglesworth et al., 2016) have been carried out, each extending the work of Durlak and colleagues and substantiating their conclusion that SEL programs are efficacious. It is clear that such programs have immediate effects, as well as long-term benefits. A final point to be made is that such interventions can be fairly simple to implement and brief in duration. For example, one group of study participants were encouraged to complete specific challenges designed to improve certain characteristics. Engaging in simple tasks, such as "When you are worried about something, write it down," or "Say 'please' and 'thank you' when asking for something," resulted in gains of up to .17 standard deviations over the course of just a few months.

Cross-disciplinary research shows that individual differences develop naturally throughout the course of life, and skills can be developed with effort – much like a physical muscle. Exercise the muscle, and it will grow. This concept is an important feature of most, if not all, definitions of SE skills and SEL.

2.4 Social and Emotional Skill Assessment

Regardless of how we conceptualize them, SE skills can develop over time. Measuring them is a key component in how we track growth. In this section, we will discuss the most common traditional methods of assessing SE skills, namely, self-report Likert items, situational judgment items, and forced-choice items. SE skills can be assessed by other methods, such as parent/teacher/other report, behavior observation, stealth assessment, etc. However, given that the goal of this chapter is to provide an overview of the most common traditional approaches to SE skills, we have chosen to focus on self-report methods.

2.4.1 Self-Report Likert

Self-report Likert items have been used for decades in SE skill and behavioral research, and their use is expected to continue into the future. As noted by Spector (2012), they are "cheap, efficient, flexible, and in appropriate settings they have good predictive validity (p. 459)." Individuals are asked to indicate their level of agreement with a number of statements (e.g., "I work hard"; see Fig. 2.4 for an example).

Although the predictive validities for Likert items are modest, they are the same or higher than other assessment methods. By making the items context specific (e.g., "I work hard **at school**"), the predictive and incremental validities increase (Bing et al., 2004; Davison & Bing, 2009; Lievens et al., 2008a). Likert items have the benefit of not requiring nearly the level or amount of reading as some other assessment methods, such as situational judgment items (discussed in the next section), thereby reducing the cognitive load on the respondent. On the other hand, these items may be particularly susceptible to response biases, such as reference effects. That is, often people answer such items by asking the question, "compared to whom?" As a consequence, it could be the case that students from very high achieving schools, for example, might rate themselves lower on their SE skills than students from low-achieving schools simply because they are using a different reference group and not because they are truly lower on these skills. This is often called the Big-Fish Little-Pond Effect (Marsh & Hau, 2003).

Please indicate the extent to which you agree or disagree with the following statements.

	Strongly disagree	Disgree	Slightly disagree	Slightly agree	Agree	Strongly agree
When I have a difference of opinion with someone, I try to understand why that person disagrees with me.	○	○	○	○	○	○

Fig. 2.4 Example Likert item

There has been considerable debate regarding the occurrence of faking on self-report measures used in high-stakes situations, such as applying for university admission or for a job. Although studies have shown that individuals are able alter their scores when instructed to appear as favorable as possible, this does not tell us the extent to which it occurs in high-stakes situations (Spector, 2012). In a study conducted by Hogan et al. (2007), they reviewed self-report scores for job applicants who were rejected for a position and reapplied six months later. They found that applicants were as likely to decrease their scores as increase them. The debate also concerns the question as to what is actually taking place when respondents misrepresent themselves on self-report SE or behavioral items. Is this faking to appear more favorably to gain admission to a school or program, which is something to be prevented? Is this actually impression management that occurs every day when individuals interact with each other (Hogan et al., 2007)? It appears that faking is not impacting criterion-related validity of self-report scales (Hogan et al., 2007; Ones et al., 1996). Even when they try to correct for faking, studies have found very little difference in criterion-related validities between uncorrected and faking corrected samples (Schmitt & Oswald, 2006). In general, correcting scores for faking is not recommended. Another option to address faking is to add warning statements indicating that faking can be detected and will result in consequences (Dwight & Donovan, 2003). Finally, other item types, such as those described below, may be effective for reducing faking.

2.4.2 Situational Judgment Test

One promising alternative to self-report items is the situational judgment test (SJT) format. In general, SJTs present scenarios describing incidents critical to effective behavior. Respondents either identify an appropriate response from a list of alternatives or indicate their level of agreement with statements concerning the appropriateness of various behaviors SJTs present (see Fig. 2.5). They have become a popular selection method in high-stakes contexts (Campion et al., 2014), with studies and meta-analyses showing that SJTs can be valid predictors of important

Instructions: Read the situation below and choose the most appropriate response.

You are taking part in a study group with classmates in preparation for a particularly difficult test. As the first review session gets underway, it becomes clear that the other members of the group have not taken good notes and are not as familiar with the material as you are.

A. Do nothing: clearly you will get better marks on the test.
B. Suggest that everyone read over the textbook in preparation for the next session.
C. Leave the group because you will be better off studying on your own.
D. Offer to use your notes as the basis for the remaining review sessions.
E. Ask the teacher for advice as to how to handle the next meeting.

Fig. 2.5 Example situational judgment item

outcomes (e.g., McDaniel et al., 2001) over and above self-report tests of personality, cognitive ability, and job experience (Chan & Schmitt, 2002; Weekley & Ployhart, 2005). They are an effective approach for assessing SE skills (e.g., interpersonal skills, procedural or tacit knowledge; Kyllonen & Lee, 2005; Lievens & Sackett, 2012; MacCann & Roberts, 2008; Oswald et al., 2004).

Some research suggests that SJTs are advantageous because they are more difficult to fake than self-report scale items (Hooper et al., 2006; Lievens et al., 2008b). A comparative evaluation of three response formats' (rate effectiveness, rank responses, and choose most/least effective) construct validity showed that the rate-SJT displayed stronger correlations with hypothesized SE skills and weaker correlations with general mental ability (Arthur Jr et al., 2014). Further, SJTs tend to score well in terms of respondent reactions due to high face validity (Bauer & Truxillo, 2006). Students report they are engaging and worth completing (Lipnevich et al., 2013), which better supports multiple administrations and retains student "buy-in" to the ongoing process of SE skill assessment. Finally, SJTs can be repurposed as formative assessments so as to provide a student with feedback on his or her competencies in the domain of interest.

One concern noted in the literature is that they are not as effective at reducing racial/ethnic subgroup differences as other noncognitive methods. In particular, the amount of reduction appears to be driven by the reading load of the SJT (Ployhart & Holtz, 2008; Whetzel et al., 2008). One way to address this concern is to reduce reading load by producing video SJTs or providing voiceover for text that should be read by test takers. Some studies, however, show that SJTs may be associated with less adverse impact (e.g., Schmitt et al., 2009). Another concern has to do with the reliability of SJTs. In a meta-analysis of the reliability of SJTs, Kasten and Freund (2015) noted that "the reliability of SJT scores is typically rather low and below recommended levels for high-stakes applications" (p. 230). However, a recent report by Sorrel et al. (2016) addressed this issue by creating a Q-matrix using cognitive diagnosis models that coded the attributes of each SJT item to a multidimensional matrix. Calculating the reliabilities (as classification accuracies) based on the matrix increased reliability of SJTs substantially.

2.4.3 Forced Choice

Forced choice (FC) methods are a family of related methods of presenting assessment content, where examinees are presented with two or more statements and asked to choose which statement is most like them. Variants of this method will have examinees choose the statements that are most and least like them or rank order the statements by degree of similarity to themselves (see Fig. 2.6). These methods contrast with single statement methods of presenting assessment content, which ask examinees to rate agreement with the statement content using Likert scales. FC methods have been most commonly used in the context of personality assessment, which will be the focus of the research discussed here. As an aside, FC

Item	Most Like Me	Least Like Me
I do more than what my teachers expect of me.		
I am concerned about other students' well-being.		
I cope well with stressful assignments.		

Fig. 2.6 Example forced choice item
Instructions: Choose the item that is most like you and least like you. Do nothing with the third item.

methods are often discussed in reference to item response theory (IRT)-based scoring methods. It should be noted that the topic of scoring method is related to, although distinct from, the topic of assessment presentation method. Scoring methods relevant to personality assessment include classical test theory (CTT) and various IRT-based methods. Both single statement and FC presentation methods can be scored using either of these scoring methods. However, it is typically recommended to use IRT-based pairwise-preference scoring models for FC presentation methods. Comparing the effectiveness of the various IRT scoring methods is beyond the scope of this chapter, and the discussion below includes research that used both unidimensional and multidimensional pairwise-preference models.

FC methods have only become popular in the personality assessment literature in the past twenty years or so, mostly due to limitations of computing power necessary to run the algorithms for IRT parameter estimation. Generally, FC assessments are complicated and time intensive to develop (see Stark et al., 2011, for a sample description of the process). One, they require several rounds of data collection and analysis to develop. After an initial item pool has been developed, test ratings using Likert scoring must be gathered. The number of ratings must be substantial (generally at least 1000 people) in order to provide accurate estimation of the item parameters. Once the parameters have been estimated, data must also be collected on the social desirability of each item, so that items of roughly equal desirability can be paired. Once the test form has been developed using these methods, the test must be validated with additional data collection efforts. However, these methods result in a test that is generally shorter than traditional single statement methods (Stark et al., 2011), can be converted into an adaptive test, and can be converted into many different test forms due to the many different combinations of statements that can be created from the statement pool (Stark et al., 2012). It should be noted that applicants typically have negative reactions to the FC presentation method, as they feel artificially limited by the statements presented and are forced to choose a statement that may not represent a completely accurate description of themselves (Converse et al., 2008).

In terms of reliability and validity, the literature has been somewhat contradictory on how FC presentation methods compare to more traditional single statement presentation methods. Although some have noted that it is more difficult to evaluate the reliability of FC assessments (Heggestad et al., 2006; Stark et al., 2011), they are generally found to have comparable reliability to single statement assessments scored using Likert methods (Chernyshenko et al., 2009; Geldhof et al., 2015). Issues of validity, especially in regards to faking, are more complicated. Most research comparing single statement and FC methods have focused on comparing construct validity and evaluating the IRT-based scoring methods. In this regard, FC methods do well, as the two presentation methods show convergence in construct validity in personality assessment (Chernyshenko et al., 2009; Christiansen et al., 2005; Geldhof et al., 2015). Fewer studies have examined the predictive validity of FC assessments, with some showing comparable validity to single statement methods (Bowen et al., 2002; Christiansen et al., 2005; White & Young, 1998), some showing more predictive validity (Drasgow et al., 2012), and some showing less predictive validity for FC assessments (Geldhof et al., 2015). Other concerns that have been raised in regard to FC methods include that given the differences in test construction across different studies, it is difficult to know exactly which features and development procedures contribute to test validity (Stark et al., 2012). Additionally, the process of responding to FC items may be more cognitively taxing than single statement items, and may therefore lead to greater fatigue, errors in responding, and diminishing gains in incremental validity of the personality assessment over cognitive ability tests (Chernyshenko et al., 2009; Vasilopolous et al., 2006). Finally, even after using the complicated methods to develop a FC test, it is no more effective in providing normative standing on a trait than Likert scoring methods (Heggestad et al., 2006).

Some studies have found that FC methods may reduce faking (Christiansen et al., 2005; Walton et al., 2021), especially when paired with the multidimensional pairwise-preference scoring method (Cao & Drasgow, 2016). However, other research has shown that there is no difference in faking between single statement and FC methods (Heggestad et al., 2006), and that FC methods don't appear to affect examinees' response processes (Guan & Carter, 2016). Furthermore, given that no scales are used in these items, FC tests eliminate scale response effects. Reference bias should also be minimized with FC tests because respondents conduct an internal (self vs. self) rather than an external (self vs. other) comparison when responding to the items.

2.5 Conclusion

In this chapter, we presented several frameworks, assessment methods, and developmental courses of SE skills in order to provide a foundation of the traditional thinking and approaches in this area and serve as a starting point/contrast to the rest of the book. The next chapter presents an overview of current research on learning analytics as applied to SE skills, while the rest of the book presents novel, groundbreaking research in this area.

References

Arthur, W., Jr., Glaze, R. M., Jarrett, S. M., White, C. D., Schurig, I., & Taylor, J. E. (2014). Comparative evaluation of three situational judgment test response formats in terms of construct-related validity, subgroup differences, and susceptibility to response distortion. *Journal of Applied Psychology, 99*(3), 535–545.

Ashton, M. C., & Lee, K. (2007). Empirical, theoretical, and practical advantages of the HEXACO model of personality structure. *Personality and Social Psychology Review, 11*(2), 150–166.

Aspen Institute. (2018). *From a nation at risk to a nation at hope: Recommendations from the National Commission on Social, Emotional, and Academic Development*. The Aspen Institute.

Barrick, M. R., & Mount, M. K. (1991). The Big Five personality dimensions and job performance: A meta-analysis. *Personnel Psychology, 44*, 1–26.

Barrick, M. R., Mount, M. K., & Judge, T. A. (2001). Personality and performance at the beginning of the new millennium: What do we know and where do we go next? *International Journal of Selection and Assessment, 9*(1–2), 9–30.

Bauer, T. N., & Truxillo, D. M. (2006). Applicant reactions to situational judgment tests: Research and related practical issues. In J. A. Weekley & R. E. Ployhart (Eds.), *SIOP organizational series. Situational judgment tests: Theory, measurement, and application* (pp. 233–249). Lawrence Erlbaum Associates Publishers.

Berg, J., Osher, D., Same, M. R., Nolan, E., Benson, D., & Jacobs, N. (2017). *Identifying, defining, and measuring social and emotional competencies*. Washington, DC: American Institutes for Research.

Bing, M. N., Whanger, J. C., Davison, H. K., & VanHook, J. B. (2004). Incremental validity of the frame-of-reference effect in personality scale scores: a replication and extension. *Journal of Applied Psychology, 89*(1), 150–157.

Bowen, C.-C., Martin, B. A., & Hunt, S. T. (2002). A comparison of ipsative and normative approaches for ability to control faking in personality questionnaires. *The International Journal of Organizational Analysis, 10*(3), 240–259.

Camara, W., O'connor, R., Mattern, K., & Hanson, M. A. (2015). *Beyond academics: A holistic framework for enhancing education and workplace success* (ACT research report series) (4th ed.). ACT.

Campion, M. C., Ployhart, R. E., & MacKenzie, W. I., Jr. (2014). The state of research on situational judgment tests: A content analysis and directions for future research. *Human Performance, 27*(4), 283–310.

Cao, M., & Drasgow, F. (2016). A meta-analytic review of forced-choice personality measures in high-stakes situations. In J. Seybert & A. L. Guan (Co-Chairs), *Recent advances in forced choice personality assessment*. Symposium conducted at the 30th annual meeting of the society of industrial/organizational psychology conference, Anaheim, CA.

CASEL. (2020). *CASEL's SEL framework: What are the core competence areas and where are they promoted?* Retrieved from: https://casel.org/wp-content/uploads/2020/12/CASEL-SEL-Framework-11.2020.pdf

Casillas, A., Way, J., Burrus, J., Allen, J., & Hanson, M. A. (2015). Behavioral skills. In *Beyond academics: A holistic framework for enhancing education and workplace success* (pp. 25–38).

Chan, D., & Schmitt, N. (2002). Situational judgment and job performance. *Human Performance, 15*(3), 233–254.

Chernyshenko, O. S., Stark, S., Prewett, M. S., Gray, A. A., Stilson, F. R., & Tuttle, M. D. (2009). Normative scoring of multidimensional pairwise preference personality scales using IRT: Empirical comparisons with other formats. *Human Performance, 22*, 1–23.

Chernyshenko, O., Kankaraš, M., & Drasgow, F. (2018). *Social and emotional skills for student success and wellbeing: Conceptual framework for the OECD study on social and emotional skills* (OECD Education Working Papers, No. 17). OECD Publishing. https://doi.org/10.1787/db1d8e59-en

Christiansen, N. D., Burns, G. N., & Montgomery, G. E. (2005). Reconsidering forced-choice item formats for applicant personality assessment. *Human Performance, 18*(3), 267–307.

Converse, P. D., Oswald, F. L., Imus, A., Hendricks, C., Roy, R., & Butera, H. (2008). Comparing personality test formats and warnings: Effects on criterion-related validity and test-taker reactions. *International Journal of Selection and Assessment, 16*(2), 155–169.

Davison, H. K., & Bing, M. N. (2009). Content validity does matter for the criterion-related validity of personality tests. *Industrial and Organizational Psychology, 2*, 501–503.

Drasgow, F., Stark, S., Chernyshenko, O. S., Nye, C. D., Hulin, C. L., & White, L. A. (2012). *Development of the tailored adaptive personality assessment system (TAPAS) to support Army selection and classification decisions* (Tech. Rep. No. 1311). US Army Research Institute for the Behavioral and Social Sciences.

Durlak, J. A., Weissberg, R. P., Dymnicki, A. B., Taylor, R. D., & Schellinger, K. B. (2011). The impact of enhancing students' social and emotional learning: A meta-analysis of school-based universal interventions. *Child Development, 82*(1), 405–432.

Dwight, S. A., & Donovan, J. J. (2003). Do warnings not to fake reduce faking? *Human Performance, 16*(1), 1–23.

Geldhof, G. J., Gestsdottir, S., Stefansson, K., Johnson, S. K., Bowers, E. P., & Lerner, R. M. (2015). Selection, optimization, and compensation: The structure, reliability, and validity of forced-choice versus Likert-type measures in a sample of late adolescents. *International Journal of Behavioral Development, 39*(2), 171–185.

Guan, A. L. & Carter, N. (2016). Personality test faking as a shift in response process: Theoretical and practical implications. In J. Seybert & A. L. Guan (Co-Chairs), *Recent advances in forced choice personality assessment*. Symposium conducted at the 30th annual meeting of the society of industrial/organizational psychology conference, Anaheim, CA.

Hampson, S. E., & Goldberg, L. R. (2006). A first large cohort study of personality trait stability over the 40 years between elementary school and midlife. *Journal of Personality and Social Psychology, 91*, 763–779.

Heggestad, E. D., Morrison, M., Reeve, C. L., & McCloy, R. A. (2006). Forced-choice assessments of personality for selection: Evaluating issues of normative assessment and faking resistance. *Journal of Applied Psychology, 91*(1), 9–24.

Hogan, J., Barrett, P., & Hogan, R. (2007). Personality measurement, faking, and employment selection. *Journal of Applied Psychology, 92*(5), 1270–1285.

Hooper, A. C., Cullen, M. J., & Sackett, P. R. (2006). Operational threats to the use of SJTs: Faking, coaching, and retesting issues. In J. A. Weekley & R. E. Ployhart (Eds.), *SIOP organizational series. Situational judgment tests: Theory, measurement, and application* (pp. 205–232). Lawrence Erlbaum Associates Publishers.

John, O. P., & DeFruyt, F. (2015). *Education and social progress: Framework for the longitudinal study of social and emotional skills in cities* (Organization for Economic Cooperation and Development, EDU/CERI/CD (2015)13). Retrieved from: http://www.oecd.org/officialdocuments/publicdisplaydocumentpdf/?cote=EDU/CERI/CD(2015)13&docLanguage=En

John, O. P., & Srivastava, S. (1999). The Big Five trait taxonomy: History, measurement, and theoretical perspectives. In *Handbook of personality: Theory and research* (2nd ed., pp. 102–138). Guilford.

Jones, S., Bailey, R., Brush, K., Nelson, B., & Barnes, S. (2016). *What is the same and what is different? Making sense of the "noncognitive" domain: Helping educators translate research into practice*. Harvard Graduate School of Education. Retrieved from: https://easel.gse.harvard.edu/files/gse-easel-lab/files/words_matter_paper.pdf

Kasten, N., & Freund, P. A. (2015). A meta-analytical multilevel reliability generalization of Situational Judgment Tests (SJTs). *European Journal of Psychological Assessment, 32*, 230–240.

Klimstra, T. A., Hale, W. W., III, Raaijmakers, Q. A., Branje, S. J., & Meeus, W. H. (2009). Maturation of personality in adolescence. *Journal of Personality and Social Psychology, 96*, 898–912.

Kyllonen, P. C., & Lee, S. (2005). Assessing problem solving in context. In O. Wilhelm & R. W. Engle (Eds.), *Handbook of understanding and measuring intelligence* (pp. 11–25). Sage.
Kyllonen, P. C., Lipnevich, A. A., Burrus, J., & Roberts, R. D. (2014). *Personality, motivation, and college readiness: A prospectus for assessment and development* (ETS research report series) (pp. 1–48).
Lievens, F., & Sackett, P. R. (2012). The validity of interpersonal skills assessment via situational judgment tests for predicting academic success and job performance. *Journal of Applied Psychology, 97*(2), 460–468.
Lievens, F., De Corte, W., & Schollaert, E. (2008a). A closer look at the frame-of-reference effect in personality scale scores and validity. *Journal of Applied Psychology, 93*(2), 268–279.
Lievens, F., Peeters, H., & Schollaert, E. (2008b). Situational judgment tests: A review of recent research. *Personnel Review, 37*(4), 426–441.
Lipnevich, A. A., MacCann, C., & Roberts, R. D. (2013). Assessing noncognitive constructs in education: A review of traditional and innovative approaches. In D. H. Saklofske, C. R. Reynolds, & V. L. Schwean (Eds.), *Oxford handbook of child psychological assessment* (pp. 750–772). Oxford University Press.
Lipnevich, A. A., Preckel, F., & Roberts, R. D. (Eds.). (2016). *Psychosocial skills and school systems in the 21st century: Theory, research, and practice*. Springer.
MacCann, C., & Roberts, R. D. (2008). New paradigms for assessing emotional intelligence: Theory and data. *Emotion, 8*(4), 540–551.
Marsh, H. W., & Hau, K. T. (2003). Big-fish--little-pond effect on academic self-concept: A cross-cultural (26-country) test of the negative effects of academically selective schools. *American Psychologist, 58*(5), 364–376.
McCrae, R. R., & Costa, P. T. (1996). Toward a new generation of personality inventories: Theoretical contexts for the five-factor model. In J. S. Wiggins (Ed.), *The five-factor model of personality: Theoretical perspectives* (pp. 51–87). Guilford Press.
McCrae, R. R., & Terracciano, A. (2005). Universal features of personality traits from the observer's perspective: Data from 50 cultures. *Journal of Personality and Social Psychology, 88*, 547–561.
McDaniel, M. A., Morgeson, F. P., Finnegan, E. B., Campion, M. A., & Braverman, E. P. (2001). Predicting job performance using situational judgment tests: A clarification of the literature. *Journal of Applied Psychology, 86*, 730–740.
OECD (2021). *Beyond academic learning: First results from the survey of social and emotional skills*. OECD Publishing, Paris. https://doi.org/10.1787/92a11084-en.
Ones, D. S., Viswesvaran, C., & Reiss, A. D. (1996). Role of social desirability in personality testing for personnel decisions: The red herring. *Journal of Applied Psychology, 81*, 660–679.
Oswald, F. L., Schmitt, N., Kim, B. H., Ramsay, L. J., & Gillespie, M. A. (2004). Developing a biodata measure and situational judgment inventory as predictors of college student performance. *Journal of Applied Psychology, 89*(2), 187–207.
Ployhart, R. E., & Holtz, B. C. (2008). The diversity–validity dilemma: Strategies for reducing racioethnic and sex subgroup differences and adverse impact in selection. *Personnel Psychology, 61*(1), 153–172.
Poropat, A. E. (2009). A meta-analysis of the five-factor model of personality and academic performance. *Psychological Bulletin, 135*, 322–338.
Prinzie, P., & Deković, M. (2008). Continuity and change of childhood personality characteristics through the lens of teachers. *Personality and Individual Differences, 45*, 82–88.
Pullman, H., Raudsepp, L., & Allik, J. (2006). Stability and change in adolescents' personality: A longitudinal study. *European Journal of Personality, 20*, 447–459.
Robbins, S. B., Lauver, K., Le, H., Davis, D., Langley, R., & Carlstrom, A. (2004). Do psychosocial and study skill factors predict college outcomes? A meta-analysis. *Psychological Bulletin, 130*(2), 261.
Roberts, B. W., Walton, K. E., & Viechtbauer, W. (2006). Patterns of mean-level change in personality traits across the life course: A meta-analysis of longitudinal studies. *Psychological Bulletin, 132*, 1–25.

Roberts, R. D., Martin, J. E., & Olaru, G. (2015). A Rosetta stone for noncognitive skills: Understanding, assessing, and enhancing noncognitive skills in primary and secondary education. Asia Society and Professional Examination Service.

Schmitt, N., & Oswald, F. L. (2006). The impact of corrections for faking on the validity of noncognitive measures in selection settings. *Journal of Applied Psychology, 91*(3), 613–621.

Schmitt, D., Allik, J., McCrae, R. R., & Benet-Martinez, V. (2007). The geographic distribution of Big Five personality traits: Patterns and profiles of human self-description across 56 nations. *Journal of Cross-Cultural Psychology, 38*, 173–212.

Schmitt, N., Keeney, J., Oswald, F. L., Pleskac, T. J., Billington, A. Q., Sinha, R., & Zorzie, M. (2009). Prediction of 4-year college student performance using cognitive and noncognitive predictors and the impact on demographic status of admitted students. *Journal of Applied Psychology, 94*(6), 1479–1497.

Sklad, M., Diekstra, R., De Ritter, M., Ben, J., & Gravesteijn, C. (2012). Effectiveness of school-based universal social, emotional, and behavioral programs. Do they enhance students' development in the area of skill, behavior, and adjustment? *Psychology and Schools, 49*, 892–909.

Sorrel, M. A., Olea, J., Abad, F. J., de la Torre, J., Aguado, D., & Lievens, F. (2016). Validity and reliability of situational judgement test scores a new approach based on cognitive diagnosis models. *Organizational Research Methods, 19*(3), 506–532.

Spector, P. E. (2012). Self-reports for employee selection. In N. Schmitt (Ed.), *The Oxford handbook of personnel assessment and selection* (pp. 443–461). Oxford University Press.

Stark, S., Chernyshenko, O. S., Drasgow, F., Lee, W. C., White, L. A., & Young, M. C. (2011). Optimizing prediction of attrition with the US Army's Assessment of Individual Motivation (AIM). *Military Psychology, 23*(2), 180.

Stark, S., Chernyshenko, O. S., & Drasgow, F. (2012). *Constructing fake-resistant personality tests using item response theory: High stakes personality testing with multidimensional pairwise preferences. New perspectives on faking in personality assessments*. Oxford University Press.

Taylor, R., Oberle, E., Durlak, J. A., & Weissberg, R. P. (2017). Promoting positive youth development through school-based social and emotional learning interventions: A meta-analysis of follow-up effects. *Child Development, 88*, 1156–1171.

Vasilopoulos, N. L., Cucina, J. M., Dyomina, N. V., Morewitz, C. L., & Reilly, R. R. (2006). Forced-choice personality tests: A measure of personality and cognitive ability?. *Human Performance, 19*(3), 175–199.

Walton, K. E., & Billera, K. A. (2016). Personality development during the school-aged years: Implications for theory, research, and practice. In A. A. Lipnevich, F. Preckel, & R. D. Roberts (Eds.), *Psychosocial skills and school systems in the 21st century* (pp. 93–111). Springer.

Walton, K. E., Radunzel, J., Moore, R., Burrus, J., Anguiano-Carrasco, C., & Murano, D. (2021). Adjectives vs. statements in forced choice and Likert item types: Which is more resistant to impression Management in Personality Assessment?. *Journal of Personality Assessment*. Online publication Feb 3, 2021.

Weekley, J. A., & Ployhart, R. E. (2005). Situational judgment: Antecedents and relationships with performance. *Human Performance, 18*(1), 81–104.

Weissberg, R. P., Durlak, J. A., Domitrovich, C. E., & Gullotta, T. P. (2015). Social and emotional learning: Past, present, and future. In J. A. Durlak, C. E. Domitrovich, R. P. Weissberg, & T. P. Gullotta (Eds.), *Handbook of social and emotional learning: Research and practice* (pp. 3–19). The Guilford Press.

Whetzel, D. L., McDaniel, M. A., & Nguyen, N. T. (2008). Subgroup differences in situational judgment test performance: A meta-analysis. *Human Performance, 21*, 291–309.

White, L. A., & Young, M. C. (1998). *Development and validation of the Assessment of Individual Motivation (AIM)*. Paper presented at the annual meeting of the American Psychological Association, San Francisco, CA.

Wiglesworth, M., Lendrum, A., Oldfield, J., Scott, A., ten Bokkel, I., Tate, K., & Emery, C. (2016). The impact of trial stage, developer involvement and international transferability on universal social and emotional learning programme outcomes: A meta-analysis. *Cambridge Journal of Education, 46*(3), 347–376.

Chapter 3
Mapping the Landscape of Social and Emotional Learning Analytics

Srećko Joksimović, Shane Dawson, Abhinava Barthakur, Oleksandra Poquet, Yuan 'Elle' Wang, Fernando Marmolejo-Ramos, and George Siemens

Abstract Educational research is increasingly implementing and studying new approaches for assessing attributes that go beyond conventional assessments of students' cognitive ability. Despite decades of research, there remains a lack of consensus in describing these skills or attributes, variously termed "non-cognitive skills", "21st century competencies", "personal qualities", "social and emotional learning skills", and "soft/core skills". Regardless, these skills and qualities reflect dimensions of learning that are broader than conventional curriculum knowledge. The importance of such skills has been well established in contemporary literature as highly relevant for success in school, university, the workplace, and engaged

S. Joksimović (✉) · S. Dawson
Education Futures, University of South Australia, Adelaide, SA, Australia

Centre for Change and Complexity in Learning, University of South Australia, Adelaide, SA, Australia
e-mail: srecko.joksimovic@unisa.edu.au; shane.dawson@unisa.edu.au

A. Barthakur · F. Marmolejo-Ramos
Centre for Change and Complexity in Learning, University of South Australia, Adelaide, SA, Australia
e-mail: abhinava.barthakur@mymail.unisa.edu.au; Fernando.Marmolejo-Ramos@unisa.edu.au

O. Poquet
Learning Planet Institute, Universite Paris Cite, Paris, France
e-mail: sasha.poquet@unisa.edu.au

Y. E. Wang
EdPlus Action Lab, Arizona State University, Scottsdale, AZ, USA
e-mail: elle.wang@asu.edu

G. Siemens
Centre for Change and Complexity in Learning, University of South Australia, Adelaide, SA, Australia

Department of Psychology, University of Texas at Arlington, Arlington, TX, USA

Y. E. Wang et al. (eds.), *Social and Emotional Learning and Complex Skills Assessment*, Advances in Analytics for Learning and Teaching,
https://doi.org/10.1007/978-3-031-06333-6_3

citizenship more broadly. The relatively new fields of learning analytics and educational data mining have introduced numerous novel methodologies to education research. This work has served to advance assessment models for social and emotional learning skills. Building on one of the most referenced social and emotional learning frameworks, this chapter provides a comprchensive overview of learning analytics methods for measuring skills such as creativity, critical thinking, or emotional regulation, among others. We recognize that the potential of learning analytics to measure SEL is largely under-utilized and pose possible ways to advance work in this domain.

Keywords Measuring skills · Learning analytics · Social and emotional learning · Frameworks · Psychometrics and learning analytics

3.1 Introduction

Developing skills that promote collaboration, leadership, critical thinking, and complex problem solving are foundational across a range of education settings from K-12 to corporate learning. These skills and competencies, detailed as social and emotional learning (SEL) in this chapter, are integral to engagement in complex problem solving and are increasingly seen as employment qualities. Supporting these skills is far from straightforward (Buckingham Shum & Deakin Crick, 2016). The decreasing half-life of knowledge and the growing integration of machine learning are collectively changing the nature of learning, work, and the environments where learning and work occur (de Laat et al., 2020). The uptake of advanced technologies in education settings has also generated new modes of learner data, developed by the fields of learning analytics (LA) and educational data mining (EDM). It is at this juncture where learning analytics can be leveraged to bring insights into how learning models and data mining methods can be used to evaluate SEL.

Measurement of skills and literacies has been of interest to LA since the inception of the field. Lockyer et al. (2013) and Dawson and Siemens (2014) described how checkpoint and process analytics can measure complex multiliteracies and learning design models. Further, a collective set of learning analytics work in 2016, defined and proposed novel methods for assessing a complex array of competencies included under the SEL domain, forming a special issue in the Journal of Learning Analytics – titled *Learning Analytics for 21st Century Competencies* (Buckingham Shum & Deakin Crick, 2016). The topic of the special issue was further developed through a series of workshops at the Learning Analytics and Knowledge conference in 2018 and 2019. LA research on 21st Century skills and SEL evolved into another special issue published in the Journal of Learning Analytics in 2020 (Joksimović et al., 2020). Despite this continued interest among LA researchers investigating SEL, there remains a lack of a coherent and system-level narrative that can guide the development of assessment models for measuring and understanding such a complex array of skills and competencies (Dawson et al., 2019).

In Chap. 2 of this book, Way and colleagues provide an extensive overview of several commonly used SEL frameworks. The authors review the Collaborative for Academic, Social, and Emotional Learning (CASEL) framework that defines SEL across five competency areas: self-awareness, self-management, responsible decision making, relationship skills, and social awareness. Then, Way and colleagues continue to review the ACT holistic framework that, similarly to CASEL, is structured around four broad skill areas – (i) core academic skills; (ii) cross-cutting capabilities; (iii) behavioral skills; and (iv) education and career navigation skills. Another framework discussed by Way and Colleagues is that by the Organization for Economic Cooperation and Development (OECD). OECD framework centers on the Big Five Personality traits alongside 6 broad domains: collaboration; open-mindedness; engaging with others; compound skills; task performance; and emotional regulation (see Fig. 3.1). While the frameworks discussed by Way and colleagues (Chap. 2) are well established, none of them have considered the role LA and EDM methods can play in advancing this work. This chapter fulfills this gap. Drawing on the OECD framework, this chapter discusses how LA can measure and

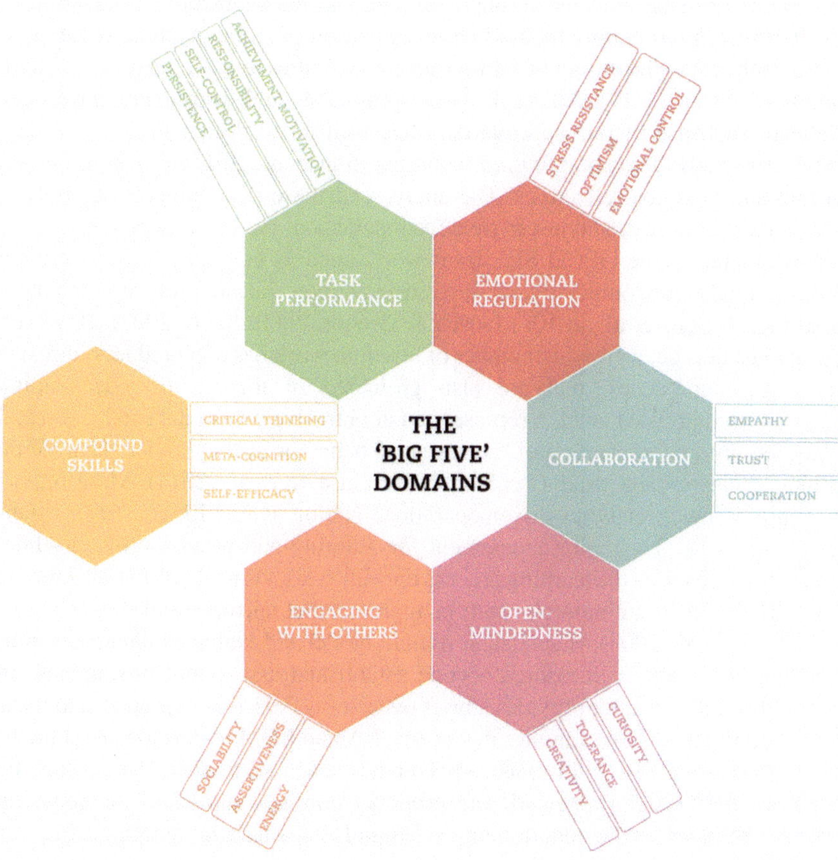

Fig. 3.1 OECD's framework for measuring social and emotional skills

promote SEL skills. In so doing, we demonstrate how LA and EDM methods can support and advance SEL research and measurement of complex skills and competencies.

3.2 Measuring Social and Emotional Learning Skills

The OECD framework is grounded in the Big Five Personality traits (Raad & Perugini, 2002). As can be seen in Fig. 3.1, SEL skills within the OECD framework are closely linked to the assessment of the five personality traits – conscientiousness, emotional stability, agreeableness, openness to experience, and extraversion, using well-established instruments (De Raad, 2000; Raad & Perugini, 2002; Rammstedt et al., 2017). Among the most commonly used instruments are the Big Five Factor Markers (Goldberg, 1992), which comes in long (100 indicators) and short formats (50 indicators), and the NEO-PI Big Five Questionnaire (Ashton, 2013). The majority of research exploring SEL outlined within the OECD framework relies on data collected through surveys and questionnaires. Evaluation of SEL in research has mainly focused on investigation of learners' perceived experiences or informing the design of future learning activities (e.g., Hannay et al., 2010; Rammstedt et al., 2017). Although these approaches and instruments have been validated, such use of retrospective data has well-noted design limitations (e.g., Brutus et al., 2013). In short, current attempts to measure SEL are over reliant on staccato and post hoc reflections. The analysis of these data hampers attempts to provide more proactive and just in time feedback about the learning process.

Subdomains of the OECD SEL framework, such as empathy, trust, or critical thinking are also predominantly measured using survey data and self-reported measures (e.g., Gerdes et al., 2010; Madsen & Gregor, 2000; Staib, 2003). However, over the last decade, a growing number of research work has explored less intrusive data-driven approaches, that are also grounded in theory, to understanding SEL. Research in EDM and LA represent an unobtrusive way of data collection and do not require an interruption of the learners' behavior as they undertake specific learning or assessment tasks (Bergner, 2017; Gašević et al., 2014). More importantly, providing an analytics-driven operationalization of the SEL constructs establishes a basis for developing assessment ***for*** learning and provision of real-time, formative feedback as the learning process unfolds (Gašević et al., 2017; Joksimović et al., 2019). This is an important shift in approach that mirrors trends in social sciences (Lazer et al., 2009) where data is used to model and evaluate human behavior, expanding the range of measures beyond established assessment instruments. In what follows, this chapter provides a brief overview of existing approaches to measure SEL and reflects on examples that utilize LA and EDM to develop novel methods for measuring SEL skills and competencies. Each section explores a specific domain of the OECD framework and proposes how LA and EDM methods can measure and promote the underpinning skills and characteristics defining SEL.

3.2.1 Collaboration

Collaboration is arguably among the most researched areas of SEL (Schneider et al., 2021). Early works investigating collaboration tended to measure the outcomes of groups or teams working together. For example, the product of group work is often the assessed artefact to demonstrate effective group or teamwork (e.g., Chikersal et al., 2017; Dillenbourg et al., 1996; Lacerenza et al., 2018). More recently, a growing number of studies have examined the process of collaboration through the use of learning analytics and data mining (D. C. Gibson, 2018; Hernández-García et al., 2018). This shift towards the use of learner trace data mirrors the growth in uptake of digital technologies to support the learning process. The analysis of digital data with varied assessment models have led to the use of multiple metrics to understand different aspects of collaboration and cooperation such as team or group work (Griffin & Care, 2015).

Griffin and Care (2015), present one of the most comprehensive frameworks for the assessment of collaborative problem solving across two broad categories of social and cognitive competencies. In so doing, Griffin and Care (2015), provided a mapping between trace data and complex skills measures, such as participation (e.g., action or interaction), perspective taking (e.g., adaptive responsiveness), or social negotiation (e.g., negotiation or self-evaluation). In developing an alternative model for collaborative problem solving, Gibson (2018) provided definitions of key dimensions of SEL (i.e., collaboration, creativity, personal learning, problem solving, and global sustainability) and metrics of specific team attributes (e.g., establishing and maintaining shared understanding or exploring and understanding) collected automatically within a digital learning space. Hernández-García et al. (2018) also demonstrate how log data can be used to measure communication, cooperation, monitoring, and coordination in group work. More recently, Martinez-Maldonado et al. (2021) proposed a conceptual model of collaboration analytics. This is, arguably, one of the first conceptual models that integrates LA, theory, and design to provide unobtrusive and data-informed evidence of collaboration, across diverse settings, providing collaboration related feedback at the individual and group level. Martinez-Maldonado et al. (2021) also provide several use-cases that operationalized the proposed model, showing the potential of LA and EDM for measuring SEL skills and competencies in the collaboration domain.

In addition to log or trace data, LA and EDM approaches draw heavily on the analysis of textual artefacts that are generated through collaborative learning tasks such as discussion forums or peer exchanges. Discourse analysis has been one of the most applied approaches to measure various aspects of collaboration as it allows for understanding concrete discursive actions and practices (Kovanović et al., 2017). Tarmazdi et al. (2015) operationalized Dickinson and McIntyre's teamwork model to provide a dashboard that depicts the quality of teamwork across the seven dimensions: leadership, orientation, monitoring, coordination, communication, feedback, and backup behavior. The approach proposed by Tarmazdi et al. (2015) builds on the combination of natural language processing, information retrieval, and sentiment analysis methods to operationalize metrics of team collaboration. Dowell and colleagues introduced Group Communication Analysis, a computational linguistic framework for automated text analysis of the sequential interactions of online team communication (Dowell et al., 2019). The framework introduced by Dowell et al. (2019) allows for measuring social impact, participation, internal cohesion, sharing or negotiating ideas, among several other dimensions. The recent advances in artificial intelligence, such as GPT-3 (Brown et al., 2020) or Google's BERT – Bidirectional Encoder Representations from Transformers (Devlin et al., 2018) – warrant further investigation into the role discourse analysis can play in measuring and supporting SEL.

In this book, our contributors focus on studying *online collaboration* (Chap. 8) and *empathy* (Chap. 4), as perhaps the most challenging concepts to be replicated by machines (Bollmer, 2017). In Chap. 8, Cukurova and colleagues broaden the discussion on providing analytics-based assessment of SEL, proposing a framework for measuring collaboration as a process. In so doing, Cukurova and colleagues build on collaborative cognitive load theory and social network analysis to examine the relationship between the interactivity gains and coordination costs. Some of the metrics used to evaluate online collaboration, as revealed by Cukurova and colleagues (Chap. 8) include number of posts viewed by a learner or entire group, number of comments each learner made or the amount of time learners spent in participating in discussions. In addition to those shallow proxies of learners' collaboration, educators also tend to opt for more complex insights, such as quality of learners' posts from the subject domain, dialogic, or learning perspective (Chap. 8). In Chap. 4, on the other hand, Knezek, Christensen and Gibson further discuss a very complex relationship between technology and empathy, noting that "real-world empathy is six times more strongly related to feelings than virtual empathy". Nevertheless, as Knezek and colleagues further posit, deploying learning analytics models for measuring empathy remains a stretch goal. Even with the long history of psychometric development regarding empathy, it is still questionable to what extent existing instruments measure what they claim to be measuring (Chap. 4). It then becomes even more challenging to develop machine learning models that would grasp the understanding of empathy in technology mediated environments.

3.2.2 Open-Mindedness

Open-mindedness is among the five domains of social and emotional skills within the OECD framework. Open-mindedness, or openness to experience, has been commonly labeled as one of the critical skills in generalizing learning across different contexts and learning tasks, in individual, group, and organizational learning (Dale et al., 2021; Lord, 2015). Although primarily measured within the Big Five inventory, alternative measures exist such as: open-minded thinking scale (Stanovich & West, 2007) or the Intellectual Humility Scale (Alfano et al., 2017). As a subdomain of open-mindedness, ***curiosity*** has been linked to a wide variety of work and learning related processes, such as learning performance, job satisfaction, socialization and forming an identity, psychological health, and decision making, among other dimensions (for a review see Wagstaff et al., 2021). The concept is generally defined as "the desire for knowledge and sensory experiences that sparks exploratory behavior, [and] represents one of the most fundamental and pervasive aspects of humanity" (Wagstaff et al., 2021, p. 363). It is noteworthy, that since 2000, more than 15 measures of curiosity have been developed and validated in contemporary literature (Wagstaff et al., 2021). The volume of measures developed in this short time span indicated both the importance of curiosity in SEL and the drive to overcome the limitations of existing evaluation approaches. Finally, ***tolerance***, another sub-domain of open-mindedness, also has been studied using self-reported measures, and is commonly seen as a context-dependent skill (Bobo & Licari, 1989; Hjerm et al., 2020; Sztejnberg & Jasiński, 2014). Nevertheless, to the best of our knowledge, in contemporary literature there are no attempts that utilize LA and EDM driven methods in operationalizing and measuring dimensions of open-mindedness or it's subdomains of curiosity and tolerance. This remains a gap in LA and EDM research and is an area of investigation that could be promoted through various means of discourse analysis or social and network analysis (Chen & Poquet, 2020; Kovanović et al., 2017; Poquet & de Laat, 2021).

In contrast to the other skills in the domain of open-mindedness, ***creativity*** received considerable attention in LA and EDM research and practice. The relationship between creativity and learning analytics is twofold – first, the use of learning analytics to support creative processes (e.g., Britain et al., 2020; Saleeb, 2021); and second in the use of learning analytics to measure creativity (e.g., Gal et al., 2017). Britain et al. (2020), for instance, proposed a platform that integrates support for design ideation and a learning analytics dashboard. The use of data-driven methods in measuring creativity are a focus of this book. Gal et al. (2017) provide means for automated measurement of creativity, using log data collected from learners' interactions with various problems in the Kodetu game-based learning environment for teaching programming skills. For instance, learners would get additional points for each solution attempt, use while statements instead of if, or using both while and if statements. Gibson (2018) further proposed how to measure creativity within a framework for unobtrusive observation of team learning within the Curtin Challenge platform. Creativity is measured via log data indices that represent proxies for idea generation, design and refinement, openness, and exploration, working creatively, and creative production.

The topic of creativity and learning analytics is further discussed in Chap. **5** of this book. Specifically, Marrone and Cropley (Chap. 5) operationalize creativity and position it as a core competence that will delineate humans from machines in the age of Artificial Intelligence. Marrone and Cropley (Chap. 5) propose the Innovation Phase Model to measure creativity and suggest how LA applications can support various dimensions of the proposed model. Further LA applications to reflect creativity can be obtained via social network analysis in online settings. Such applications build on Ronald Burt's seminal research (Burt, 2004) to identify idea generation and propagation (creativity) within and across groups (e.g., Lee & Tan, 2017).

3.2.3 Engaging with Others

Engaging with others as a SEL domain has been studied in educational and learning analytics research from two aspects: (i) observed from the original notion of extraversion or (ii) as a part of the collaboration domain (Sect. 3.2.1). On the other hand, the three subdomains – i.e., energy, assertiveness, and sociability – have been usually observed in somewhat different contextualization than defined within the OECD framework (Fig. 3.1). For instance, although the existing literature does not necessarily cover ***energy*** as a construct, learners' motivation and self-efficacy that link to energy from the conceptual point of view, have been explored in the learning analytics domain. In that sense, Ahmad Uzir et al. (2020), de Barba et al. (2020), as well as Jovanovic et al. (2019), among others, rely on trace data to extract various time effort or regulatory indicators (e.g., entropy of weekly session counts). ***Assertiveness***, on the other hand, have been primarily studied in the context of understanding factors that shape team effectiveness, using self-reported measures (e.g., Pearsall & Ellis, 2006). The cotemporary literature does not recognize LA or EDM driven approaches in studying assertiveness. Finally, the notion of ***sociability*** has been primarily contextualized in studying group and teamwork, predominantly utilizing social network analysis (SNA) techniques (e.g., Lancieri, 2017). Being a central methodology in learning analytics, SNA offers multiple opportunities for automated data collection and analysis (Dowell et al., 2015; Joksimović et al., 2018a; Skrypnyk et al., 2015). This topic has been further discussed in Chap. 14 of this book, where Benson and Houtti propose several social interaction metrics (e.g., number of likes or number of views) to disentangle learners' engagement with a case-based online learning platform.

In the domain of computer supported collaborative learning, engaging with others is a dimension where various (learning analytics) algorithms have been developed to foster social matching and collaboration between learners. For instance, Adamson et al. (2014) introduced Bazaar – a platform for multi-dimensional multi-party dialog, where an artificial agent matches learners based on their topic of interest. This idea of fostering social interactions has been further discussed in Chap. 13, where Wang, Camacho and Goel explore the design of AI-based social matching systems to enhance social interactions between online learners. The proposed social matching platform utilizes natural language processing to extract various information about learners – such as hobbies, city, or country of birth – to build profiles of online learners. Profiles of learners are further matched based on learners' preferences (e.g., geographically or based on their age). Nevertheless, in the evaluation of the proposed solution, considerable concerns have been raised by learners about the transparency of the proposed solution as well as about losing agency in building social connections. Finally, in Chap. 15, Li and Krasny present a study that explores how a MOOC can foster participants taking climate actions and helping spread those actions through their social networks.

3.2.4 Compound Skills

Compound skills have been extensively researched in education and learning analytics. The concept includes 3 sub-domains namely: critical thinking, self-efficacy, and metacognition. This grouping of characteristics very much aligns with theories of self-regulated learning – one of the most cited theories in LA research. ***Critical thinking*** has been valued as a crucial factor that characterizes students who are prepared for increasingly complex life and work environments (Paul & Elder, 1990). The assessment of critical thinking in learning analytics is primarily driven by the work in the Community of Inquiry (CoI) and cognitive presence as one of the CoI domains (cite, cite). For instance, Kovanovic and colleagues (2016) rely on linguistic properties extracted from learners' discussion forum posts, providing an overview of the classification features most indicative of the different phases of cognitive presence. A similar approach has been applied by Neto and colleagues (2018) to develop a classification method for automated analysis of cognitive presence in discussion messages written in Portuguese or earlier by Paul Corrich (2011). ***Metacognition*** (i.e., thinking about thinking) is regarded as a critical skill in students' learning, problem solving, reasoning, development, school transitions, and achievement of longer-term life outcomes in general (Veenman et al., 2006). As one of the most promising approaches, trace-based self-reports, as proposed by Zhou and Winne (2012) – meaning, each self-report was embedded into a task – can be used to provide analytics-driven insights into metacognitive skills. Additionally, recent literature in LA and EDM proposed a variety of analytics methods that rely on sequence or process mining have been proposed for identifying learning tactics and strategies (Matcha et al., 2019; Van Laer & Elen, 2018), as well as reflective writing and related automated text analysis methods for understanding learners' metacognition (Gibson et al., 2016). Finally, Jovanovic et al. (2019) utilized students' traced self-reports on cognitive load and ***self-efficacy*** to investigate how a predictive model for self-reports can be made based on log data.

Compound skills have been addressed in several chapters in this book. Specifically, Gabriel and colleagues (Chap. 6) discusses novel, learning analytics driven methods for measuring motivational and affective processes related to self-regulated learning, across various learning environments. In so doing, Gabriel and colleagues recognize a shift in how LA and EDM researchers define and measure motivation and affect utilizing multimodal data. In Chap. 7, Rakovic and Winne provide a summary of research accessing and synthesizing content across multiple sources, writing processes and self-regulated learning. As a result, Rakovic and Winne propose a two-dimensional typology of cognitive and metacognitive processes in self-regulated writing using multiple sources spanning two problem spaces in writing tasks, rhetorical and content. In Chap. 10, Hillaire and colleagues further introduce an automated method for measuring reading comprehension skills. The proposed solution is primarily based on measuring valence interpretation from learners' self-reports and discussion comments. Finally, San Pedro and colleagues (Chap. 11) leverage recent methodological advances in measurement, LA and EDM to develop automated detectors of learners' academic emotions, and engagement, using data collected by a middle school mathematics software. Such detectors are further used to predict which students will go to selective colleges several years later.

3.2.5 Task Performance

As defined in the OECD framework (Fig. 3.1), conscientiousness or task performance includes a range of skills "that determine the propensity to be self-controlled, responsible towards others, hardworking, persistent, orderly, virtuous and rule abiding" (OECD, 2017, p. 17). The construct itself is rather intriguing. As Gamez (2014) explains, in its original definition, consciousness "can only be measured through first person reports" (*ibid.*, p.1). However, in the domain of LA and EDM research, task performance has been commonly linked to academic performance and various forms of summative and formative assessment. For instance, Xing et al. (2021) utilized Bayesian Network model to analyze log data of learners' interactions with an engineering challenge. Additionally, Janssen et al. (2016) measure the task performance in a Minecraft based environment that captures parameters such as time, number of errors, or number of used rails. More recently, a considerable body of research emerged that utilizes multimodal learning analytics to measure task performance. In that sense Bitner and Le (2021) rely on EEG-devices to differentiate attention values between incorrect and correct solutions in a problem solving task, Abrahamson et al. (2015) used eye-tracking devices, whereas Larmuseau et al. (2020) relied on a variety of metrics (e.g., heart rate variability, skin temperature, or galvanic skin response) to explore the association between cognitive load and task performance.

The concept of persistence, as defined in the OECD framework (Fig. 3.1) primarily draws on several interrelated concepts: (i) concept of mastery oriented behaviors, and (ii) self-constructs such as self-efficacy beliefs, self-concept or confidence (Stankov et al., 2014). However, of particular importance for understanding learners' persistence in their beliefs that they can succeed is the theory of situational and individual interest (Renninger & Hidi, 2019). Persistence, commonly defined as a continuous engagement with a difficult task, is one of the four mastery behaviors being critical for success in school and life in general. Persistence has been usually captured using self-reported instruments (Christensen & Knezek, 2014). For example, Computer Attitude Questionnaire is a Likert-scale self-report instrument that consists of two subscales – Motivation/Persistence and Study Habits (Christensen & Knezek, 2002). The instrument has been validated across various formal and informal settings for almost 30 years. However, this subdomain can be captured using a combination of self-reported (i.e., learners' self-efficacy beliefs) and trace data. Additionally, the indicator closely relates to learners' capacity to adapt or attempt a new strategy after receiving information that the current strategy is not successful. Specifically, providing feedback during the task performance, learners may enhance their effort, persistence, and self-efficacy through prompting of meta-motivational (Miele & Scholer, 2018) and emotion regulation strategies (Gross, 2015).

3.2.6 Emotional Regulation

Affect and motivation play a crucial role in learners' ability to monitor and regulate their learning, particularly when studying and engaging with STEM content (Azevedo et al., 2017; Efklides et al., 2018). Although the initial approaches to measuring learners' motivation and affect were predominantly centered around the utilization of self-reported instruments, recent advances in LA and EDM showed comparable results employing trace data (D'Mello et al., 2017; Porter et al., 2020). In one of the recent studies, Bosch and D'Mello (2017), for example made considerable advances in mapping affective states, such as anger, anxiety, boredom, confusion, curiosity, disgust, fear, frustration, flow/engagement, happiness, sadness, and surprise to the traces of learner interactions in online settings. Triangulating data from learners' face recordings, self-reports, and trace data, (Bosch & D'Mello, 2017) detected certain behaviours (e.g., reading, coding) that trigger specific affective states (e.g., boredom, engagement, curiosity, frustration). Despite still being its infancy, such research provides a sound basis for more salient operationalisation and measurement of complex constructs, such as motivation and affect (Lajoie et al., 2019).

Although *stress resistance*, *optimism*, and *emotional control* are somewhat specific constructs and are not being addressed as such (Fig. 3.1), their various manifestations in learning have been extensively studied in educational research, learning analytics, and educational data mining. Specifically, negative emotions, such as frustration, have been linked to less use of self-regulatory learning strategies and lower levels of engagement with learning tasks (Efklides et al., 2018). For example, utilizing sequence mining (Kinnebrew et al., 2013) and coherence analysis (Segedy et al., 2015), the nature of learners' interactions with the task (e.g., the strategies applied) before and after the feedback can be explored to provide automated means for the assessment of ***emotional control***. Additionally, Epistemic Network Analysis (Shaffer, 2018), can be used to quantify differences in the nature of learners' interactions with the task before and after receiving negative feedback. The ***stress resistance*** subdomain has been primarily linked to the notion of managing anxiety and performing well in high-pressure situations (OECD, 2017). Numerous studies showed a detrimental effect of anxiety on learning and performance (e.g., Gabriel et al., 2020). However, it has been only recently that considerable advances have been made in measuring anxiety and stress resistance utilising various learning analytics methods and approaches. One of the most comprehensive approaches represents a work by Smets et al. (2018) who developed an approach for ambulatory stress detection, relying on physiological and contextual measurements, collected through wearable devices and smartphones. Specifically, using two wearable devices, Smets et al. (2018) captured three physiological signals unobtrusively – the electrocardiogram, skin conductance, and skin temperature, namely – and revealed "significant differences between physiological features for ECG, SC, and ST between different stress levels" (*ibid.*, p.67). Ramakrishnan et al. (2019), on the other hand, showed that the combination of several physiological factors – namely,

EEG, heart rate variability, heart rate, and galvanic skin response – show promising results in differentiating stressful from non-stressful states.

Despite the complexity of measuring various concepts, significant advances have been made over the last several years in capturing proxies of emotional control, trust, critical thinking, or creativity, among other dimensions. WIth more data being available, various methods gained more prominence in measuring SEL. Some of the most utilized approaches include sequence and process mining (Jovanovic et al., 2019; Matcha et al., 2019), Bayesian network models (Xing et al., 2021), combination of self-reflective prompts and trace data (Jovanović et al., 2019; Zhou & Winne, 2012), as well as various methods of natural language processing and discourse analysis in general (Dowell et al., 2019; Gibson et al., 2016). Multimodal learning analytics further brought a plethora of opportunities, including unprecedented sources of data that allow for capturing various psycho-physiological measures as proxies for SEL. Further adoption of the proposed methods, however, to great extent will depend on the ability to develop reliable and valid assessment of SEL dimensions, such as those proposed in the OECD framework (Fig. 3.1).

3.3 Bridging Psychometrics and Learning Analytics

While the pedagogical shift from academic knowledge to competence in curricula is gaining widespread attention (Milligan, 2020), valid and replicable assessment of these competencies (and SEL skills more broadly) has been of increasing concern. This is primarily due to the complex nature of these constructs, making it hard to quantify and measure them (Joksimović et al., 2020). As such, educators and policymakers argue the need for requisite changes in the methods of assessments from those used to measure the learning within the cognitive domain. This, along with the confluence of digitalisation and advancement in contemporary education research have paved the way for interdisciplinary approaches to assess the acquisition of social and emotional skills. One such promising area of blended methodology research includes the intersection of established psychometric theories and approaches and learning analytics methodologies (e.g., Milligan & Griffin, 2016).

Assessments from the psychometrics perspective is a process of evaluating students' knowledge and learning based on their performance in various assessment situations. The field of psychometrics follows a top-down approach, starting from theory to identifying and collecting behavioural evidence and observable indicators eliciting various skills and other learner attributes (Drachsler & Goldhammer, 2020). With respect to learning analytics, the field is concerned with understanding and optimising learning and providing personalized learning experiences (Joksimović et al., 2019). Following a bottom-up methodology, LA researchers collect fine-grained student learning data to analyse learner behaviour and draw inferences about the learning process (Drachsler & Goldhammer, 2020). Overall, although LA and psychometrics are different in their theoretical and methodological assumptions, they fundamentally share a similar goal of positively impacting learning (Mislevy, 2019).

Drawing on the advantages from psychometrics and LA, researchers have adopted more blended methodologies for measuring and sensemaking of various SEL skills which were otherwise beyond our grasp. Milligan & Griffin (2016) used a similar methodology to quantify a 21st century competency – crowd-sourced learning capability – associated with effective MOOC based learning. In their study, the authors demonstrate how blended methodologies can be used to operationalise an assessment rubric and categorise learners into different groups based on these rubric measures. In another study, Hu et al. (2017) used a three step methodology marrying psychometrics and state of the art learning analytics techniques to investigate online problem-solving capabilities of learners in primary school. Similarly, He and colleagues (2017) adopted a psychometric approach of topic modelling to examine students' online forum postings. Therefore, these studies have consolidated the narrative raised by several researchers of developing blended methodologies as the future for assessment of SEL skills.

This amalgamation of methods has the potential to provide fine-grain data to evidence complex skills and support assessment. A goal that would not be achieved by using only one of the aforementioned approaches. This is well demonstrated in the work of Barthakur and colleagues (Barthakur et al., 2022) (forthcoming). The authors utilise a blended methodology, combining state-of-the-art learning analytics techniques with psychometrics, to measure complex skills development within a workplace learning MOOC. The authors illustrate a systematic and scalable approach to assess leadership skills based on learners' self-reflective writings. Rather than evaluating learners based on their final course grade, this study provides empirical evidence of measuring leadership skills that have previously been beyond our grasp. This blend of psychometrics and learning analytics, besides utilising various data sources (e.g., trace-data, assessments, and survey data as well as from aggregated information), is specifically suited for near real-time assessment for learning processes and formative feedback purposes. Such an approach would be fundamentally different from standardized assessments, such as PISA, that focus primarily on providing assessment of learning, without the intention to optimize learning or the environments in which learning occurs – a holy grail of learning analytics.

3.4 The Role of Design

Learning analytics has a tremendous, yet underutilized, potential to help address challenges associated with measuring SEL skills and competencies (Gašević et al., 2016; Knight et al., 2013; Mangaroska & Giannakos, 2019). Being utilized either as an assessment ***of*** learning or as providing the means for assessment ***for*** learning. Learning analytics provides tools and methods for assessing complex skills and competencies in a timely and formative manner (Gašević et al., 2014; Knight et al., 2013; Pardo & Siemens, 2014). However, it is questionable whether the evidence for skills and competencies we aim to measure is present in the data we collect (Joksimović et al., 2019).

Despite technological advances and recent proliferation of various learning management systems, two approaches stand out when it comes to the assessment ***of*** complex skills. First, PISA standardized assessment provides frameworks for measuring collaborative problem solving (2015), creativity (2021/2022) and learning in the digital age (2024/2025). Another line of research in complex skills assessment is structured around intelligent tutoring systems (Anderson et al., 1985). Both approaches are characterized by the closed and context-specific nature of the technology used, that allows for very specific and structured interactions to occur. More importantly, both streams include a set of tasks specifically ***designed*** so learners could elicit specific skills and competencies. Similar approaches have been adopted in several chapters in this book. Specifically, in Chap. 10, Hillaire and colleagues relied on Udio – an online reading environment – to capture various measures of reading comprehension. In Chap. 12 Prieto and colleagues introduce LAPills platform, whereas in Chap. 13, Wang and colleagues presented an AI-based social matching system.

Joksimovic and colleagues (2018b) provided a comprehensive overview of the approaches to measuring learning at scale. Although contextualized in MOOC settings, conclusions derived from the work presented by Joksimovic and colleagues (2018b), to a great extent, hold in more traditional fully online or blended learning settings. In their work, Joksimovic and colleagues (2018b) provided a conceptualization and operationalisation of engagement in online learning, consisting of cognitive, behavioral, academic, and affective engagement. The study further showed that contemporary literature primarily provides instrumentation for measuring behavioral and academic engagement, with some crude proxies for measuring cognitive engagement. Very little has been done in the context of affective engagement. The reason for this is rather straightforward, as Joksimovic and colleagues (2018b) further posit – existing online courses are not designed for learners to elicit behaviors that would allow for capturing various aspects of affective or even cognitive engagement. Moreover, the evidence for behavioral and academic engagement is not necessarily driven by learning activities designed for, but rather by the underlying platform used to deliver learning experience.

For instance, measuring curiosity does not seem feasible using standard log data collected by contemporary learning management systems that are being a primary data source for most learning analytics applications to date. However, building on the work by Zhou and Winne (2012) or Jovanovic et al. (2019), among others, and providing means for gauging curiosity with specific activities incorporated in the course and assessment design, could yield valid indicators for measuring curiosity. Likewise, tasks that allow students to elicit persistence differ from effort or challenge-seeking ones, in a sense that persistence tasks require continuous interaction with complex, difficult problems (Porter et al., 2020). Persistence is also one of the defining aspects of successful problem solving (van Horik & Madden, 2016; van Horik & Madden, 2016). However, while the anticipation of interest commonly leads to engagement, persistence depends on "the quality of the interest experience"

(Renninger & Hidi, 2019, p. 168). In such settings, a combination of students' self-efficacy belief, confidence, and judgments of learning, along with trace data will be used to provide the evidence for their ability to persist in completing learning activities.

The notion of ***design for learning*** (Goodyear & Carvalho, 2014; Jones, 2015), therefore, becomes instrumental in the fulfilment of the full potential of learning analytics. Drawing on the research in learning networks where the paradigm of design for learning has been commonly applied, design for learning assumes that the focus of the analysis of learning networks is always activity-centered (Goodyear & Carvalho, 2014). However, "activity cannot be designed: it is emergent" (Goodyear & Carvalho, 2014, p. 18). In operationalizing focal dimensions necessary for understanding learning and providing means for assessment for learning, the concept of engagement, as an overarching construct in the field of education that brings together "many separate lines of research under one conceptual model" (Appleton et al., 2006, p. 427), should be revisited. Engagement, in this context, is also emergent and cannot be designed. We can design environments and activities to foster learners' engagement. As such, the concept of learner engagement complements Goodyear and Carvalho (2014) notion of activity, which is being recognized as a main focus in design for learning in networks. Thus, engagement here is also viewed as emergent (i.e., cannot be designed), encapsulating measurable evidence of learners' activities in diverse learning environments.

3.5 Conclusion

As Martin et al. (2016) stated, "even with increasing attention to the importance of 21st century skills, there is still relatively little known about how to measure these sorts of competencies effectively" (*ibid.*, p. 37). A half a decade later, this statement still holds. This is by no means due to a lack of research in this domain. A simple search on Scopus for "21st century skills", "social and emotional learning" or "non-cognitive skills" shows rapid growth in the number of articles on the topic in the same period. The lack of evidence about "how to measure these sorts of competencies effectively" (Martin et al., 2016, p. 37) stems from the fact that there is a little consensus among researchers and practitioners on what to call this cluster of concepts that represents "21st century skills", "social and emotional learning" or "non-cognitive skills" (Matteson et al., 2016). In Joksimovic and colleagues (2020) we argued that none of the terms being used is complete enough, as well as that a "range of terms will be needed, with limited chance of a unifying term in the near future" (Joksimović et al., 2020, p. 1). This is further supported by the "Explore SEL" list,[1] which provides 40 related frameworks used to measure SEL.

[1] http://exploresel.gse.harvard.edu/

While challenges exist in defining SEL, more reliable approaches can be achieved through careful integration of the methods from psychometrics into LA through advanced assessment approaches (see Sect. 3.3). Real-time feedback on SEL development requires insights into learners' proficiency in different skills and competencies. The majority of the existing conceptual frameworks outlining SEL components do not envision the link between analytics and psychometrics. With forecast demand accelerating for SEL skills in the future of work, it's increasingly important for LA researchers to create constructs of SEL through integrating psychometric models and practices. This approach, while still not fully defining SEL, will serve to advance the research domain sufficiently to begin interacting with the actual outputs of assessments. These assessments, in turn, serve to inform, reframe, and revise SEL models to ensure greater accuracy over time.

References

Abrahamson, D., Shayan, S., Bakker, A., & Van Der Schaaf, M. (2015). Eye-tracking Piaget: Capturing the emergence of attentional anchors in the coordination of proportional motor action. *Human Development, 58*(4–5), 218–244.

Adamson, D., Dyke, G., Jang, H., & Rosé, C. P. (2014). Towards an agile approach to adapting dynamic collaboration support to student needs. *International Journal of Artificial Intelligence in Education, 24*(1), 92–124. https://doi.org/10.1007/s40593-013-0012-6

Ahmad Uzir, N., Gašević, D., Matcha, W., Jovanović, J., & Pardo, A. (2020). Analytics of time management strategies in a flipped classroom. *Journal of Computer Assisted Learning, 36*(1), 70–88. https://doi.org/10.1111/jcal.12392

Alfano, M., Iurino, K., Stey, P., Robinson, B., Christen, M., Yu, F., & Lapsley, D. (2017). Development and validation of a multi-dimensional measure of intellectual humility. *PLOS ONE, 12*(8), 1–28. https://doi.org/10.1371/journal.pone.0182950

Anderson, J. R., Boyle, C. F., & Reiser, B. J. (1985). Intelligent tutoring systems. *Science, 228*(4698), 456–462. https://doi.org/10.1126/science.228.4698.456

Appleton, J. J., Christenson, S. L., Kim, D., & Reschly, A. L. (2006). Measuring cognitive and psychological engagement: Validation of the student engagement instrument. *Journal of School Psychology, 44*(5), 427–445. https://doi.org/10.1016/j.jsp.2006.04.002

Ashton, M. C. (2013). Chapter 2—Personality traits and the inventories that measure them. In M. C. Ashton (Ed.), *Individual differences and personality* (2nd ed., pp. 27–55). Academic. https://doi.org/10.1016/B978-0-12-416009-5.00002-5

Azevedo, R., Mudrick, N., Taub, M., & Wortha, F. (2017). Coupling between metacognition and emotions during STEM learning with advanced learning technologies: A critical analysis, implications for future research, and design of learning systems. *Teachers College Record, 119*(13), n13.

Barthakur, A., Kovanovic, V., Joksimovic, S., Zhang, Z., Richey, M., & Pardo, A. (2022). Measuring leadership development in workplace learning using an automated system. *British Journal of Educational Technology.*

Bergner, Y. (2017). Measurement and its uses in learning analytics. In C. Lang, G. Siemens, A. F. Wise, & D. Gaševic (Eds.), *The handbook of learning analytics* (1st ed., pp. 34–48). Society for Learning Analytics Research (SoLAR). http://solaresearch.org/hla-17/hla17-chapter1

Bitner, R. A., & Le, N.-T. (2021). Can EEG-devices differentiate attention values between incorrect and correct solutions for problem-solving tasks? *Journal of Information and Telecommunication, 0*(0), 1–20. https://doi.org/10.1080/24751839.2021.1950319

Bobo, L., & Licari, F. C. (1989). Education and political tolerance: Testing the effects of cognitive sophistication and target group affect. *Public Opinion Quarterly, 53*(3), 285–308. https://doi.org/10.1086/269154

Bollmer, G. (2017). Empathy machines. *Media International Australia, 165*(1), 63–76.

Bosch, N., & D'Mello, S. (2017). The affective experience of novice computer programmers. *International Journal of Artificial Intelligence in Education, 27*(1), 181–206.

Britain, G., Jain, A., Lupfer, N., Kerne, A., Perrine, A., Seo, J., & Sungkajun, A. (2020). Design is (A)Live: An environment integrating ideation and assessment. In *Extended Abstracts of the 2020 CHI conference on human factors in computing systems*, pp. 1–8. https://doi.org/10.1145/3334480.3382947

Brown, T. B., Mann, B., Ryder, N., Subbiah, M., Kaplan, J., Dhariwal, P., Neelakantan, A., Shyam, P., Sastry, G., Askell, A., et al. (2020). Language models are few-shot learners. *ArXiv Preprint ArXiv:2005.14165*.

Brutus, S., Aguinis, H., & Wassmer, U. (2013). Self-reported limitations and future directions in scholarly reports: Analysis and recommendations. *Journal of Management, 39*(1), 48–75. https://doi.org/10.1177/0149206312455245

Buckingham Shum, S., & Deakin Crick, R. (2016). Learning analytics for 21st century competencies. *Journal of Learning Analytics, 3*(2), 6–21. https://doi.org/10.18608/jla.2016.32.2

Burt, R. S. (2004). Structural holes and good ideas. *American Journal of Sociology, 110*(2), 349–399. 10.1.1.388.2251.

Chen, B., & Poquet, O. (2020). Socio-temporal dynamics in peer interaction events. In *Proceedings of the tenth international conference on learning analytics & knowledge*, pp. 203–208. https://doi.org/10.1145/3375462.3375535

Chikersal, P., Tomprou, M., Kim, Y. J., Woolley, A. W., & Dabbish, L. (2017). Deep structures of collaboration: Physiological correlates of collective intelligence and group satisfaction. In *Proceedings of the 2017 ACM conference on computer supported cooperative work and social computing*, pp. 873–888. https://doi.org/10.1145/2998181.2998250

Christensen, R., & Knezek, G. (2002). Advancing teachers through stages of adoption of technology in the classroom. In D. A. Willis, J. Price, & N. Davis (Eds.), *Proceedings of society for information technology & teacher education international conference 2002* (pp. 1983–1987). Association for the Advancement of Computing in Education (AACE). https://www.learntechlib.org/p/17561

Christensen, R., & Knezek, G. (2014). Comparative measures of grit, tenacity and perseverance. *International Journal of Learning, Teaching and Educational Research, 8*(1).

D'Mello, S., Dieterle, E., & Duckworth, A. (2017). Advanced, Analytic, Automated (AAA) measurement of engagement during learning. *Educational Psychologist, 52*(2), 104–123. https://doi.org/10.1080/00461520.2017.1281747

Dale, G., Cochrane, A., & Green, C. S. (2021). Individual difference predictors of learning and generalization in perceptual learning. *Attention, Perception, & Psychophysics, 83*(5), 2241–2255. https://doi.org/10.3758/s13414-021-02268-3

Dawson, S., & Siemens, G. (2014). Analytics to literacies: The development of a learning analytics framework for multiliteracies assessment. *International Review of Research in Open and Distance Learning, 15*(4), 284–305.

Dawson, S., Joksimović, S., Poquet, O., & Siemens, G. (2019, March 4). Increasing the impact of learning analytics. In *Proceedings of the international conference on learning analytics and knowledge*. https://doi.org/10.1145/3303772.3303784

de Barba, P. G., Malekian, D., Oliveira, E. A., Bailey, J., Ryan, T., & Kennedy, G. (2020). The importance and meaning of session behaviour in a MOOC. *Computers & Education, 146*, 103772. https://doi.org/10.1016/j.compedu.2019.103772

de Laat, M., Joksimovic, S., & Ifenthaler, D. (2020). Artificial intelligence, real-time feedback and workplace learning analytics to support in situ complex problem-solving: A commentary. *The International Journal of Information and Learning Technology*, in-print. https://doi.org/10.1108/IJILT-03-2020-0026

De Raad, B. (2000). *The big five personality factors: The psycholexical approach to personality* (pp. vii, 128). Hogrefe & Huber Publishers.

Devlin, J., Chang, M.-W., Lee, K., & Toutanova, K. (2018). Bert: Pre-training of deep bidirectional transformers for language understanding. *ArXiv Preprint ArXiv:1810.04805*.

Dillenbourg, P., Baker, M., Blaye, A., & O'Malley, C. (1996). The evolution of research on collaborative learning. In *Learning in humans and machine: Towards an interdisciplinary learning science* (pp. 189–211).

Dowell, N., Skrypnyk, O., Joksimović, S., Graesser, A. C., Dawson, S., Gašević, D., de Vries, P., Hennis, T., & Kovanović, V. (2015, June 26). Modeling learners' social centrality and performance through language and discourse. In *Proceedings of the 8th international conference on educational data mining*.

Dowell, N. M., Nixon, T. M., & Graesser, A. C. (2019). Group communication analysis: A computational linguistics approach for detecting sociocognitive roles in multiparty interactions. *Behavior Research Methods, 51*(3), 1007–1041.

Drachsler, H., & Goldhammer, F. (2020). Learning analytics and eAssessment—Towards computational psychometrics by combining psychometrics with learning analytics. In D. Burgos (Ed.), *Radical solutions and learning analytics: Personalised learning and teaching through big data* (pp. 67–80). Springer. https://doi.org/10.1007/978-981-15-4526-9_5

Efklides, A., Schwartz, B. L., & Brown, V. (2018). *Motivation and affect in self-regulated learning: Does metacognition play a role?*

Gabriel, F., Buckley, S., & Barthakur, A. (2020). The impact of mathematics anxiety on self-regulated learning and mathematical literacy. *Australian Journal of Education, 64*(3), 227–242. https://doi.org/10.1177/0004944120947881

Gal, L., Hershkovitz, A., Morán, A. E., Guenaga, M., & Garaizar, P. (2017). Suggesting a log-based creativity measurement for online programming learning environment. In *Proceedings of the fourth (2017) ACM conference on learning @ scale*, pp. 273–277. https://doi.org/10.1145/3051457.3054003

Gamez, D. (2014). The measurement of consciousness: A framework for the scientific study of consciousness. *Frontiers in Psychology, 5*, 714. https://doi.org/10.3389/fpsyg.2014.00714

Gašević, D., Dawson, S., & Siemens, G. (2014). Let's not forget: Learning analytics are about learning. *TechTrends, 59*(1), 64–71. https://doi.org/10.1007/s11528-014-0822-x

Gašević, D., Dawson, S., Rogers, T., & Gašević, D. (2016). Learning analytics should not promote one size fits all: The effects of instructional conditions in predicting academic success. *The Internet and Higher Education, 28*, 68–84. https://doi.org/10.1016/j.iheduc.2015.10.002

Gašević, D., Kovanović, V., & Joksimović, S. (2017). Piecing the learning analytics puzzle: A consolidated model of a field of research and practice. *Learning: Research and Practice, 3*(1), 63–78. https://doi.org/10.1080/23735082.2017.1286142

Gerdes, K. E., Segal, E. A., & Lietz, C. A. (2010). Conceptualising and measuring empathy. *British Journal of Social Work, 40*(7), 2326–2343.

Gibson, D. C. (2018). Unobtrusive observation of team learning attributes in digital learning. *Frontiers in Psychology, 9*, 834. https://doi.org/10.3389/fpsyg.2018.00834

Gibson, A., Kitto, K., & Bruza, P. (2016). Towards the discovery of learner metacognition from reflective writing. *Journal of Learning Analytics, 3*(2), 22–36. https://doi.org/10.18608/jla.2016.32.3

Goldberg, L. R. (1992). The development of markers for the Big-Five factor structure. *Psychological Assessment, 4*(1), 26.

Goodyear, P., & Carvalho, L. (2014). Introduction: Networked learning and learning networks. In L. Carvalho & P. Goodyear (Eds.), *The architecture of productive learning networks* (pp. 3–22). Routledge.

Griffin, P., & Care, E. (2015). *Assessment and teaching of 21st century skills. Methods and approach*. Springer.

Gross, J. J. (2015). Emotion regulation: Current status and future prospects. *Psychological Inquiry, 26*(1), 1–26.

Hannay, J. E., Arisholm, E., Engvik, H., & Sjoberg, D. I. K. (2010). Effects of personality on pair programming. *IEEE Transactions on Software Engineering, 36*(1), 61–80. https://doi.org/10.1109/TSE.2009.41

Hernández-García, Á., Acquila-Natale, E., Chaparro-Peláez, J., & Conde, M. Á. (2018). Predicting teamwork group assessment using log data-based learning analytics. *Computers in Human Behavior, 89*, 373–384. https://doi.org/10.1016/j.chb.2018.07.016

Hjerm, M., Eger, M. A., Bohman, A., & Fors Connolly, F. (2020). A new approach to the study of tolerance: Conceptualizing and measuring acceptance, respect, and appreciation of difference. *Social Indicators Research, 147*(3), 897–919. https://doi.org/10.1007/s11205-019-02176-y

Hu, Y., Wu, B., & Gu, X. (2017). Learning analysis of K-12 students' online problem solving: A three-stage assessment approach. *Interactive Learning Environments, 25*(2), 262–279.

Janssen, D., Tummel, C., Richert, A., & Isenhardt, I. (2016). *Towards measuring user experience, activation and task performance in immersive virtual learning environments for students*. ILRN.

Joksimović, S., Dowell, N., Poquet, O., Kovanović, V., Gašević, D., Dawson, S., & Graesser, A. C. (2018a). Exploring development of social capital in a CMOOC through language and discourse. *The Internet and Higher Education, 36*, 54–64. https://doi.org/10.1016/j.iheduc.2017.09.004

Joksimović, S., Poquet, O., Kovanović, V., Dowell, N., Mills, C., Gašević, D., Dawson, S., Graesser, A. C., & Brooks, C. (2018b). How do we model learning at scale? A systematic review of research on MOOCs. *Review of Educational Research, 88*(1), 43–86. https://doi.org/10.3102/0034654317740335

Joksimović, S., Kovanović, V., & Dawson, S. (2019). The journey of learning analytics. *HERDSA Review of Higher Education, 6*, 27–63.

Joksimović, S., Siemens, G., Wang, Y. E., San Pedro, M. O. Z., & Way, J. (2020). Beyond cognitive ability. *Journal of Learning Analytics, 7*(1), 1–4.

Jones, C. (2015). *Networked learning: An educational paradigm for the age of digital networks*. Springer. https://books.google.co.uk/books?id=NKyRngEACAAJ

Jovanović, J., Gašević, D., Pardo, A., Dawson, S., & Whitelock-Wainwright, A. (2019). Introducing meaning to clicks: Towards traced-measures of self-efficacy and cognitive load. In *Proceedings of the 9th international conference on learning analytics & knowledge*, pp. 511–520. https://doi.org/10.1145/3303772.3303782

Jovanovic, J., Mirriahi, N., Gašević, D., Dawson, S., & Pardo, A. (2019). Predictive power of regularity of pre-class activities in a flipped classroom. *Computers & Education, 134*, 156–168. https://doi.org/10.1016/j.compedu.2019.02.011

Kinnebrew, J., Mack, D., & Biswas, G. (2013). Mining temporally-interesting learning behavior patterns. *Educational Data Mining 2013*.

Knight, S., Buckingham Shum, S., & Littleton, K. (2013). Epistemology, pedagogy, assessment and learning analytics. In *ACM international conference proceeding series*, pp. 75–84. https://doi.org/10.1145/2460296.2460312

Kovanović, V., Joksimović, S., Gašević, D., Hatala, M., & Siemens, G. (2017). Content analytics: The definition, scope, and an overview of published research. In C. Lang, G. Siemens, A. F. Wise, & D. Gaševic (Eds.), *The handbook of learning analytics* (1st ed., pp. 77–92). Society for Learning Analytics Research (SoLAR). http://solaresearch.org/hla-17/hla17-chapter1

Lacerenza, C. N., Marlow, S. L., Tannenbaum, S. I., & Salas, E. (2018). Team development interventions: Evidence-based approaches for improving teamwork. *The American Psychologist, 73*(4), 517–531. https://doi.org/10.1037/amp0000295

Lajoie, S. P., Pekrun, R., Azevedo, R., & Leighton, J. P. (2019). Understanding and measuring emotions in technology-rich learning environments. *Learning and Instruction, 101272*. https://doi.org/10.1016/j.learninstruc.2019.101272

Lancieri, L. (2017). *Asymmetry in the perception of friendship in students groups*. International Association for Development of the Information Society.

Larmuseau, C., Cornelis, J., Lancieri, L., Desmet, P., & Depaepe, F. (2020). Multimodal learning analytics to investigate cognitive load during online problem solving. *British Journal of Educational Technology, 51*(5), 1548–1562. https://doi.org/10.1111/bjet.12958

Lazer, D., Pentland, A., Adamic, L., Aral, S., Barabási, A.-L., Brewer, D., Christakis, N., Contractor, N., Fowler, J., Gutmann, M., Jebara, T., King, G., Macy, M., Roy, D., & Van Alstyne, M. (2009). Computational social science, *Science, 323*(5915), 721–723. https://doi.org/10.1126/science.1167742

Lee, A. V. Y., & Tan, S. C. (2017). *Promising ideas for collective advancement of communal knowledge using temporal analytics and cluster analysis*.

Lockyer, L., Heathcote, E., & Dawson, S. (2013). Informing pedagogical action: aligning learning analytics with learning design. *American Behavioral Scientist, 57*(10), 1439–1459. https://doi.org/10.1177/0002764213479367

Lord, M. (2015). Group learning capacity: The roles of open-mindedness and shared vision. *Frontiers in Psychology, 6*, 150. https://doi.org/10.3389/fpsyg.2015.00150

Madsen, M., & Gregor, S. (2000). Measuring human-computer trust. *11th Australasian Conference on Information Systems*, *53*, 6–8.

Mangaroska, K., & Giannakos, M. (2019). Learning analytics for learning design: A systematic literature review of analytics-driven design to enhance learning. *IEEE Transactions on Learning Technologies, 12*(4), 516–534. https://doi.org/10.1109/TLT.2018.2868673

Martin, C. K., Nacu, D., & Pinkard, N. (2016). Revealing opportunities for 21st century learning: An approach to interpreting user trace log data. *Journal of Learning Analytics*, *3*(2), 37–87. https://doi.org/10.18608/jla.2016.32.4

Martinez-Maldonado, R., Gašević, D., Echeverria, V., Fernandez Nieto, G., Swiecki, Z., & Buckingham Shum, S. (2021). What do you mean by collaboration analytics? A conceptual model. *Journal of Learning Analytics*, *8*(1), 126–153. https://doi.org/10.18608/jla.2021.7227

Matcha, W., Gašević, D., Jovanović, J., Pardo, A., Maldonado-Mahauad, J., Pérez-Sanagustín, M., et al.. (2019). Detection of learning strategies: A comparison of process, sequence and network analytic approaches. In *European conference on technology enhanced learning*, pp. 525–540.

Matteson, M. L., Anderson, L., & Boyden, C. (2016). 'Soft skills': A phrase in search of meaning. *Portal: Libraries and the Academy, 16*(1), 71–88.

Miele, D. B., & Scholer, A. A. (2018). The role of metamotivational monitoring in motivation regulation. *Educational Psychologist, 53*(1), 1–21.

Milligan, S. (2020). Standards for developing assessments of learning using process data. In M. Bearman, P. Dawson, R. Ajjawi, J. Tai, & D. Boud (Eds.), *Re-imagining university assessment in a digital world* (pp. 179–192). Springer. https://doi.org/10.1007/978-3-030-41956-1_13

Milligan, S., & Griffin, P. (2016). Understanding learning and learning design in MOOCs: A measurement-based interpretation. *Journal of Learning Analytics*, *3*(2), 88–115. https://doi.org/10.18608/jla.2016.32.5

Mislevy, R. J. (2019). Advances in measurement and cognition. *The ANNALS of the American Academy of Political and Social Science, 683*(1), 164–182.

OECD. (2017). *Social and emotional skills: Well-being, connectedness and success*. OECD.

Pardo, A., & Siemens, G. (2014). Ethical and privacy principles for learning analytics. *British Journal of Educational Technology, 45*(3), 438–450. https://doi.org/10.1111/bjet.12152

Paul, R., & Elder, L. (1990). *Critical thinking*. Sonoma State University.

Pearsall, M. J., & Ellis, A. P. J. (2006). The effects of critical team member assertiveness on team performance and satisfaction. *Journal of Management, 32*(4), 575–594. https://doi.org/10.1177/0149206306289099

Poquet, O., & de Laat, M. (2021). Developing capabilities: Lifelong learning in the age of AI. *British Journal of Educational Technology, 52*(4), 1695–1708. https://doi.org/10.1111/bjet.13123

Porter, T., Molina, D. C., Blackwell, L., Roberts, S., Quirk, A., Duckworth, A. L., & Trzesniewski, K. (2020). Measuring mastery behaviors at scale: The persistence, effort, resilience and challenge-seeking task (PERC). *Journal of Learning Analytics, 7*(1), 5–18.

Raad, B. D. E., & Perugini, M. E. (2002). *Big five factor assessment: Introduction*. Hogrefe & Huber Publishers.

Ramakrishnan, A., Pardes, A., Lynch, W., Molaro, C., & Platt, M. L. (2019). A machine learning approach to identifying objective biomarkers of anxiety and stress. *BioRxiv*. https://doi.org/10.1101/745315

Rammstedt, B., Danner, D., & Lechner, C. (2017). Personality, competencies, and life outcomes: Results from the German PIAAC longitudinal study. *Large-Scale Assessments in Education, 5*(1), 2. https://doi.org/10.1186/s40536-017-0035-9

Renninger, K. A., & Hidi, S. (2019). *Interest development and learning*.

Saleeb, N. (2021). Closing the chasm between virtual and physical delivery for innovative learning spaces using learning analytics. *The International Journal of Information and Learning Technology, 38*(2), 209–229. https://doi.org/10.1108/IJILT-05-2020-0086

Schneider, B., Dowell, N., & Thompson, K. (2021). Collaboration analytics—Current state and potential futures. *Journal of Learning Analytics, 8*(1), 1–12. https://doi.org/10.18608/jla.2021.7447

Segedy, J. R., Kinnebrew, J. S., & Biswas, G. (2015). Using coherence analysis to characterize self-regulated learning behaviours in open-ended learning environments. *Journal of Learning Analytics, 2*(1), 13–48.

Shaffer, D. W. (2018). Epistemic network analysis. In *International handbook of the learning sciences* (pp. 520–531). https://doi.org/10.4324/9781315617572-50

Skrypnyk, O., Joksimović, S., Kovanović, V., Gašević, D., & Dawson, S. (2015). Roles of course facilitators, learners, and technology in the flow of information of a CMOOC. *International Review of Research in Open and Distance Learning, 16*(3), 188–217.

Smets, E., Rios Velazquez, E., Schiavone, G., Chakroun, I., D'Hondt, E., De Raedt, W., Cornelis, J., Janssens, O., Van Hoecke, S., Claes, S., Van Diest, I., & Van Hoof, C. (2018). Large-scale wearable data reveal digital phenotypes for daily-life stress detection. *Npj Digital Medicine, 1*(1), 67. https://doi.org/10.1038/s41746-018-0074-9

Staib, S. (2003). Teaching and measuring critical thinking. *Journal of Nursing Education, 42*(11), 498–508.

Stankov, L., Morony, S., & Lee, Y. P. (2014). Confidence: The best non-cognitive predictor of academic achievement? *Educational Psychology, 34*(1), 9–28.

Stanovich, K. E., & West, R. F. (2007). Natural myside bias is independent of cognitive ability. *Thinking & Reasoning, 13*(3), 225–247. https://doi.org/10.1080/13546780600780796

Sztejnberg, A., & Jasiński, T. L. (2014). Measurement of the tolerance general level in the higher education students. *International Journal of Latest Research in Humanities and Social Science (IJLRHSS), 1*(4), 01–07.

Tarmazdi, H., Vivian, R., Szabo, C., Falkner, K., & Falkner, N. (2015, July). *Using learning analytics to visualise computer science teamwork*. https://doi.org/10.1145/2729094.2742613

van Horik, J. O., & Madden, J. R. (2016). A problem with problem solving: Motivational traits, but not cognition, predict success on novel operant foraging tasks. *Animal Behaviour, 114*, 189–198.

Van Laer, S., & Elen, J. (2018). Towards a methodological framework for sequence analysis in the field of self-regulated learning. *Frontline Learning Research, 6*(3), 228–249.

Veenman, M. V., Van Hout-Wolters, B. H., & Afflerbach, P. (2006). Metacognition and learning: Conceptual and methodological considerations. *Metacognition and Learning, 1*(1), 3–14.

Wagstaff, M. F., Flores, G. L., Ahmed, R., & Villanueva, S. (2021). Measures of curiosity: A literature review. *Human Resource Development Quarterly, 32*(3), 363–389. https://doi.org/10.1002/hrdq.21417

Xing, W., Li, C., Chen, G., Huang, X., Chao, J., Massicotte, J., & Xie, C. (2021). Automatic assessment of students' engineering design performance using a Bayesian network model. *Journal of Educational Computing Research, 59*(2), 230–256. https://doi.org/10.1177/0735633120960422

Zhou, M., & Winne, P. H. (2012). Modeling academic achievement by self-reported versus traced goal orientation. *Learning and Instruction, 22*(6), 413–419.

Part I
Key SEL Attributes

Chapter 4
Empathy: How Can Technology Help Foster Its Increase Rather Than Decline in the 21st Century?

Gerald Knezek, Rhonda Christensen, and David Gibson

Abstract Empathy (a caring concern for the thoughts and feelings of others) has been a focus of studies at the national and international level since the late 1980s when Japan delayed introduction of microcomputers into elementary schools in part due to concerns that technology would turn young children into "non-thinking machines." (Children & Computers in Schools, 1996). In the early 1990s, three years of study by the USA authors of the current chapter, in conjunction with colleagues in Japan and Mexico, found no tangible evidence of those specific concerns at the early primary school level, but did uncover evidence of a construct they labeled Computer Seclusion emerging at the middle school level (Computers in the Schools, 1996). Since those early studies, many other trends based on the original IEA items customized by Japan have emerged. Among these are the extensive gender gap regarding empathy that is now known to exist for young learners from the first grade through completion of secondary education. Another is that empathy is positively related to many other characteristics, such as self-concept, study habits, and creative tendencies. In this chapter findings from empathy data gathered from 5000 middle school students in 2009, compared with previous studies using the same item set and more recent findings in the literature, will be used to present implications for best practices regarding the current strong interest in empathy as a contributor to twenty-First century skills that appear to be declining among today's youth in our society. Conjectured reasons for these declines are also included in the chapter, as well as coaching historical findings in the context of the newly-emerging importance of social emotional intelligence.

Keywords Empathy · Measures of empathy · Empathy norms · Educational technology

G. Knezek (✉) · R. Christensen
Learning Technologies, University of North Texas, Denton, TX, USA
e-mail: Rhonda.christensen@unt.edu

D. Gibson
UNESCO Chair for Data Sciences, Curtin University, Perth, Australia

Y. E. Wang et al. (eds.), *Social and Emotional Learning and Complex Skills Assessment*, Advances in Analytics for Learning and Teaching,
https://doi.org/10.1007/978-3-031-06333-6_4

Empathy has been defined as the ability to sense other people's emotions, coupled with the ability to imagine what someone else might be thinking or feeling and compassionately take appropriate action (Goleman, 2008). Empathy is a social phenomenon that requires us to consider capacities of both social and emotional intelligence. It has been a focus of studies at the national and international level since the late 1980s and is believed to be impacted by more than social and emotional intelligence, including the use of technology as well as technology-mediated communication. Newer constructs of empathy, for example, include virtual empathy (Rosen et al., 2012) and digital empathy which is defined as "care and concern for others expressed through computer-mediated communications" (Terry & Cain, 2016, p. 1).

For many nations, concerns about empathetic side effects of information technology have emerged at the national policy level. For example, Japan delayed the introduction of microcomputers into elementary schools in part due to concerns that technology would turn young children into "non-thinking machines" (Collis et al., 1996). In the early 1990s, 3 years of study in conjunction with colleagues in Japan and Mexico found no tangible evidence of those specific concerns at the early primary school level. Since those early studies, many other trends based on the original International Association for the Evaluation of Educational Achievement (IEA) items have emerged. Among these trends are the extensive gender gap regarding empathy as well as a relationship to other characteristics such as self-concept, study habits, and creative tendencies.

The study of empathy and its relationship to technology is not new. However, with the ubiquitous nature of the use of technology by students and new approaches in learning analytics, the use of sophisticated means to collect and represent capacities of empathy can now go beyond self-reports to include unobtrusive observation of behavior in naturalistic settings. Since more recently empathy has been elevated to the role of a desirable outcome measure (Knezek & Christensen, 2018) that can be incorporated into learning analytics systems, much research is needed to see how such data might be included in learning analytics systems of the future. Accuracy and privacy, for example, are important matters to be carefully considered. Is it ethical to have socio-emotional characteristics given to an entity without the knowledge of where that data will be used? These are important considerations in the context of how learning analytics systems can best capture and represent individual learner empathy. This theme will be addressed from several perspectives.

The chapter will include findings from two decades of empathy studies, focusing on the relationship of empathy to technology and exploring how learning analytics may be applied to provide feedback to help improve empathy in students and teachers. The goal of the chapter is to propose research-informed ideas about ways in which learning analytics systems can build on quality measures to capture and represent individual learner empathy, thereby enabling learning analytics systems to better support the fostering of empathy in learners in the future.

4.1 Literature Review

Empathy is a construct representing people's compassionate awareness, knowledge, attitudes and behaviors towards others (Ekman, 2003; Goleman, 2008); a lack of empathy is associated with more negative attitudes and behaviors (Belman & Flanagan, 2010). The construct is a complex entity that appears to have at least three major components: (1) a cognitive mechanism (awareness, recognition, understanding), (2) an affective component that allows the matching of emotions with other people and (3) a compassionate component or behavioral response that is part of the self and social management of emotions (Ekman, 2003; Goleman, 2008; Powell & Roberts, 2017). Observable social behaviors may indicate one's capacity for empathy, in particular, behaviors that indicate cognitive awareness of the emotional states of others and social skills that indicate compassionate empathy toward others.

How do the different types of empathy relate to the uses of technology? Cognitive empathy may in fact be enhanced by technology. However, affective and compassionate empathy, feeling what another person feels and taking action, may be hampered by many of the social media tools common in the twenty-first century, and lead to byproducts already emerging, such as bullying via social media.

Large multidimensional data sets from digital learning tools (e.g. digital games, simulations, communication tools, social media) combined with self-report measures may be useful in understanding an individual's empathy status, capabilities and propensities as well as the causes and remedies of the lack of empathy. Feedback made possible by digital learning tools can inform both teachers and students before, during and after engagements with technology-enhanced learning experiences and point the way toward improvements for learning (Ifenthaler et al., 2017). In addition, given the ubiquitous nature of technology in every facet of our lives, understanding the relationship of technology and empathy is important for understanding social and cultural challenges facing society.

4.1.1 Relationship of Technology to Empathy

Is technology changing the way we empathize with others? According to Turkle (2016), technology affects how we communicate with others and therefore impacts our relationships. Technology may impede the ability to read facial expressions and identify emotions (Uhls et al., 2014). Research studies have suggested that empathy is essential in the development of resilience, ability to adapt to change and the recognition and appreciation of the perspectives of others (Gordon, 2012). The healthcare industry has understood the importance of empathy in patient-provider relationships for decades (Terry & Cain, 2016). Research in the field of medicine has indicated that the expression of empathy is reduced in digital telemedicine encounters (Terry & Cain, 2016), leading the authors to believe the expression of empathy is also likely to be reduced in other digital encounters.

Many researchers and practitioners in the field of education have noted a decline in empathy over the last decade that could be related to the rise in personal technology use. For example, Konrath et al. (2011) have speculated that technology has reduced our ability to be empathetic to others. In a study of empathy between 1979 and 2009 with more than 13,700 university students, researchers found major decreases in empathy over that time. Their hypothesis based on the longitudinal data was that one likely contributor to the declining empathy was the rising prominence of personal technology and media use in everyday life (Konrath et al., 2011). In that context, Shaffer identified the rise of multi-subculturalism (Shaffer, 2005), an increasing stance of isolation into smaller and smaller subcultures made possible by global communication technologies and practices (e.g. blocking others, recommendation systems). Ribble and Miller (2013) conveyed the concern that the misuse of technology is related to the lack of empathy being displayed by students. These authors went on to state that other factors, such as the lack of emotional connection on social networks, increased violent media exposure, and the lack of face-to-face interaction that technology provides also contribute to lower empathy (Ribble & Miller, 2013). If increasing isolation is partially driving a decrease in empathy, as the data suggests, then educators might want to consider how to introduce learning opportunities for modeling and reflecting on empathetic responses that re-build the bridges from self to others. Learning analytics, if created in near real-time and embedded into learning experiences, has the potential to provide information and feedback for the journey toward increased empathy.

The relationship between technology use and empathy appears to not be simple. Rosen et al. (2012) termed a phrase "virtual empathy" to assess the empathetic responses via virtual environments from their findings that virtual empathy has real impacts on relationships especially in the areas of social networking and instant messaging. Those who were able to practice virtual empathy actually increased their ability to express empathy face-to-face (Rosen et al., 2012). Perhaps not so surprising, playing violent video games was the only form of media use that was directly related to real-world empathy in a negative manner (Spradlin et al., 2012). The same researchers (Carrier et al., 2015) noted that real-world empathy is six times more strongly related to feelings than virtual empathy, which suggests that as immersion and emotional valence are increased in digital experiences, the impacts on empathy may be increased. The example provided was that "a person would need five to six supportive comments on a social network to get the same feeling as receiving a hug from a friend to feel better" (Carrier et al., 2015, p. 46).

There appears to be a relationship between online communication tools and empathy. Collins (2014) found a positive correlation between empathic concern and using chat, time spent using and connecting on Facebook similar to a study that found a positive relationship between conversing with others online and empathic communication (Ivcevic & Ambady, 2012). Carrier et al. (2015) reported a positive impact on empathy from online chatting, video chatting and social networking that led to face-to-face interactions. The Pew Internet and American Life Project survey stated that 48% of teens who participate in online communication feel that it strengthens their relationships with peers (Madden & Rainie, 2003). Konrath and colleagues (2011) found that those who were better able to express virtual empathy were also better able to express real-world empathy and that practicing virtual empathy actually led to people being better at expressing empathy face to face. However, Carrier et al. (2015) found that increased time spent in video gaming online was associated with lower real-world empathy in males and females, and suggested that perhaps the lack of nonverbal cues in the online world contributed to findings of lower virtual empathy compared to real-word empathy among an anonymous sample of more than 1000 young adults.

Certain uses of technology may aid in improving empathy in some students. According to Belman and Flanagan (2010), digital games are well-suited to support educational programs that foster empathy because games allow players to inhabit the roles and perspectives of other people in a uniquely, immersive way. A digital simulation study of high school students showed that those who played the simulation game expressed more global empathy and greater interest in learning about other countries (Bachen et al., 2012).

Contrary to these findings, there are also negative impacts on empathy from the use of technology depending on how much and in what way it is used. Goleman (2008) indicated that technology makes disconnecting from others easier to do. Carrier et al. (2015) found going online decreases empathy, specifically cognitive empathy. These researchers speculated the decline may be due to the lack of nonverbal cues in most computer-mediated communication (Carrier et al., 2015). The increasing practicality of learning context data acquisition through systems such as learning analytics, offers hope that a deeper understanding of the relationship between technology and empathy may someday emerge.

One learning analytics approach to studying the relationship of technology to empathy is to intensely localize a particular digital experience. For example, the digital traces might be studied from a digital game or simulation designed to elicit cognitive, emotional or compassionate empathetic interactions. Inherent in the design of the experience is the assumption that someone could think, feel and behave from the perspective of a role in the setting. The data from the enactment would then supply streams of information for the process of constructing classifiers and testing their validity within a model of the user, the task and the evidence (Mislevy et al., 1999). Another learning analytics approach would start with existing sources of data that are best if fine grained and from multiple sources, and are time-stamped for coordination with the production of evidence of the thinking, feeling and action that is expected in a given situation. In either approach, a variety of analytic methods might make use of network theory (e.g. Epistemic Network Analysis, Shaffer et al., 2009), or concepts from the physical properties of networks such as entropy-as-an-indicator of order (Feng & Kirkley, 2020). It is also possible that machine learning in conjunction with any number of statistical methods can, in what might be considered a rigorous new conception of mixed methods research, bridge not only qualitative with quantitative, but also computational and theoretical modeling (Gibson & Ifenthaler, 2017).

4.1.2 Gender and Empathy

There has been a consistent pattern in the research literature showing that females report higher empathy than males (Christov-Moore et al., 2014; Eisenberg & Lennon, 1983; Lennon & Eisenberg, 1987). Cassels et al. (2010) reported their gender findings in the context of reconfirming that females were higher ($p < .001$) than males in two measures of affective empathy, *personal distress* and *empathetic concern*. Carrier et al. (2015) conducted a study with over 1700 members from the "Net Generation" (those born after 1980) comparing real-world empathy to virtual empathy. Their findings regarding gender included similarities for males and females. These researchers found that while cognitive empathy scores were higher than affective empathy scores, cognitive empathy scores were lowered more in the online world than were affective empathy scores (Carrier et al., 2015). Perhaps it is the type of technology used that impacts empathy differently in males and females. Learning analytics applied to large sets of data can help answer these questions.

4.1.3 Impacts in the Classroom

Empathy of both teachers and students impacts learning in the classroom. More than two decades ago Lewis et al. (1996) reported that "students work harder, achieve more, and attribute more importance to schoolwork in classes in which they feel liked, accepted, and respected by the teacher and fellow students" (p. 20). They summarized their findings by reporting that their project "… has shown that when kids care about one another—and are motivated by important, challenging work—they're more apt to care about learning" (p. 16). Mendes (2003) described steps that teachers can take to let students know that teachers empathize with student crises, and care about individual student learning.

In the twenty-first century empirical research is emerging that socio-emotional attributes such as empathy can be transmitted from teachers to their students. For example, one study conducted by Christensen (2002) used a time-lag regression technique called panel analysis to demonstrate that, over the course of a school year, positive teacher attitudes tend to imbue positive attitudes in their students. Thus, both modeling empathy for students and teaching empathy to students – so that students tend to possess a caring concern for the thoughts and feelings of others – can be envisioned to contribute to desirable learner outcomes. Several studies completed since the 1990s are relevant to this topic.

Empathetic teachers create classroom environments that impact their students' lives, as well as learning, because they understand the social nature of teaching effectively (Darling-Hammond, 2000; McAllister & Irvine, 2002). As the divergence between the economic and ethnic background of teachers and students grows, the need increases for teachers to understand and empathize with their students. Warner (1984) emphasizes the importance of teachers' ability to empathize with their students and believes empathy can be taught to teachers. It is important for teacher education programs to focus on the need to educate teachers in understanding the perspectives of diverse learners (Darling-Hammond, 2000) as well as how to instill these traits in their students.

Teaching empathy in Denmark has been compulsory since 1993 for 1 h each week for ages 6–16 (Morning Future, 2019). Many believe the required empathy focus contributes to the high level of happiness in the country (Helliwell et al., 2018). In the U.S., the *Making Caring Common Project* at Harvard's Graduate School of Education has created recommendations for educators in building empathy in their school communities (Jones et al., 2020). These recommendations include five steps: (1) Model empathy, (2) Explain and give importance for empathy, (3) Practice empathy, (4) Set clear ethical expectations, and (5) Make school culture a priority (Jones et al., 2020).

Near-real time learning analytics may have a role to play, along with validated report-measures, in providing data for timely, accurate feedback contextualized as closely as possible to one's responses in settings that elicit empathetic emotions and compassionate care for others. Such uses of learning analytics methods have begun to accumulate research on digital games, simulation-based learning and the application of machine learning to large data sets created by social platforms, learning management systems and other digital learning platforms (Sin & Muthu, 2015; Tlili & Chang, 2019). We next review some key validated measures and suggest ways of developing indicators and classifiers for learning analytics.

The unobtrusive collection of data through simulations and collaborative learning has potential for the application of learning analytics to provide useful feedback to users regarding empathy as well as assess the effectiveness of possible empathy curricula introduced to students. A framework has been introduced with the complementary methods of social network analysis (SNA) and epistemic network analysis (ENA) which illustrates how indicators of collaborative learning emerge from social ties and content analysis of discourse when modeled as overlapping networks (Shaffer & Ruis, 2015).

4.2 Measures of Empathy

In this section we review several validated measures and make suggestions about possible learning analytics situations and approaches that may 1 day expand the types of indicators of empathy that can be studied rigorously with both machine learning and psychometrics. We begin with a discussion of types of empathy, then describe several validated instruments measuring empathy constructs and end with a description of how machine-based learning could be used to produced indicators of empathy based on data within a learning analytics system.

There are at least two dimensions of empathy, namely understanding the emotions of others (cognitive empathy) and actually experiencing their emotional states (affective empathy) (Zych et al., 2019, p. 88). There is also possibly a third

Table 4.1 Examples of empathy measures

Scale	Source	Type of empathy
Interpersonal Reactivity Index (IRI)	Davis (1980, 1983) and Cassels et al. (2010)	Cognitive and affective
Basic Empathy Scale	Jolliffe and Farrington (2006)	Cognitive and affective
Computer Attitude Questionnaire Empathy Scale	Christensen and Knezek (2001, 2008) and Knezek and Christensen (1996)	Affective
Compassion Scale of Dispositional Positive Emotion Scales (DPES)	Shiota et al. (2006)	Compassionate

dimension of empathy, called compassionate empathy (Goleman, 2008), that is related to willingness to take appropriate action based on experiencing an emotional empathetic state. As shown in Table 4.1, this compassionate dimension of empathy seems to be missing in the literature of both psychometrics and learning analytics.

4.2.1 Indicators of Empathy Based on Traditional Psychometrics

Self-report measures for non-cognitive constructs such as empathy are generally developed according to a four-stage process in classical psychometrics (DeVellis, 2011). The first stage is that items are selected or written based on an accepted blueprint for a construct of interest. In stage 2, pilot-test survey items are then administered to a representative sample of the target audience and checked for alignment with each other to ensure acceptable internal consistency reliability (Cronbach's alpha) for a scale to be produced by summing or averaging across these items in the future. In the third stage, data collected from a large survey sample are subjected to factor analysis or another dimension confirmation technique such as multidimensional scaling (Dunn-Rankin et al., 2004) to ensure that responses from the survey participants can be translated into independent, reliable scales. In the fourth stage, the validated instrument is administered to a group of participants and scale scores are produced to represent the extent of possession of an attribute, such as empathy, by each individual. The goal of this process is to find items that indirectly infer characteristics of a person that are not directly observable. The theoretical basis is that the level of agreement with items on a scale representing a construct are caused by the degree of possession of a construct (such as empathy) by the individual.

4.2.2 Examples of Validated Empathy Instruments

One instrument that has been used in studies of empathy is *The Interpersonal Reactivity Index (IRI)* (Davis, 1980), with a description reported by Cassels et al. (2010) as a 28-item index used to assess both cognitive and affective components of dispositional empathy. It is a self-report measure of empathy and has been shown to be valid (Davis, 1980, 1983). The IRI includes four different areas of empathy, two of which are consistent with indictors of affective empathy referenced in this chapter:

1. Personal distress: the tendency to experience distress and/or discomfort when witnessing another person's distress (e.g., "Being in a tense emotional situation scares me.").
2. Empathic concern: the tendency to feel sympathy and/or concern for others in negative situations (e.g., "I often have tender, concerned feelings for people less fortunate than me.)

(Cassels et al., 2010, p. 314)

Many studies of empathy have been based on measures of general empathy specified in the *Basic Empathy Scale* by Jolliffe and Farrington (2006) (Zych et al., 2019). Examples of affective empathy items on the Basic Empathy Scale include "After being with a friend who is sad about something, I usually feel sad" and "I tend to feel scared when I am with friends who are afraid." Sample cognitive empathy items include "When someone is feeling 'down' I can usually understand how they feel" and "I can usually work out when people are cheerful" (Jolliffe & Farrington, 2006). The initial survey included 40 items and was administered to 363 adolescents and exploratory factor analysis revealed that two constructs could be measured using 20 items (Jolliffe & Farrington, 2006). The 20-item version was then administered to a separate group of 357 adolescents from the same schools and confirmatory factor analysis validated the two-factor solution. Scale scores produced for this second group were used to produce a large number of findings, including the following (Jolliffe & Farrington, 2006):

(a) females were found to be higher than males on both affective and cognitive empathy;
(b) empathy was positively related to parental supervision and socioeconomic status;
(c) empathy was positively correlated with the personality trait of openness, for both males and females;
(d) adolescents who would help victims of bullying had greater empathy;
(e) cognitive empathy was positively associated with the personality trait of extraversion;
(f) affective empathy was positively associated with personality trait of neuroticism;
(g) for females, empathy was positively correlated with intelligence;
(h) for males, empathy was positively correlated with the personality traits of agreeableness and conscientiousness.

Cognitive empathy items of the *Basic Empathy Scale* focus on recognition, awareness, and understanding of feelings in others while affective empathy items identify the respondent as being involved with (caught up in) the feelings themselves. Sanchez-Perez et al. (2014) described the difference between the two as the ability to construct the mental state of another (cognitive empathy) versus experiencing an emotionally concordant response aligned with the affective state of another (affective empathy). Cassels et al. (2010) have pointed out that affective empathy, as an emotional response to others' distress, can take on two forms: empathetic concern or personal distress. Distinctions between these two forms have been previously described, as they emerged from the work of Davis (1980, 1983) and others.

Compassionate empathy scales are less frequently found in social sciences literature. One scale that would appear to align with the third component of empathy as defined by Goleman (2008) is the *Compassion Scale of the Dispositional Positive Emotion Scales (DPES)* by Shiota et al. (2006). The five items on this scale are: (1) *It's important to take care of people who are vulnerable*; (2) *When I see someone hurt or in need, I feel a powerful urge to take care of them*; (3) *Taking care of others gives me a warm feeling inside*; (4) *I often notice people who need help*; and (5) *I am a very compassionate person* (Shiota et al., 2006, p. 71). Shiota et al. (2006) reported reliability to be Cronbach's Alpha = .80 for this scale.

4.2.3 Prospects for Indicators of Empathy via Machine Learning

Psychometric instruments may not be capturing the full scope of established components of empathy (for example compassionate actions), so this is one area where learning analytics systems could possibly add a richer context in the future. The development of classifiers in machine learning and learning analytics often begins with a hypothesis (e.g. an educated guess) or a model (e.g. an existing useful computational system), and follows a process similar to the blueprint for a psychometric construct. Expert knowledge at the early stage is presumed to be aligned or complementary to if not fully mapping of what Cattell called the 'nomological network' of the construct (Cattell, 1957). Experts have a role in codifying critical observations and evidence or signals in data streams from authentic professional practice environments, including digital and simulated environments. Their knowledge is used to build structured entities (e.g. metatagged and labeled data) for a training file. Then algorithms are developed to automate the experts' observations for massive new sets of data, in a process that includes exploration, pattern recognition, data transformation and norming, and then used with a reward function to train and optimize the algorithms for effective predictions. The algorithms that constitute the detector (or sets of detectors) are trained via data examples and a reward function for correct predictions, based on how well the detector assigns a correct categorical class label to particular data points.

For example, in one study, the state of a student in "engaged concentration" was constructed from evidence of both "concentration" and "being on-task" based on a distillation of expert observations. Automated detectors were then created and validated on the data using ten-fold participant-level cross-validation. For example, ten groups were randomly formed, the detector was trained on nine of the groups, then tested on the tenth, and this process was iterated until a strong predictability is reached. In this study several detectors were created using a number of computational methods. Then, the best-performing method was selected for each state based on performance using standard tools of statistics such as Cohen's Kappa and A' of Wilcoxin (Paquette et al., 2015). A' is equivalent to the area under the ROC curve in signal detection theory (Hanley & McNeil, 1982). A model with an A' of 0.5 performs at chance, and a model with an A' of 1.0 performs perfectly.

4.2.4 Selected Findings from Studies by the Authors

The authors of the current chapter began refining instruments for learning dispositions believed to be related to or influenced by information technologies in the early 1990s. This refinement continued into the early twenty-first Century (Christensen & Knezek, 2001) with research using these instruments continuing for two decades beyond (Christensen & Knezek, 2008; Knezek & Christensen, 1996, 2008, 2018). The empathy scale used for these studies was originally developed for the Computers in Education (CompEd) 20-nation international study led by the International Association for the Evaluation of Educational Achievement (IEA) headquartered in the Hague, Netherlands (Collis et al., 1996). The instrument has remained intact over the past three decades as one of the scales of the Computer Attitude Questionnaire (Christensen & Knezek, 2001, 2008; Knezek & Christensen, 1996). Individual items are listed in Table 4.2.

Internal consistency reliability for the CAQ Empathy Scale formed from these nine items, for children from upper elementary school through secondary education, typically fall in the range of alpha = .76 to alpha = .77 (Collis et al., 1996, p. 78). These Empathy scale items shown in Table 4.2 appear to be primarily in the realm of affective empathy as defined by researchers including Jolliffe and Farrington (2006). Findings presented in the following sections are based on the CAQ Empathy Scale.

Table 4.2 Computer Attitude Questionnaire (CAQ) empathy scale

1. I feel sad when I see a child crying
2. I sometimes cry when I see a sad play or movie
3. I get angry when I see a friend who is treated badly
4. I feel sad when I see old people alone
5. I worry when I see a sad friend
6. I feel very happy when I listen to a song I like
7. I do not like to see a child play alone, without a friend
8. I feel sad when I see an animal hurt
9. I feel happy when I see a friend smiling

4.2.5 Empathy Norms Are Related to Culture

In the early 1990s an international study was able to confirm through focusing on attitudes toward computers of young children in the USA, Mexico and Japan that empathy was an important consideration in distinguishing among norms in different cultures in the broader context of learning dispositions (Knezek et al., 1993). Evidence also emerged that learning dispositions (including those related to technology) may shift in children who move away from their original cultural homeland. The implication is that young children residing in a country away from their homeland, and attending school in that new environment, evolve somewhat toward the cultural norm of their new environment. This is consistent with the more recent findings of Cassels et al. (2010) as reported in their study of 190 adolescents and young adults in Canada: "The bicultural individuals' scores fell in between the East Asian and Western groups, but revealed significant differences from their 'unicultural' peers, demonstrating shared influences of community and family" (p. 309).

As technology increasingly links common interest groups across nations and cultures in the twenty-first century, we might anticipate that the interaction of technology with cultural background is becoming more complex than in the past. Certainly, it is no longer simply a case of from which region of the world one's ancestors originated, or even the part of the world in which one resides. As Shaffer points out, the multitude of subcultures of practice (e.g. how doctors, lawyers and teachers ply their various trades) is at once global and multicultural, yet current educational systems are organized around highly localized and unitary conceptions of both the means and ends of education (Shaffer, 2005). Learning analytics systems of the future will need to be clear about the contexts of the activities leading to the data-as-evidence as well as the cultural impacts on both the creators of learning experiences (e.g. how the structure, timing and data gathering capabilities of an embedded sensor-net may reveal the decisions and intentions of the creators) as well as those participating in, learning from and otherwise impacted by the experiences.

4.2.6 Empathy Declines Less Than Other Learning Dispositions as Grade Level Increases

From 1999 through 2004 the authors conducted research on technology affordances and learning dispositions while serving as external evaluators for a U.S. federally-funded *Technology Innovation Challenge Grant* involving 51 public school districts. One finding by the authors was that K-12 student attitudes tend to become less positive in general as the students progress through school (Knezek & Christensen, 2005). As shown in Fig. 4.1, students in first grade appear to have positive attitudes related to all of the measured indices including: Creative Tendencies, Empathy, Computer Importance, Computer Enjoyment, Motivation, Attitudes Toward School, Study Habits, Motivation to Study, and Attitudes toward Computers.

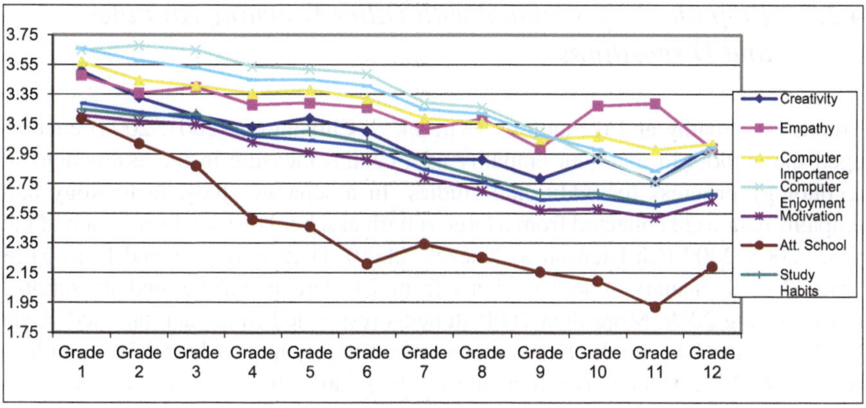

Fig. 4.1 Grade 1–12 trends in student attitudes across 14 Texas schools

The general trend from grade 1 to grade 11 is one of consistent decline, followed by a grade 11–12 trend toward more positive attitudes during the final year in school.

The seemly parallel paths of empathy and computer attitudes across level of maturation in school children that are graphically displayed Fig. 4.1, raised the question in early research by the authors as to whether these two might be related. Precisely why neither empathy nor computer importance were found to have extensive declines across years of maturation, unlike many other recorded learning dispositions, remains largely a mystery. However, the associations identified in the following section begin to shed some light on directions that future researchers may need to explore more deeply.

One previously unreported trend shown in Fig. 4.1 is that empathy appears to increase across grades 10 and 11, as a mirror image of attitudes toward school, creative tendencies, computer importance and computer enjoyment, and most other learning dispositions measured for this project. It then returns to the grades 1–12 trend line and declines at grade 12, the last year of secondary school.

A learning analytics approach taking account of this finding of dispositional decline in all measures across the years might explore 'what changes?' from grade 9–10. The goal would be to add multiple sources of new objective data (not self-reported data) collected from grade 9 through 12 and then search for potential sources of influence in patterns (e.g. patterns of curriculum, teacher expectations, social mixing of students, fluctuations in student behavior, social opportunities, movement from buses to cars, increases in autonomy, etc.). As mentioned earlier, often the first step is to find, transform and examine existing sources of data collected unobtrusively as multiple measures that stem from the common activity of 'going to school.'

4.2.7 Empathy Is Associated with Other Learning Attitudes and Dispositions

Trends in empathy and technology (Knezek & Christensen, 2001, 2005) encouraged the authors to maintain empathy as a central measure in the assessment of technology progress for additional studies. In a separate study, technology and empathy data were collected from a large suburban school district during each of the years 2005–2009 (Christensen & Knezek, 2009). Data were gathered from 5128 grade 3–5 elementary school students from 24 elementary/intermediate schools during spring 2009. More than 1600 students responded from each targeted grade level, from the 24 schools. Respondents were roughly half male (50.3%) and half female (49.7%). Associations focusing on the relationship of empathy to technology and other learning dispositions were previously examined for these data (Christensen & Knezek, 2009), but additional analyses were completed in 2020 focusing on empathy as an outcome (dependent) variable.

Analyses in 2020 began with detailed examination of associations among learning disposition indices previously reported for the 2009 data using outcomes from the 1990–2009 studies for guidance. This is consistent with the grounded theory research methodology advanced by Creswell (2013) and others where a research team "… sets out to discover or construct theory from data, systematically obtained and analyzed using comparative analysis" (Tie et al., 2019, np). Correlational analyses showed that greater Empathy was found to be most strongly associated with Creative Tendencies (r = .44), but was also significantly ($p < .05$) associated with five other learning dispositions. In the case of Computer Importance (r = .203) and Computer Enjoyment (r = .218), both types of computer attitudes were significantly ($p < .0005$) correlated with Empathy to an educationally meaningful degree (Bialo & Sivin-Kachala, 1996), with magnitudes of association comparable to Cohen's d effect sizes of .41 and .45, respectively (Cohen, 1988; Lenhard & Lenhard, 2016).

In order to more deeply explore the relationship of technology and empathy from the perspective of empathy as an outcome measure, a regression analysis was completed using Empathy as the targeted outcome measure with the other learning dispositions as predictors. The result was that Computer Enjoyment, Computer Importance, Study Habits, Motivation, Creative Tendencies and Attitudes Toward School together explained 27% of Empathy (RSQ = .270). All except Motivation were individually significant ($p < .05$). Creative tendencies (beta = .308) emerged as the strongest contributor to Empathy, followed by Attitude Toward School (beta = .180) and Study Habits (beta = .136). Computer Enjoyment (beta = .054) and Computer Importance (beta = .033) were not as strong but worthy of mention because both were positive contributors to Empathy. These relationships imply that learning analytics systems of the future will need to be capable of accounting for complex combinations of multiple contributors to empathy, in situations where nurturing empathy is deemed to be a desirable learning outcome.

4.2.8 Measurable Differences Exist in Level of Empathy Related to Gender

Do females have greater empathy than males at the elementary school level? As shown in Table 4.3, across 5082 students in grades 3–5, 50.3% male and 47.7% female, with nearly equal distribution by grade level, two of seven learning dispositions emerged as confirming sizeable differences by gender. These were Empathy ($p < .0005$, ES = .56) and Attitude Toward School ($p < .0005$, ES = .54). Both were more positive for females, with a magnitude of effect due to reported gender that would be considered moderate according to guidelines by Cohen (1988) and educationally meaningful (ES > .3) according to established research criteria (Bialo & Sivin-Kachala, 1996). Note that across all seven types of learning dispositions reported in Table 4.2, only Computer Importance) was rated more positive by males ($p = .26$, NS), and the magnitude of the effect in favor of males (Cohen's d = .03) was small according to guidelines by Cohen (1988).

The 2009 report produced by the authors on these data (Christensen & Knezek, 2009) pointed out longitudinal trends across multiple years for the same school system. Specifically reported was that the effect of gender on empathy was comparable across multiple years, with Cohen's d = 55 in 2009; .72 in 2008; .63 in 2007; .62 in 2006; and .54 in 2005 (Christensen & Knezek, 2009). In addition, the authors illustrated that differences by gender in empathy remained consistent when the same students were assessed across 3 years, in grades 3, 4, and 5. Females appear to have maintained a high level of empathy across these grade levels, while males appear to have successively declined each year in school.

Multidimensional scaling (MDS) applied to the 2009 data set illustrated that males and females differ not only in their level or magnitudes of empathy but also in their perceptions of how empathy relates to other learning dispositions known to be related to use of technology. As shown in Fig. 4.2, the MDS 2-dimensional solution (Z score standardization) is meaningful for males, as the 2-dimensional solution shown accounts for 97.3% of object proximity while the 1-dimensional solution (not shown) would account for 54.1%, a loss of more than 20% of the variance explained. This loss of explanatory power by forced reduction to a 1-dimensional solution is largely because the best-fit line would run from the top center of the display, though the cluster of the four items of empathy, motivation, study habits, creative tendencies, computer enjoyment, and computer importance – while attitude toward school would remain distant from all other items in the analysis.

The MDS 2-dimensional solution shown in Fig. 4.3 accounts for 97.8% of object proximity for females, with empathy closer to the centroid of all items than was the case for males. Note that a 1-dimensional solution with a line drawn from lower left to upper right (not shown) would account for almost as much (97.3%) for females. This implies that empathy lies near the center of a linear continuum that has attitude toward school on one extreme and computer attitudes on the other for females, with empathy in the middle. Attitude toward school appears to lie on the same continuum as empathy for females.

Table 4.3 Gender differences across six learning dispositions for 5000 students in grades 3–5

Measure	Gender	N	Mean	Std. Dev.	Effect size	Sig.
Empathy	Male	2550	3.04	0.72	0.55	0.0005
	Female	2532	3.42	0.59		
	Total	5082	3.23	0.69		
Computer enjoyment	Male	2571	3.47	0.50	0.12	0.0005
	Female	2545	3.52	0.46		
	Total	5116	3.50	0.48		
Computer importance	Male	2566	3.27	0.56	−0.03	0.2550
	Female	2540	3.25	0.55		
	Total	5106	3.26	0.56		
Motivation	Male	2553	3.02	0.63	0.05	0.0670
	Female	2531	3.05	0.63		
	Total	5084	3.04	0.63		
Study habits	Male	2553	3.07	0.63	0.18	0.0005
	Female	2531	3.19	0.59		
	Total	5084	3.13	0.61		
Attitude toward school	Male	2556	2.41	0.85	0.54	0.0005
	Female	2538	2.87	0.81		
	Total	5094	2.64	0.86		
Creative tendencies	Male	2553	3.04	0.63	0.14	0.0005
	Female	2534	3.12	0.58		
	Total	5087	3.08	0.61		

The visual placement of objects in Figs. 4.2 versus 4.3 implies that it may be attitude toward school, more so than the relationship of technology to empathy, that is more complex for males and causes MDS to require two dimensions to adequately represent the seven learning dispositions featured in this analysis. Overall, patterns displayed in Figs. 4.2 and 4.3 imply that females likely possess different mental maps for empathy than males. Learning analytics systems may need to account for these differences when considering empathy as a contributor to learning, because the impacts will likely not be the same for males as for females.

4.3 Topics Warranting Additional Study

Several topics warrant additional study as we move toward meaningful representation of empathy in learning analytics systems. Three topics deemed to be important for advancing the field are addressed in this section.

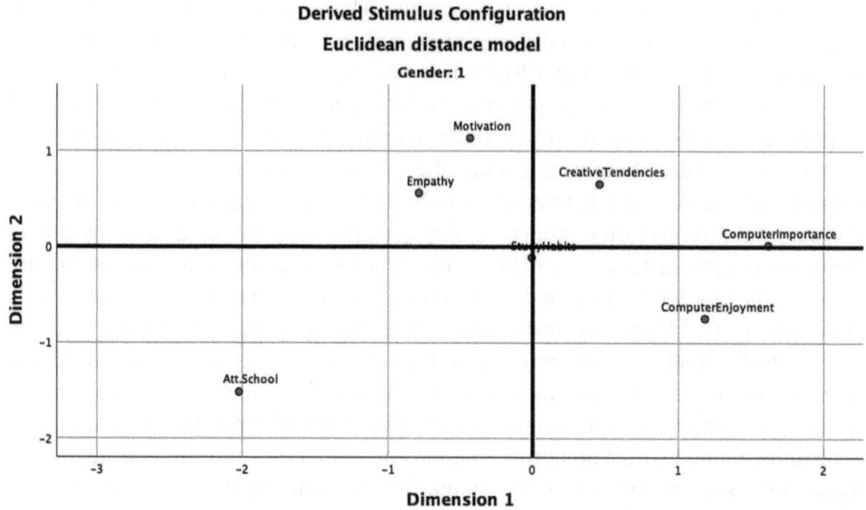

Fig. 4.2 Two-dimensional representation of perceived distances among seven learning dispositions for males

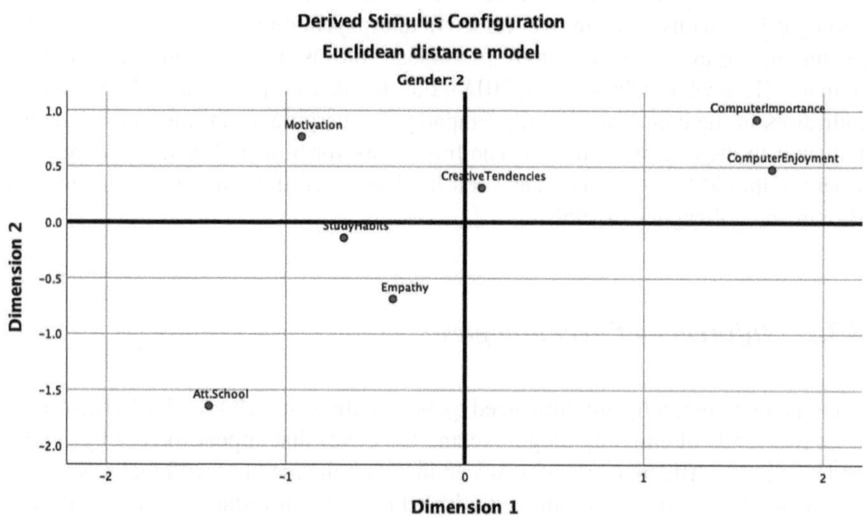

Fig. 4.3 Two-dimensional representation of perceived distances among seven learning dispositions for females

4.3.1 Placement in Learning Frameworks

Is empathy a dependent variable, and independent variable, or both, for the purposes of fostering learning and individual attribute development? Ongoing research needs to continue to address how empathy is related to currently-accepted "standard" indices gathered by learning analytics systems. These are currently focused on academic performance, grades and career readiness outcomes. If we follow the recommendations of Ifenthaler et al. (2020) about the need to put a focus on the learner and learning back into learning analytics, then a system responsive to individual learner needs rather than simply institutional management information should pay attention to gender, ethnicity, and a handful of socio-emotional variables including empathy when advising a learner of his or her status regarding satisfactory progress and suggested remediation strategies. For example, learning analytics systems of the future might regard targeted learner outcomes as a vector of desirable attributes known to be essential for productive and rewarding contributions to society among graduates of an educational institution. Currently, the primary outcome indicator appears to be academic performance in the form of grades and standardized test scores, or course completion rates. Some universities systematically follow up with alumni regarding careers and employment rates. A broader perspective might need to be implemented with a blueprint grounded in accepted twenty-first Century skills frameworks such as the Four C's: communication, collaboration, critical thinking, and creativity (NEA, 2020) – in order systematically address perceived broader needs of society beyond cognitive skills development. We suggest it has long been established that when viewing the society-contributor readiness of an individual, his or her behavior is known to be influenced by the three historically established psychological domains of cognitive, affective and psychomotor (Bloom et al., 1956). Technology plays an increasingly-prominent role as a partner in each of these domains (Knezek & Christensen, 2018). For decades we have focused on outcome indicators in the cognitive domain. Empathy is emerging as an important outcome indicator in the affective domain. The framework for how it should be viewed and where it should be positioned has been in place since the seminal publications of Bloom and colleagues in 1956.

4.3.2 Influence of Technologies

What kinds of empathy are influenced by which kinds of technologies? There are at least two kinds of empathy (cognitive and affective) that appear to vary by gender and by culture. Affective empathy, which involves engagement with feelings of others, seems to be potentially aligned with educational affordances and specifically technologies supporting two-way communications between people. Cognitive empathy, on the other hand, involves recognition and understanding of an unfortunate situation of another. This appears to be able to be conveyed by one-way technologies such as television. Is there a meaningful distinction in types of communication technologies? The research literature is not clear. If there is a meaningful distinction, then where would a modern system such as twitter be situated, and what does this imply for potential outcomes during children's schooling years? This is basic research that needs to be pursued in detail, in order for the new knowledge to be properly applied in teaching and learning systems and environments, including learning analytics systems.

4.3.3 Role of Learning Analytics Systems

How can learning analytics systems best capture and represent individual learner empathy? Since empathy has been elevated to the role of a desirable outcome measure that could be incorporated into learning analytics systems, much research is needed to see how such data might be combined in learning analytics systems of the future. Accuracy and privacy are very important matters to be carefully considered. Are we willing to have our socio-emotional characteristics given to an entity such as a university (or a company the university contracts with), the same as a mid-term or final examination test score?

In the field of data science and data analytics in particular, a difficult challenge is determining which kinds of data are important to gather, and how these data can it be transformed into variables that are useful for answering important research questions. This process is often referred to as feature engineering and selection (Kuhn & Johnson, 2020). If, as proposed by Ifenthaler et al. (2020), that a major goal of learning analytics systems should be to benefit the learner and learning, then the question of which data elements to gather, what is important to assess, and via what measures, becomes paramount. If learning analytics systems are to become tools for guiding learners and education toward attainment of 'productive future citizen' societal goals, and beyond simple completion of a degree or mastery of content, then they will have to evolve toward accommodating measures of socio-emotional attributes such as empathy as well as cognitive performance.

Yet several areas of research need to further advance before the full incorporation of empathy into the fabric of human development /education systems in general and learning analytics systems in particular are likely to take place. For example, even with the rather long history of psychometric development regarding empathy, limitations in what is being measured undoubtedly remain. For example, we should note that in the long line of instrument refinements based on *The Interpersonal Reactivity Index (IRI)* (Davis, 1980), or the *Basic Empathy Scale* (Jolliffe & Farrington, 2006), if there are no items concerning compassionate empathy (Goleman, 2008), then that factor is not going to be confirmed through procedures like confirmatory factor analysis by the original authors nor by any others who follow. Therefore, research to date may not be capturing the full scope of important aspects of empathy, and this is one area where learning analytics systems could possibly add richer context in the future.

4.4 Conclusion

Does technology influence empathy? Is technology fostering empathy at a distance, and/or empathy's decline, in the twenty-first century? The answer appears to be "yes" to each of these questions, if one focuses on the devices, applications, and technology systems that serve as communication mediums among human beings. However, it may not be the technology itself that is strengthening or reducing empathy, but rather the extensions of possibilities over distance or time to the world community through the Internet and other media channels, or the reduction of bandwidth leading to misunderstandings regarding the intended message of the sender versus the interpretation of the receiver – that lead to positive or negative outcomes. The myriad of communication technologies available to a large portion of the world's population in the twenty-first century are not all equal in their tendencies to facilitate and/or magnify positive or negative reactions in humans regarding empathy as a caring concern for the thoughts and feelings of fellow human beings. Perhaps additional research in the coming decades will help more precisely identify which systems need policy-level scrutiny and regulation, versus those that need encouragement and support for broader implementation to further desirable societal goals, one of which is fostering empathy in our society of the future. We envision a day where via embedded learning analytics, technology is better enabled to play a dynamic positive role, in line with the recommendation "…to create a mutual dependency between empathy and technology: Using technology to help people cultivate empathy among people, so empathy in the society may allow people to help each other" (Nishida, 2013, p. 277).

References

Bachen, C. M., Hernandez-Ramos, P. F., & Raphael, C. (2012). Simulating REAL LIVES: Promoting global empathy and interest in learning through simulation games. *Simulation & Gaming, 43*(4), 437–460.

Belman, J., & Flanagan, M. (2010). Designing games to foster empathy. *Cognitive Technology, 14*(2), 5–15.

Bialo, E., & Sivin-Kachala, J. (1996). The effectiveness of technology in schools: A summary of recent research. *School Library Media Quarterly, 25*(1), 51–57.

Bloom, B. S., Engelhart, M. D., Furst, E. J., Hill, W. H., & Krathwohl, D. R. (1956). *Taxonomy of educational objectives, handbook I: The cognitive domain.* David McKay.

Carrier, L. M., Spradlin, A., Bunce, J. P., & Rosen, L. D. (2015). Virtual empathy: Positive and negative impacts of going online upon empathy in young adults. *Computers in Human Behavior, 52*, 39–48.

Cassels, T. G., Chan, S., Chung, W., & Birch, S. A. J. (2010). The role of culture in affective empathy: Cultural and bicultural differences. *Journal of Cognition and Culture, 10*(3–4), 309–326. https://doi.org/10.1163/156853710X531203

Cattell, R. (1957). *Personality and motivation: Structure and measurement.* Harcourt, Brace & World.

Christensen, R. (2002). Effects of technology integration education on the attitudes of teachers and students. *Journal of Research on Technology in Education, 34*(4), 411–433.

Christensen, R., & Knezek, G. (2001). Instruments for assessing the impact of technology in education. In Assessment/evaluation in educational information technology. *Computers in the Schools, 18*(2/3/4), 5–25.

Christensen, R., & Knezek, G. (2008). Self-report measures and findings for information technology attitudes and competencies. In *International handbook of information technology in primary and secondary education* (pp. 349–359). Springer.

Christensen, R., & Knezek, G. (2009). *CFB elementary student technology report 2009*. Institute for the Integration of Technology into Teaching & Learning.

Christov-Moore, L., Simpson, E. A., Coude, G., Grigaityte, K., Iacoboni, M., & Pier, F. F. (2014). Empathy: Gender effects in brain and behavior. *Neuroscience & Biobehavioral Reviews, 46*, 604–627. https://doi.org/10.1016/j.neubiorev.2014.09.001

Cohen, J. (1988). *Statistical power analysis for the behavioral sciences*. Erlbaum.

Collins, F. (2014). *The relationship between social media and empathy*. Published masters' thesis. https://digitalcommons.georgiasouthern.edu/cgi/viewcontent.cgi?article=2189&context=etd

Collis, B., Knezek, G., Lai, K., Miyashita, K., Pelgrum, W., Plomp, T., & Sakamoto, T. (1996). *Children and computers in school*. Lawrence Erlbaum.

Creswell, J. (2013). *Research design: Qualitative, quantitative, and mixed methods approaches* (4th ed., p. 2013). SAGE.

Darling-Hammond, L. (2000). How teacher education matters. *Journal of Teacher Education, 51*(3), 166–173.

Davis, M. H. (1980). A multi-dimensional approach to individual differences in empathy. *Catalogue of Selected Documents in Psychology, 10*, 85–104.

Davis, M. H. (1983). The effects of dispositional empathy on emotional reactions and helping: A multidimensional approach. *Journal of Personality, 51*, 167–184.

DeVellis, R. F. (2011). *Scale development* (3rd ed.). Sage Publications.

Dunn-Rankin, P., Knezek, G., Wallace, S., & Zhang, S. (2004). *Scaling methods* (2nd ed.). Lawrence Erlbaum.

Eisenberg, N., & Lennon, R. (1983). Sex differences in empathy and related responding. *Psychological Bulletin, 94*, 100–131.

Ekman, P. (2003). *Emotions revealed*. Times Books.

Feng, S., & Kirkley, A. (2020). *Mixing patterns in interdisciplinary collaboration networks: Assessing interdisciplinarity through multiple lenses*. Retrieved from http://arxiv.org/abs/2002.00531

Gibson, D., & Ifenthaler, D. (2017). Preparing the next generation of education researchers for big data in higher education. In B. Daniel (Ed.), *Big data and learning analytics in higher education* (pp. 29–42). Springer International Publishing. https://doi.org/10.1007/978-3-319-06520-5_

Goleman, D. (2008). Hot to help: When can empathy move us to action? *Greater Good Magazine, March*, n/a. Retrieved from https://greatergood.berkeley.edu/article/item/hot_to_help

Gordon, M. (2012). *Roots of empathy: Changing the world, child by child*. Thomas Allen Publishers.

Hanley, J., & McNeil, B. (1982). The meaning and use of the area under a receiver operating characteristic (ROC) curve. *Radiology, 143*, 29–36.

Helliwell, J. F., Layard, R., & Sachs, J. D. (2018). *World happiness report*. Sustainable Development Solutions Network. ISBN 978-0-9968513-6-7.

Ifenthaler, D., Gibson, D., & Dobozy, E. (2017). The synergistic and dynamic relationship between learning design and learning analytics. *ASCILITE, 2017*, 1–5.

Ifenthaler, D., Gibson, D., Prasse, D., Shimada, A., & Yamada, M. (2020). Putting learning back into learning analytics: Actions for policy makers, researchers, and practitioners. *Educational Technology Research and Development*. https://doi.org/10.1007/s11423-020-09909-8

Ivcevic, Z., & Ambady, N. (2012). Personality impressions from identity claims on Facebook. *Psychology of Popular Media Culture, 1*(1), 38–45. https://doi.org/10.1037/a0027329

Jolliffe, D., & Farrington, D. P. (2006). Development and validation of the Basic Empathy Scale. *Journal of Adolescence, 29*, 589–611. https://doi.org/10.1016/j.adolescence.2005.08.010

Jones, S., Weissbourd, R., Bouffard, S., Kahn, J., & Ross Anderson, R. (2020). *For educators: How to build empathy and strengthen your school community* (Making Caring Common Project). Harvard Graduate School of Education. https://mcc.gse.harvard.edu/resources-for-educators/how-build-empathy-strengthen-school-communit

Knezek, G., & Christensen, R. (1996). *Validating the computer attitude questionnaire (CAQ)*. Paper presented at the annual meeting of the southwest educational research association (New Orleans, LA, January 1996). https://files.eric.ed.gov/fulltext/ED398243.pdf

Knezek, G., & Christensen, R. (2001). Impact of new information technologies on teachers and students. *IFIP World Conference on Computers in Education, 7*, 169–178.

Knezek, G., & Christensen, R. (2005). *Summary of KIDS project findings for 1999–2004: Research and project evaluation* (p. 6). University of North Texas Institute for the Integration of Technology into Teaching and Learning.

Knezek, G., & Christensen, R. (2008). The importance of information technology attitudes and competencies in primary and secondary education. In *International handbook of information technology in primary and secondary education* (pp. 321–328). Springer.

Knezek, G., & Christensen, R. (2018). The evolving role of attitudes and competencies in information and communication technology in education. In J. Voogt, G. Knezek, R. Christensen, & K. W. Lai (Eds.), *Second handbook of information technology in primary and secondary education* (pp. 239–253). Springer.

Knezek, G., Miyashita, K., & Sakamoto, T. (1993). Cross-cultural similarities in attitudes toward computers and the implications for teacher education. *Journal of Information Technology for Teacher Education, 2*(2), 193–204. https://doi.org/10.1080/0962029930020207

Konrath, S. H., O'Brien, E. H., & Hsing, C. (2011). Changes in dispositional empathy in American college students over time: A meta-analysis. *Personality and Social Psychology Review, 15*(2), 180–198.

Kuhn, M., & Johnson, K. (2020). *Feature engineering and selection: A practical approach for predictive models*. Taylor Francis/CRC Press.

Lenhard, W., & Lenhard, A. (2016). *Calculation of effect sizes*. Retrieved from: https://www.psychometrica.de/effect_size.html. Psychometrica. https://doi.org/10.13140/RG.2.2.17823.92329

Lennon, R., & Eisenberg, N. (1987). Gender and age differences in empathy and sympathy. In N. Eisenberg & J. Strayer (Eds.), *Empathy and its development* (pp. 195–217). Press Syndicate of the University of Cambridge.

Lewis, C., Schapps, E., & Watson, M. (1996). The caring classroom's academic edge. *Educational Leadership, 54*(1), 16–21.

Madden, M., & Rainie, L. (2003). *America's online pursuits: The changing picture of who's online and what they do*. Pew Internet & American Life Project. https://www.pewresearch.org/internet/2003/12/22/part-1-introduction-13/

McAllister, G., & Irvine, J. J. (2002). The role of empathy in teaching culturally diverse students: A qualitative study of teachers' beliefs. *Journal of Teacher Education, 53*(5), 433–443.

Mendes, E. (2003). What empathy can do. *Educational Leadership, 61*(1), 56–59.

Mislevy, R., Steinberg, L., & Almond, R. (1999). *Evidence-centered assessment design*. Educational Testing Service. Retrieved from http://www.education.umd.edu/EDMS/mislevy/papers/ECD_overview.html

Morning Future. (2019). Empathy? In *Denmark they're learning it in school*. https://www.morningfuture.com/en/article/2019/04/26/empathy-happiness-school-denmark/601/.

National Education Association (NEA). (2020). *Preparing 21st century students for a global society: An educator's guide to the "Four Cs"*. http://dl.icdst.org/pdfs/files3/0d3e72e9b873e0ef2ed780bf53a347b4.pdf

Nishida, T. (2013). Toward mutual dependency between empathy and technology. *AI & SOCIETY, 28*, 277–287. https://doi.org/10.1007/s00146-012-0403-5

Paquette, L., Rowe, J. P., Baker, R. S. J. D., Mott, B. W., Lester, J. C., DeFalco, J., … Georgoulas, V. (2015). Sensor-free or sensor-full: A comparison of data modalities in multi-channel affect detection. In *Proceedings of the 8th international conference on educational data mining* (pp. 93–100).

Powell, P. A., & Roberts, J. (2017). Situational determinants of cognitive, affective, and compassionate empathy in naturalistic digital interactions. *Computers in Human Behavior, 68*, 137–148. https://doi.org/10.1016/j.chb.2016.11.024

Ribble, M., & Miller, T. (2013). Educational leadership in an online world: Connecting students to technology responsibly, safely, and ethically. *Online Learning, 17*(1), 137–145. https://doi.org/10.24059/olj.v17i1.310

Rosen, L. D., Cheever, N. A., & Carrier, L. M. (2012). *iDisorder: Understanding our obsession with technology and overcoming its hold on us*. Palgrave Macmillan.

Sanchez-Perez, N., Fuentes, L. J., Jolliffe, D., & Gonzalez-Salinas, C. (2014). Assessing children's empathy through a Spanish adaptation of the basic empathy scale: parent's and child's report forms. *Frontiers in Psychology, 5*(1438), 1–13. https://doi.org/10.3389/fpsyg.2014

Shaffer, D. W. (2005). *Multisubculturalism: Computers and the end of progressive education.* Wisconsin Center for Education Research. Retrieved from http://coweb.wcer.wisc.edu/cv/papers/multisubculturalism-draft1.pdf%5Cn, http://coweb.wcer.wisc.edu/dws

Shaffer, D. W., & Ruis, A. R. (2015). *Epistemic network analysis: A worked example of theory-based learning analytics*. Wisconsin Center for Education Research, University of Wisconsin–Madison. https://doi.org/10.18608/hla17.015

Shaffer, D., Hatfield, D., Svarovsky, G., Nash, P., Nulty, A., Bagley, E., & Mislevy, R. (2009). Epistemic network analysis: A prototype for 21st-century assessment of learning. *International Journal of Learning and Media, 1*(2), 33–53. Retrieved from http://www.mitpressjournals.org/doi/abs/10.1162/ijlm.2009.0013

Shiota, M. N., Keltner, D., & John, O. P. (2006). Positive emotion dispositions differentially associated with Big Five personality and attachment style. *Journal of Positive Psychology, 1*, 61–71.

Sin, K., & Muthu, L. (2015). Application of big data in educational data mining and learning analytics. *ICTACT Journal on Soft Computing, 05*(04), 1035–1049. https://doi.org/10.21917/ijsc.2015.0145

Spradlin, A., Bunce, J., Rosen, L. D., Carrier, L. M., & Cheever, N. A. (2012). Virtual friendships: A study of digital media usage and empathy. In L. D. Rosen (Ed.), *iDisorder: Understanding our obsession with technology and overcoming its hold on us*. Palgrave Macmillan.

Terry, C., & Cain, J. (2016). The emerging issue of digital empathy. *American Journal of Pharmaceutical Education, 80*(4), Article 58.

Tie, Y., Birks, M., & Francis, K. (2019). *Grounded theory research: A design framework for novice researchers*. Sage Open Medicine. PMCID: PMC6318722, PMID: 30637106. https://doi.org/10.1177/2050312118822927

Tlili, A., & Chang, M. (2019). Data analytics approaches in educational games and gamification systems: Summary, challenges, and future insights. In A. Tlili & M. Chang (Eds.), *Data analytics approaches in educational games and gamification systems. Smart computing and intelligence*. Springer, Singapore. https://doi.org/10.1007/978-981-32-9335-9_13

Turkle, S. (2016). *Reclaiming conversation: The power of talk in a digital age*. Penguin Books.

Uhls, Y. T., Michikyan, M., Morris, J., Garcia, D., Small, G. W., Zgourou, E., & Greenfield, P. M. (2014). Five days at outdoor education camp without screens improves preteen skills with nonverbal emotion cues. *Computers in Human Behavior, 39*, 387–392.

Warner, R. E. (1984). Can teachers learn empathy? *Education Canada, 24*(1), 39–41.

Zych, I., Baldry, A. C., Farrington, D. P., & Llorent, V. J. (2019). Are children involved in cyberbullying low on empathy? A systematic review and meta-analysis of research on empathy versus different cyberbullying roles. *Aggression and Violent Behavior, 45*, 83–97. https://doi.org/10.1016/j.avb.2018.03.004

Chapter 5
The Role of Learning Analytics in Developing Creativity

Rebecca L. Marrone and David H. Cropley

Abstract As society transitions into Industry 4.0 (whereby industries integrate digital and physical objects) the role of educational systems is to shape students into twenty-first century STEM citizens. Modern education systems develop competencies such as creativity alongside STEM capabilities in response to this emerging workforce. The individual, social, and economic importance of creativity has encouraged scientists, educators, and psychologists to study creativity's nature, assessment, and development. On an individual level, creativity is important when assisting problem solving and decision making. At the societal level, creativity can lead to scientific breakthroughs, conflict resolutions, and inventions. Creativity is fundamental to a thriving modern economy. Educators now focus on encouraging creativity in their students, thus, equipping them with the necessary skills to become workforce ready. For a society to successfully prepare for the Future of Work, a concerted emphasis on creativity in both school and the workforce is needed. This chapter will explore the role of Learning Analytics in developing creativity in education.

Keywords Creativity · Industry 4.0 · Education · Learning analytics

Tagline: *Creativity and Innovation are integral to Learning Analytics and Data Science and going forward in this AI world, these creative competencies are the things that will give humans an edge over the machines.*

R. L. Marrone (✉)
Education Futures, University of South Australia, Adelaide, Australia
e-mail: rebecca.marrone@unisa.edu.au

D. H. Cropley
UniSA STEM, University of South Australia, Mawson Lakes, Australia
e-mail: david.cropley@unisa.edu.au

Y. E. Wang et al. (eds.), *Social and Emotional Learning and Complex Skills Assessment*, Advances in Analytics for Learning and Teaching,
https://doi.org/10.1007/978-3-031-06333-6_5

5.1 Introduction

The emergence of cyber-physical systems, artificial intelligence, big data, and automation (Industry 4.0) has focused attention on the so-called Future of Work. At the core of Industry 4.0 is the automation of many of the tasks humans currently perform. However, despite the rapid adoption of Industry 4.0 technologies, the challenges of the twenty-first century—health, security, climate, population, and food—remain, and developing uniquely human competencies beyond those of machines is more important than ever. Competencies such as creativity, complex problem solving and critical thinking are widely regarded as essential to success in the emerging, Industry-4.0-enabled, work environment (Gray, 2016; Nakano & Wechsler, 2018; Sternberg & Lubart, 1996). There is, however, a challenge: how are these competencies to be developed, efficiently and effectively, in an educational paradigm that has frequently neglected them in favour of the acquisition of declarative (i.e., factual) knowledge and "hard" skills (see, for example, Cropley, 2015).

Paradoxically, part of the solution to the rapid development of these competencies may lie in the very technologies that have drawn attention to their need. As education systems pivot towards the systematic development of what have been treated, traditionally, as *soft skills*, the discipline of Learning Analytics (LA) offers a mechanism to enhance this transition by bridging the feedback gap to learners. "Learning analytics is the measurement, collection, analysis and reporting of data about learners and their contexts, for purposes of understanding and optimising learning and the environments in which it occurs" (Society for Learning Analytics Research, 2012). In an increasingly digital world, LA enables new ways of achieving excellence in teaching and learning for both educator and student (Siemens, 2012). LA is synonymous with modern data-driven education. It provides educators with sophisticated techniques to study teaching and learning that can close the feedback loop to learners, offering more timely, precise, and actionable feedback.

Scholars, including Buckingham Shum and Fergusson (2012), have already recognised the potential of LA to support the development of creativity, and related competencies, calling for "analytics that can support the development of dispositions such as creativity…" (p. 7), even as curriculum authorities around the world (Patston et al., 2021) begin to embed these competencies into the frameworks of school education. Concurrently, an acknowledgment of the importance of creativity, innovation, and technology skills to national prosperity in the twenty-first century continues to grow (Deloitte, 2016). What is now clear is that developing these competencies is a priority for education systems at all levels, from primary to tertiary. The emerging field of Learning Analytics is well-placed to support the more rapid and successful development of these competencies.

LA will support the development of competencies such as creativity in modern, digitally enabled, learning environments in a variety of ways. These will depend, fundamentally, on an ability to extract, from current digital learning platforms and activities, data related to the development of key competencies such as creativity, and, the ability to provide teachers and learners with more rapid and *personalised* feedback. Some attempts have been made to do this (e.g., Gal et al., 2017; Britain et al., 2020) however all of these are held back by limitations in the understanding of what creativity is, and what needs to be measured. This is not an uncommon problem (see Cropley & Cropley, 2009; Patston et al., 2018), but must be addressed.

Before LA can be applied to the task of encouraging creativity, creativity must be understood not merely as a skill, but as a more complex, multifaceted *competency*. Creativity consists not only of cognitive skills (e.g., divergent thinking), but also attitudes and dispositions (e.g., openness, a tolerance for uncertainty), and environmental factors[1] (e.g., the classroom climate[2]) in addition to subject-specific knowledge (Cropley & Cropley, 2012).

Another matter that must be addressed before applying LA to encouraging creativity is the importance of dispelling the myths and misconceptions often associated with it. A common myth is that creativity is thought of as a trait that people are born with—"you either have it, or you don't" (Olken, 1964, p. 149). Similarly, it is frequently conceived of too narrowly, as exclusively concerned with aesthetics—"creativity is about art, isn't it?" This is because creativity is often associated with a lack of rigour, impulsive behaviour, free expression of ideas without regard to quality, and other "soft" factors.

As the study of creativity has developed within educational psychology, a gradual shift in our understanding of the term *creativity* has occurred (Patston et al., 2018). Many countries have adopted creativity into their school curricula (see Patston et al., 2021 for a recent review) and many industries and organisations also recognise its value (Khalili, 2016). In this chapter, we propose a roadmap on how to develop a user's creativity through LA platforms irrespective of the subject, course, or age group. Thus, ensuring the successful development of this twenty-first-century competency.

5.2 The Definition of Creativity

Two basic components are needed by LA analysts when answering *what is creativity*? These components will answer the fundamental question and remove the basic blocks to connecting creativity with LA and ensure that progress is made in both fields. Rhodes (1961) conceptualises creativity as a system of 4 interacting psychosocial dimensions, the 4P's of Person, Product, Process and Press (the environment). The second component is understanding that creativity occurs not in a single step,

[1] The environment is defined as both the social environment (influence of various aspects of society) and the institutional environment (e.g., day-to-day workplace or classroom).

[2] The climate refers to the conditions of the environment—is the environment favourable or not for creativity?

but as a series of stages or phases, each involving unique elements of the 4P's. Before exploring these components, it is important to understand that a clear, and widely accepted, definition represents the consensus that has emerged over decades of creativity research. Plucker et al. (2004) have captured all the essential factors in the following definition: creativity is "the interaction among *aptitude, process and environment* by which an individual or group produces a *perceptible product* that is both *novel and useful* as defined within a *social context*" (p. 90).

5.3 The Four Ps of Creativity

The first component needed by LA analysts to support creativity in education is to recognise that creativity is characterised in terms of 4Ps: Person, Product, Process and Press (environment). Rhodes's conceptual framework was first described in 1961 and provides a robust framework for understanding the who, what, when, where and how of creativity in LA.

5.4 Person—Who Are the Creators?

The *person* addresses the personality factors relating to the individual involved in the creation of the *product*. Research has shown that personal properties (e.g., optimism, openness, self-confidence, self-regulation), motivation (both intrinsic and extrinsic) and feelings (e.g., excitement, hope, fear) are distinct dimensions of the person that each have a bearing on creativity (Cropley & Cropley, 2013; Yeh & Lin, 2015). Furthermore, these dimensions of the person interact with each other in various ways such that different combinations have unique consequences for creativity (Baer, 2010). Research has shown how using log data to provide feedback can increase students' self-regulation (Silva et al., 2018) and increase their motivation (Aluja-Banet et al., 2019). Authors also note that they can detect and support student emotions in an online environment (Rienties & Rivers, 2014). It is therefore suggested that the traits required for being creative can be measured and supported through LA.

5.5 Product—What Do They Create?

The *product* addresses the output of the creative activity. It is widely accepted that an essential core of creativity, whether in art and poetry or Engineering and science, is the tangible artefact. This definition of *product* can be extended to any product, process, system or service that is *both novel and useful* (Plucker et al., 2004).

The product is often viewed as an objective approach to creativity, as it deals with tangible objects available for measurement (Kozbelt et al., 2010). While more recent definitions of the creative product debate the existence of higher-order characteristics (Cropley & Cropley, 2005), definitions from as far back as Stein (1953) suggest that the product is a combination of novelty and usefulness.

For example, for an object to be regarded as creative, it must be original and surprising, and it must solve a real problem or satisfy a real need. When examining LA, the product is the most obvious of the 4P's and can vary greatly. The product could include students' outputs such as a history essay, PhD thesis, or an individual art piece. Four criteria define a product's creativity (Cropley & Kaufman 2012; Cropley & Cropley 2011): relevance and effectiveness; novelty; elegance, and genesis (ability to shift paradigms). Products can be classified using these four dimensions arranged in a hierarchy ranging from "routine" products (characterised by effectiveness alone) to "innovative" products (characterised by effectiveness, novelty, elegance and genesis), with "original" and "elegant" products falling between the poles of routine and innovative. Whilst each field favours different characteristics, we propose that in education, the hierarchy of importance should be effectiveness, novelty, elegance, and then genesis in education.

To promote creativity through LA, the product should be designed to assist the user in developing their creativity. To empirically measure the creativity of a product, feedback can be given to students on three key factors.

1. ***Is* my product (essay, proof, artefact) creative?** Research by Gal et al., (2017) highlights that fluency (defined as a total number of responses) can be automatically measured through a LA platform such as the coding game Kodetu. The authors analysed log data and determined how creative a student's attempted solution was. The authors use standard definitions of creativity (fluency), and the results can highlight to teachers how creative both an attempted solution and a correct solution is. Whilst fluency is a measure of the *process* of decision making, it can still be used to indicate "how much" overall creativity the solution possesses (e.g. expressed as a percentage—80%—or as a qualitative descriptor, the product is *very* creative). A teacher could use this feedback as a summative assessment of the student's work. Whilst fluency is only a small subset of the definition of creativity and is an output of the process that students went through; it is a starting point for LA researchers to measure this competency.
2. ***Why* is my product creative** (or not creative)? If a student completes a task in an online environment, a teacher will provide concrete feedback and examples of what specific characteristics contribute (or fail to contribute) to the product's creativity. This approach highlights the particular actions a student can take to address their products shortcomings. Using the results in this manner resembles formative evaluation.
3. ***How* is my product creative?** Using creative production activities, including cycles of feedback and revision, skills such as creativity, technical self-efficacy, and working through complexity can be nurtured (Blikstein, 2013; Vossoughi &

> Bevan, 2014; Barron et al., 2014). Martin et al., in their 2016 study, suggest that analysis of log data through a social learning analytics approach may highlight opportunities to enhance creative production. These authors suggest that a teacher could use log data to observe *how* students have edited their work. This approach allows insight into every stage of the creative process and provides feedback on *how* students can adapt their approach to be more creative. The teacher could use the feedback to indicate the categories and strengths that contribute to the product's creativity and to what extent.

In addition to feedback, data collected in online environments can also be used to comment on the related aspects of the person, process and press. This addresses the fourth level of concrete and differentiated feedback for fostering creativity in the classroom—How do I go about *being* more creative? For example, Dooley and Lindner (2002) demonstrated that the competency catalogue approach could be used both to develop curricula and other teaching materials (i.e., elements of press) and also as a coaching, counselling, and mentoring tool (i.e., addressing the person). As an example of the latter, D. H. Cropley and Cropley (2000, p. 209) proposed "creativity counselling," which focused on process and person. These authors suggested that feedback might be along the lines, "You generated a high level of novelty, but expressed it only in a rough and unfinished way". This advice also informs teacher practice—how can the student be supported in achieving these changes? Research highlights that the approach and the provision of near real-time feedback can encourage support and engagement in teaching and learning and can foster creativity in the user (Britain et al., 2020).

5.6 Process—How Do They Create It?

Process identifies the thinking style that results in an innovative product (Rhodes, 1961). Guilford (1957) first described two main thinking styles associated with creativity: divergent and convergent thinking. While divergent thinking is solely associated with creativity, it is essential to recognise that convergent thinking also plays a critical role, particularly when creativity is considered in problem-solving and LA. Characteristics of divergent thinking include thinking unconventionally, producing multiple answers, or seeing new possibilities. Typical results of this process involve developing alternative or numerous solutions, encountering surprising answers, deviating from the usual, and opening new and exciting opportunities.

In any discussion of process, it is also essential to recognise that creativity does not come from nowhere. It rests on a foundation of knowledge and requires effort. To be a creative student, you first need to be a capable student! The characteristics of convergent thinking vital in supporting the overall process include thinking logically and homing in on the single best answer. Typical results of this process have a quick, "correct" answer and greater familiarity with what already exists (Cropley,

2006a, b). It is vital that both convergent and divergent thinking can be used interchangeably depending on the context in which they are used (Cropley & Cropley, 2009).

Within LA, a product could be designed to encourage the process of both divergent and convergent thinking at different stages of an exercise or activity. For example, an educator could employ a LA feedback tool through Moodle to help students engage in a problem-solving task (*e.g., how to grow food in space*). Wang, in their 2014 study, suggests that Sentiment Analysis could be used to monitor how students are tracking in their divergent thinking phase. Based on this text analysis, real-time feedback could be provided (*i.e., you have only come up with three alternate solutions for growing food, and they are all carbohydrate-based, have you considered fats or proteins?*) before allowing students to proceed with the convergent thinking phase. By providing real-time feedback, the user is encouraged to move through both phases, potentially in a more in-depth manner, thus contributing to a more creative outcome for the task. Additionally, the use of LA in this context allows data to be generated beyond 'how long a user was logged in for' and contributes to an ever-growing data bank of 'what is creative' and 'what isn't creative'. Therefore, LA can objectively demonstrate how to best support creativity dependent on the task. Research highlights how LA dashboards provide insight into student work and guide it as a task is unfolding (Arnold & Pistilli, 2012; Verbert et al., 2013). This real-time feedback has been demonstrated to increase the creative output (Britain et al., 2020).

5.7 Press—Where Does the Creativity Happen?

Press examines the role of environmental and social factors on creativity and is defined loosely as the level of 'pressure' applied or perceived in an environment. Press can be used to address both: (a) how the 'climate'[2] can either facilitate or inhibit creativity; and (b) how the 'environment'[1] reacts to the production of creativity (Cropley, 2015). The positive or negative influence does not directly shape the creative outcome, but rather, these influences mediate or moderate the outcome by affecting variables related to the creative process or person. Press, therefore, highlights elements such as support for creativity (e.g., encouraging risk-taking through a forum), and how the physical environment may foster creativity (e.g., through the provision of appropriate feedback and support) (Cropley, 2015). Press also highlights how society tolerates radical deviations from norms (e.g., 'are creative LA systems or lessons ridiculed or hailed?').

In a LA context, press would relate to how the user interacts with the digital learning environment and the analytics, i.e., through a dashboard such as a LMS. Again, in an education context, whether it be K-12 or higher education, a LA platform must be designed to promote favourable press conditions. As an example, evidence from a Massive Online Open Course (MOOCS) on 'teaching creativity at scale' highlighted that data collected through the MOOC provided insight into how to foster community and engagement amongst students (Tahirsylaj et al., 2018). By analysing such vast amounts of data these authors were able to adapt their course to best suit the students to develop and enhance their creative skills. That is, the authors used analytics to foster the right press conditions for creativity to flourish. Further evidence of facilitating the right press conditions include Clinnin who, in her 2014 study, found that users who engaged with forum discussions were more likely to meet learning objectives. Additionally, users could also be prompted to take risks and develop their willingness to fail without reprimand or embarrassment if a LMS provides feedback on their progress without observation from a teacher (Tahirsylaj, 2012). This may build confidence and stimulate creativity in the user as they can receive non-judgemental feedback.

5.8 Phase—The Stages of Creativity

Whilst the 4P's are a comprehensive framework to understanding what impacts creativity, an extension of this framework is required when recognising and fostering creativity through LA. Creativity and LA are concerned with solving problems; however, the solutions do not emerge in a single step and one must be prompted to understand that there is a sequence of stages that is followed. The first stage is starting with the recognition that there is a problem to be solved. Next is determining the possible ways of solving that problem, narrowing these down to a few probable solutions. Finally, the best option for development and implementation is selected. To understand how to develop creativity through LA, it is first necessary to understand how the 4Ps *intersect* with the stages that we know characterise problem-solving.

The answer to this issue is, therefore, a fifth P—*Phases*. These are the steps involved in generating novel and effective LA products, platforms, or solutions. Guilford (1959) described creativity as problem-solving and defined it as having four stages:

1. *recognition* that a problem exists;
2. *production* of a variety of relevant ideas;
3. *evaluation* of the various possibilities produced;
4. drawing of appropriate conclusions that lead to the *solution* of the problem.

Guilford's model corresponds closely to Wallas's (1926) well-known four-phase model. Wallas explains that in the initial phase of *preparation,* a person becomes thoroughly familiar with a content area. In the *incubation* phase, the person "moves through" or "stews over" the information obtained in the previous phase. In the phase of *illumination,* a solution emerges. The last phase is *verification*. The person tests the solution that has emerged from the phases of *incubation* and *illumination*. More recently, the Wallas model has been refined by adding three additional phases

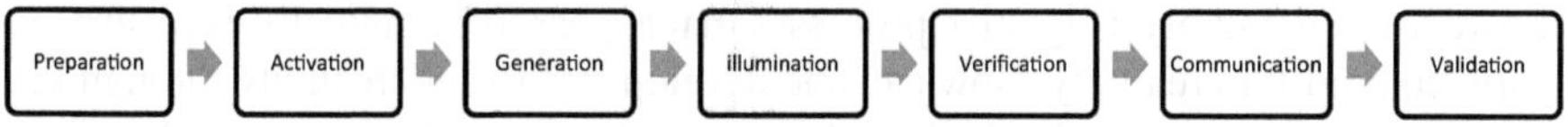

Fig. 5.1 The extended phase model of the creative process

(*activation*, *communication*, *validation*) by (Cropley & Cropley, 2008; Cropley, 2006a, b), who have conceptualised creativity as involving *seven* consecutive *phases* (Fig. 5.1).

The phases of creativity captured in the Extended Phase Model shown in Fig. 5.1 and the fundamental movement between convergent and divergent thinking link strongly to LA's steps as the mechanism by which products and systems are realised. There are several important and recurrent themes both in creative and LA problem-solving. These include the non-linear progression that the process frequently follows. Here is where creativity and Learning Analytics come together. As users of the LA systems move through a series of stages, these involve either convergent or divergent thinking.

The LA system must be designed to support the user to move through these seven stages based on interaction with a task. Take the food challenge task mentioned above. A user could receive creativity training and engage in modules relevant to the phase, i.e., you are in the preparation phase. It would be best if you did X. Then educational data mining techniques could be employed as they are designed as 'an engine to make decisions or guide actions' (Campbell et al., 2007 as cited in Buckingham Shum & Ferguson, 2012 p. 4). This feedback could then allow the educator to reshape the activity. Research shows that students exhibit more creativity and are more engaged when receiving feedback and when courses are adapted to suit their needs (Tahirsylaj et al., 2018).

5.9 Paradoxes of Creativity

To understand the interaction of creativity and LA, one critical factor must be acknowledged. Each of the 4Ps described in previous sections is not *always* good or *always* bad for Creativity (Cropley, 1997; Cropley & Cropley, 2008). Aldowah et al. (2019), in their review on LA for twenty-first-century education, highlight that various LA techniques are required across educational sectors when solving specific academic problems. These authors note that previous LA reviews such as Romero and Ventura (2010), Sacin et al. (2009), Schrire (2004), and Van Barneveld et al. (2012) fail to consider the association between different LA techniques when addressing specific educational problems. Aldowah et al. (2019) propose that to solve various problems, various techniques must be considered. Therefore, sometimes it is necessary to think analytically and sometimes synthetically. This suggests a paradox in creativity. Cognitive processes that appear to be mutually exclusive are *both* required for creativity. How can one develop and foster creativity through LA if one must simultaneously think both convergently *and* divergently? Therefore, discussions of creativity are confronted by several apparent *paradoxes*: Aspects of the processes of creativity, the personal properties associated with it, the conditions that foster its emergence and the products it yields seem to be mutually incompatible.

Similarly, a lack of structure and leadership pressure in any environment may encourage creativity sometimes but hinder it at other times. Properties of the individual—a risk-taker, for example—may be favourable to creativity at some points in the process but unfavourable at different times. The solution to this paradox lies in the fifth P—*Phases*. LA and creativity take place across distinct phases. It is possible to build a model of creativity through LA that identifies the relationships between the person, the process, the product and the press *at each phase* and specifies exactly what conditions favour or inhibit creativity, at each point in the problem-solving process.

5.10 The Innovation Phase Model

The Innovation Phase Model is a framework that addresses the paradoxes of creativity and specifies the thinking skills, personal properties, classroom climate and outcomes associated with each phase of a process of creative problem solving (Cropley & Cropley, 2012; Cropley et al., 2011). As a roadmap for the development of LA to support creativity in education, the IPM outlines what an individual needs to do/be at any stage of creative problem-solving.

The IPM consists of 7 phases and five dimensions and produces 42 nodes, as highlighted in Table 5.1. Phase by phase, Table 5.1 shows the conditions that foster creativity and innovation *change*. What is good for creativity and innovation in, for example, the *activation* phase, may hinder Innovation in the *verification* phase. The 42 nodes in the IPM are 42 things that LA can measure in real-time or near-real-time. The nodes are relevant to creativity irrespective of the task, subject or course. The key to successful creativity and innovation in any context is adjusting and adapting to the favourable conditions at each stage of the IPM.

5.11 How Can LA Use the IPAI to Encourage Creativity?

The Innovation Phase Model has been repeatedly empirically tested through the Innovation Phase Assessment Instrument (IPAI) described in (Cropley & Cropley, 2012; Cropley et al., 2011). The IPAI highlights the relationship between creativity and innovation and demonstrates how teams or organisations may be well-aligned

Table 5.1 The Innovation Phase Model (IPM)

	Phase	Invention					Exploitation	
Dimension	Poles	Preparation knowledge, problem recognition	Activation problem definition, refinement	Generation many candidate solutions	Illumination a few promising solutions	Verification a single optimal solution	Communication a working prototype	Validation a successful 'product'
Process	Convergent versus divergent	Convergent	Divergent	Divergent	Convergent	Convergent	Mixed	Convergent
Person (motivation)	Reactive versus proactive	Mixed	Proactive	Proactive	Proactive	Mixed	Reactive	Reactive
Person (properties)	Adaptive versus innovative	Adaptive	Innovative	Innovative	Innovative	Adaptive	Adaptive	Adaptive
Person (feelings)	Conserving versus generative	Conserving	Generative	Generative	Generative	Conserving	Conserving	Conserving
Product	Routine versus creative	Routine	Creative	Creative	Creative	Routine	Routine	Routine
Press	High demand versus low demand	High	Low	Low	Low	High	High	High

or misaligned to the different conditions that favour creativity at each stage of the innovation process.

To provide a concrete example, let's hypothesise that an educator wants to use LA to encourage creativity through a STEM problem-based learning task. The first phase in the IPAI is *preparation*. In this phase, the goal is to develop knowledge and recognise the problem. From a process point of view, an educator wants a task at this phase to support convergent thinking i.e., an educator will want to see a user spending a considerable amount time researching the *same* topic before moving onto a *new* topic. Therefore, the user logs into their Learning Management System (LMS) and explores different elements of the proposed challenge. Whilst logged into the LMS, the user leaves trace data based on their actions. An indication of convergent thinking may be spending an extended amount of time on a webpage or watching a video on a complex topic. Blomberg et al., in their 2014 study showed that watching a video on a complex topic enables students to better evaluate and integrate their learning, or in other words demonstrate convergent thinking. This trace data can then be used to provide meaningful feedback and encourage the user to remain in the convergent phase. For example, if an educator had OnTask[3] embedded into their LMS, the student would receive personalised and actionable feedback throughout their participation in this phase and the course. Research shows that students are generally dissatisfied with the quality of the feedback they receive (P. Ferguson, 2011). Similarly, educators are under pressure to provide consistent and relevant feedback. By using trace data, tailored and specific feedback could be provided to students. Tailored and specific feedback has been shown to positively impact student perception of feedback quality (Pardo et al., 2019). Therefore, LA can support user creativity and minimise the time needed for educators to provide feedback.

When considering the person element of the 4P's, in this phase the user needs to be motivated. Therefore, videos could be placed on the Moodle, and the educator could receive feedback if the user is watching the videos or not (i.e., through log data) (Khan, & Pardo, 2016). Through this data, the educator could determine whether the user is 'prepared' enough to move onto the next phase. This feedback would allow educators to ascertain where more support is needed or determine if users can be extended into the next phase. This data and feedback would highlight how well aligned or misaligned an individual is, and therefore it can also predict how creative a final solution will be. This example demonstrates how a digital learning environment (DLE) can set up the means to develop creativity more effectively and that LA then uses this DLE to measure outputs that provide the feedback and personalised information that makes learning more effective. This IPAI could be implemented in any DLE irrespective of the task, course, or subject. The IPAI could

[3] OnTask integrates data from any online learning tool and lets the instructor define progress indicators.

provide a holistic picture of how creative one is across different classes and where they could use support. It could be used in K-12, higher education, or an organisation and is not subject or task-specific.

5.12 Benefits

A LA-based approach to developing creativity in users can transform this competency and support the successful transition of individuals into Industry 4.0. LA can assist the development of creativity by minimising the time educators take to provide accurate and objective feedback on how creatively competent their students are. LA can provide near-real-time to real-time feedback, which can be easily distributed to all stakeholders. This process also encourages personalised learning as changes can be made to support development as a user moves through a task. This could potentially positively increase engagement and motivation in both user and educator. Additionally, users and educators learn to work alongside developing technologies such as machine learning or artificial intelligence that could be implemented through LA. Therefore, they could create the view of these technologies as 'colleagues' and consequently develop positive attitudes that will help the transition into Industry 4.0.

LA can also develop the objectives of creativity researchers to objectively determine what factors relate to student grade, motivation, and other factors across various populations. Therefore, researchers can create tasks that support these connections. Whilst modern and sophisticated approaches to the measurement of twenty-first-century skills such as creativity have been proposed (see Wilson & Scalise, 2015), there has been less focus on authentic learning and working environments. For example, the measurement of (complex and collaborative) problem-solving has been measured through the Organisation for Economic and Co-operation and Developments (OECD) Programme for International Student Assessment (PISA). However, PISA is undertaken in highly controlled conditions and does not represent everyday learning conditions (Rosen & Foltz, 2014). However, Learning Analytics allows the measurement of these skills to occur in authentic settings with minimal external interference (Buckingham Shum & Crick, 2016). Recent research has offered promising improvements in LA measurement validity to provide reliable means for developmental assessment of twenty-first-century skills such as creativity (Gašević, 2019). Finally, it is suggested that utilising LA can develop more creative students and consequently contribute to more valued employees.

5.13 Limitations

Whilst there is a push to embed creativity into school curriculums, there is a lack of empirical research connecting creativity and LA. In a LA context, creativity is measured almost solely through fluency tasks, and this is only one tiny aspect of creativity. Fluency is not a measure of the product's creativity; it is a measure of the ideational capacity of the person generating the solution. Additionally, fluency is the total number of responses and we need to consider other elements to determine overall creativity, such as originality, i.e. not just how many responses but also how many unique ideas does each response provide? Currently, the field of LA is very basic in measuring creativity, but current data analysis methods should be applied to extend this competency, and researchers should be encouraged to develop this connection.

5.14 Conclusion

In conclusion, this chapter has provided a summary of how LA can develop creativity in students of all ages across digital learning environments that support various courses and tasks. As society progresses into Industry 4.0, using data to drive the development of competencies such as creativity becomes crucial. We have provided a roadmap of how LA can utilise the theories of the 4P's and the phases of creativity to support the development of creativity.

References

Aldowah, H., Al-Samarraie, H., & Fauzy, W. M. (2019). Educational data mining and learning analytics for 21st century higher education: A review and synthesis. *Telematics and Informatics, 37*, 13–49.

Aluja-Banet, T., Sancho, M. R., & Vukic, I. (2019). Measuring motivation from the virtual learning environment in secondary education. *Journal of Computational Science, 36*, 100629.

Arnold, K. E., & Pistilli, M. D. (2012, April). Course signals at Purdue: Using learning analytics to increase student success. In *Proceedings of the 2nd international conference on learning analytics and knowledge* (pp. 267–270).

Baer, J. M. (2010). Is creativity domain specific? In J. C. Kaufman & R. J. Sternberg (Eds.), *The Cambridge handbook of creativity* (pp. 321–341). Cambridge University Press.

Barron, B., Gomez, K., Pinkard, N., & Martin, C. K. (2014). *The digital youth network: Cultivating digital media citizenship in urban communities*. MIT Press.

Blikstein, P. (2013). Digital fabrication and "making" in education: The democratization of invention. In J. Walter-Herrmann & C. Büching (Eds.), *FabLabs: Of machines, makers and inventors* (pp. 1–21). Transcript Publishers.

Blomberg, G., Sherin, M. G., Renkl, A., Glogger, I., & Seidel, T. (2014). Understanding video as a tool for teacher education: Investigating instructional strategies to promote reflection. *Instructional Science, 42*(3), 443–463.

Britain, G., Jain, A., Lupfer, N., Kerne, A., Perrine, A., Seo, J., & Sungkajun, A. (2020, April). Design is (A) live: An environment integrating ideation and assessment. In *Extended Abstracts of the 2020 CHI conference on human factors in computing systems* (pp. 1–8).

Buckingham Shum, S., & Crick, R. D. (2016). Learning analytics for 21st century competencies. *Journal of Learning Analytics, 3*(2), 6–21.

Buckingham Shum, S., & Ferguson, R. (2012). Social learning analytics. *Journal of Educational Technology & Society, 15*(3), 3–26. Cambridge, UK: Cambridge University Press.

Campbell, J. P., DeBlois, P. B., & Oblinger, D. G. (2007). Academic analytics: A new tool for a new era. *EDUCAUSE Review, 42*(4), 40.

Cropley, A. (1997). Creativity: A bundle of paradoxes. *Gifted and Talented International, 12*(1), 8–14.

Cropley, A. J. (2006a). In praise of convergent thinking. *Creativity Research Journal, 18*(3), 391–404.

Cropley, D. H. (2006b, June). The role of creativity as a driver of innovation. In *2006 IEEE international conference on management of innovation and technology* (vol. 2, pp. 561–565). IEEE.

Cropley, D. H. (2015). *Creativity in engineering: Novel solutions to complex problems*. Academic Press.

Cropley, D. H., & Cropley, A. J. (2000). Fostering creativity in engineering undergraduates. *High Ability Studies, 11*(2), 207–219.

Cropley, D. H., & Cropley, A. J. (2005). Engineering creativity: A systems concept of functional creativity. In J. C. Kaufman & J. Baer (Eds.), *Faces of the muse: How people think, work and act creatively in diverse domains* (pp. 169–185). Lawrence Erlbaum.

Cropley, A. J., & Cropley, D. H. (2008). Resolving the paradoxes of creativity: An extended phase model. *Cambridge Journal of Education, 38*(3), 355–373.

Cropley, A. J., & Cropley, D. (2009). *Fostering creativity: A diagnostic approach for higher education and organizations*. Hampton Press.

Cropley, D. H., & Cropley, A. J. (2011). Understanding value innovation in organisations: A psychological framework. *International Journal of Creativity and Problem Solving, 21*(1), 17–36.

Cropley, D. H., & Cropley, A. J. (2012). A psychological taxonomy of organizational innovation: Resolving the paradoxes. *Creativity Research Journal, 24*(1), 29–40.

Cropley, D. H., & Cropley, A. J. (2013). *Creativity and crime: A psychological analysis*. Cambridge University Press.

Cropley, D. H., & Kaufman, J. C. (2012). Measuring functional creativity: Non-expert raters and the creative solution diagnosis scale. *The Journal of Creative Behavior, 46*(2), 119–137.

Cropley, D. H., Kaufman, J. C., & Cropley, A. J. (2011). Measuring creativity for innovation management. *Journal of Technology Management & Innovation, 6*(3), 13–30.

Deloitte. (2016). *Talent for survival: Essential skills for humans working in the machine age*. Retrieved from https://www2.deloitte.com/content/dam/Deloitte/uk/Documents/Growth/deloitte-uk-talent-for-survival-report.pdf

Dooley, K. E., & Lindner, J. R. (2002). Competency-based behavioral anchors as authentication tools to document distance education competencies. *Journal of Agricultural Education, 43*(1), 24–35.

Ferguson, P. (2011). Student perceptions of quality feedback in teacher education. *Assessment & Evaluation in Higher Education, 36*(1), 51–62.

Gal, L., Hershkovitz, A., Morán, A. E., Guenaga, M., & Garaizar, P. (2017, April). Suggesting a log-based creativity measurement for online programming learning environment. In *Proceedings of the fourth (2017) ACM conference on learning@ scale* (pp. 273–277).

Gašević, D. (2019, August 05). *Using learning analytics to measure 21st-century skills* [Paper presentation]. Research Conference 2019 – Preparing students for life in the 21st century: Identifying, developing and assessing what matters. Retrieved from https://research.acer.edu.au/research_conference/RC2019/5august/3

Gray, A. (2016). *The 10 skills you need to thrive in the Fourth Industrial Revolution*. World Economic Forum. Retrieved from https://www.weforum.org/agenda/2016/01/the-10-skills-you-need-tothrive-in-the-fourth-industrial-revolution/

Guilford, J. P. (1957). Creative abilities in the arts. *Psychological Review, 64*(2), 110.
Guilford, J. P. (1959). Traits of creativity. In H. H. Anderson (Ed.), *Creativity and its cultivation* (pp. 142–161). Harper.
Khalili, A. (2016). Linking transformational leadership, creativity, innovation, and innovation-supportive climate. *Management Decision.*
Khan, I., & Pardo, A. (2016, April). Data2U: Scalable real time student feedback in active learning environments. In *Proceedings of the sixth international conference on learning analytics & knowledge* (pp. 249–253).
Kozbelt, A., Beghetto, R. A., & Runco, M. A. (2010). Theories of creativity. In J. C. Kaufman & R. J. Sternberg (Eds.), *The Cambridge handbook of creativity* (pp. 20–47). Cambridge University Press.
Martin, C. K., Nacu, D., & Pinkard, N. (2016). Revealing opportunities for 21st century learning: An approach to interpreting user trace log data. *Journal of Learning Analytics, 3*(2), 37–87.
Nakano, T. D. C., & Wechsler, S. M. (2018). Creativity and innovation: Skills for the 21st century. *Estudos de Psicologia (Campinas), 35*(3), 237–246.
Olken, H. (1964). Creativity training for engineers its past, present, and future. *IEEE Transactions on Education, 4*, 149–161.
Pardo, A., Jovanovic, J., Dawson, S., Gašević, D., & Mirriahi, N. (2019). Using learning analytics to scale the provision of personalised feedback. *British Journal of Educational Technology, 50*(1), 128–138.
Patston, T. J., Cropley, D. H., Marrone, R. L., & Kaufman, J. C. (2018). Teacher implicit beliefs of creativity: Is there an arts bias? *Teaching and teacher education, 75*, 366–374.
Patston, T. J., Kaufman, J. C., Cropley, A. J., & Marrone, R. (2021). What is creativity in education? A qualitative study of international curricula. *Journal of Advanced Academics, 32*, 207–230.
Plucker, J. A., Beghetto, R. A., & Dow, G. T. (2004). Why isn't creativity more important to educational psychologists? Potentials, pitfalls, and future directions in creativity research. *Educational Psychologist, 39*(2), 83–96.
Rhodes, M. (1961). An analysis of creativity. *The Phi Delta Kappan, 42*(7), 305–310.
Rienties, B., & Rivers, B. A. (2014). Measuring and understanding learner emotions: Evidence and prospects. *Learning Analytics Review, 1*(1), 1–27.
Romero, C., & Ventura, S. (2010). Educational data mining: A review of the state of the art. *IEEE Transactions on Systems, Man, and Cybernetics, Part C (Applications and Reviews), 40*(6), 601–618.
Rosen, Y., & Foltz, P. W. (2014). Assessing collaborative problem solving through automated technologies. *Research and Practice in Technology Enhanced Learning, 9*(3), 389–410.
Sacin, C. V., Agapito, J. B., Shafti, L., & Ortigosa, A. (2009). *Recommendation in higher education using data mining techniques*. Paper presented at the Educational Data Mining 2009.
Schrire, S. (2004). Interaction and cognition in asynchronous computer conferencing. *Instructional Science, 32*(6), 475–502.
Siemens, G. (2012, April). Learning analytics: Envisioning a research discipline and a domain of practice. In *Proceedings of the 2nd international conference on learning analytics and knowledge* (pp. 4–8).
Silva, J. C. S., Zambom, E., Rodrigues, R. L., Ramos, J. L. C., & de Souza, F. D. F. (2018). Effects of learning analytics on students' self-regulated learning in flipped classroom. *International Journal of Information and Communication Technology Education (IJICTE), 14*(3), 91–107.
Society for Learning Analytics Research. About, 2012. Available from http://www.solaresearch.org/about/. Accessed 5 Jan 2021.
Stein, M. I. (1953). Creativity and culture. *Journal of Psychology: Interdisciplinary and Applied, 36*, 311–322.
Sternberg, R. J., & Lubart, T. I. (1996). Investing in creativity. *American Psychologist, 51*(7), 677. https://doi.org/10.1037/0003-066X.51.7.677
Tahirsylaj, A. S. (2012). Stimulating creativity and innovation through Intelligent Fast Failure. *Thinking Skills and Creativity, 7*(3), 265–270.

Tahirsylaj, A., Mann, B., & Matson, J. (2018). Teaching creativity at scale: Overcoming language barriers in a MOOC. *International Journal of Innovation, Creativity and Change, 4*(2), 1–19.
Van Barneveld, A., Arnold, K. E., & Campbell, J. P. (2012). Analytics in higher education: Establishing a common language. *EDUCAUSE Learning Initiative, 1*(1), 1–11.
Verbert, K., Duval, E., Klerkx, J., Govaerts, S., & Santos, J. L. (2013). Learning analytics dashboard applications. *American Behavioral Scientist, 57*(10), 1500–1509.
Vossoughi, S., & Bevan, B. (2014). Making and tinkering: A review of the literature. *National Research Council Committee on Out of School Time STEM, 67*, 1–55.
Wallas, G. (1926). *The art of thought*. Harcourt, Brace and Company.
Wang, S. (2014). A module-based learning analytics system for Facebook supported collaborative creativity learning. In *2014 IEEE 14th international conference on advanced learning technologies* (pp. 495–496). https://doi.org/10.1109/ICALT.2014.146
Wilson, M., & Scalise, K. (2015). Assessment of learning in digital networks. In *Assessment and teaching of 21st century skills* (pp. 57–81). Springer.
Yeh, Y. C., & Lin, C. F. (2015). Aptitude-treatment interactions during creativity training in E-learning: How meaning-making, self-regulation, and knowledge management influence creativity. *Educational Technology & Society, 18*(1), 119–131.

[illegible]

Chapter 6
Using Learning Analytics to Measure Motivational and Affective Processes During Self-Regulated Learning with Advanced Learning Technologies

Florence Gabriel, Elizabeth B. Cloude, and Roger Azevedo

Abstract Self-regulated learning is an important predictor of students' academic achievement, employability, and career progression. Cognitive, affective, metacognitive, and motivational processes play a crucial role in students' ability to effectively monitor and regulate their learning while using advanced learning technologies (ALTs). This chapter focuses primarily on motivational and affective processes related to self-regulated learning with different types of ALTs such as serious games, intelligent tutoring systems, simulations and immersive technologies. While initial approaches to measuring students' motivation and affect have been predominantly centered around self-reported instruments, recent advances in learning analytics and educational data mining show significant benefits in using multimodal data as they reveal the dynamics of learning processes as they unfold with ALTs. As such, our chapter focuses on the use of novel techniques aimed at detecting, tracking, modeling, and fostering students' motivational and affective processes during learning, problem solving, and reasoning with various ALTs. We discuss implications for measuring motivational and affective processes using multimodal data for researchers, students, and educators by combining both objective and subjective methodologies.

F. Gabriel (✉)
Centre for Change and Complexity in Learning, Education Futures, University of South Australia, Adelaide, Australia
e-mail: Florence.Gabriel@unisa.edu.au

E. B. Cloude
Department of Learning Sciences and Educational Research, University of Central Florida, Orlando, FL, USA
e-mail: elizabeth.cloude@soartech.com

R. Azevedo
School of Modeling Simulation and Training, University of Central Florida, Orlando, FL, USA
e-mail: roger.azevedo@ucf.edu

Y. E. Wang et al. (eds.), *Social and Emotional Learning and Complex Skills Assessment*, Advances in Analytics for Learning and Teaching,
https://doi.org/10.1007/978-3-031-06333-6_6

Keywords Self-regulated learning · Motivation · Affect · Learning analytics · Trace data

6.1 Introduction

Advanced learning technologies (ALTs) such as serious games, intelligent tutoring systems, simulations, and immersive technologies can be powerful tools for fostering students' learning about complex topics. However, they require students to maintain increased levels of self-regulation to be successful (Schunk & Greene, 2018). Self-regulated learning (SRL) can be described an active cyclical process which includes goal setting, strategic planning, monitoring strategies, and self-evaluation (Schunk & DiBenedetto, 2020; Winne, 2018) and is a strong predictor of academic success (Schunk & Greene, 2018). SRL has recently been a focus of attention in the learning analytics community (see Emara et al., 2021) and driven new research and methods in SRL and ALTs such as serious games (e.g., Cloude et al., 2020a). Learning analytics is the measurement, collection, analysis and reporting of data about students and their contexts, for purposes of understanding and optimizing learning and the environments in which it occurs. It sits at the convergence of learning (e.g., educational research, learning and assessment sciences, educational technology), analytics (e.g., statistics, visualization, computer/data sciences, artificial intelligence), and human-centered design (e.g., usability, participatory design, sociotechnical systems thinking; see Lang et al., 2017). Cognitive, affective, metacognitive and motivational factors play an important role in SRL (Azevedo et al., 2018). Researchers have access to a wide array of techniques to collect large amounts of fine-grained multimodal data on cognitive and metacognitive processes, including log files, eye tracking, and think-alouds (Azevedo et al., 2019; Jarvela et al., 2020). However, techniques for measuring motivation and affect have been more challenging to develop and are still in their infancy (Ainley & Ainley, 2019). This chapter will introduce innovative measures of motivational and affective processes that rely primarily on learning analytics methods and approaches and present implications of measuring multimodal motivational and affective processes for researchers, students, and educators using a combination of both objective and subjective methodologies.

6.1.1 Motivation and Affect

Motivational and affective processes are both critical for successful learning and are part of a complex system of interdependently connected SRL processes. They drive initial engagement, perseverance and ultimately performance on learning tasks by influencing when, how, why, and which learning strategies are used (Efklides, 2011; Schumacher & Ifenthaler, 2018). Positive affect (e.g., enjoyment, hope, pride) promotes motivation, enabling students to control their engagement with a learning task, and ultimately enhancing their commitment to achieving their learning goals (Pekrun, 2006).

Motivation is a goal-oriented process that allows individuals to initiate, direct and maintain behavior (Schunk et al., 2014). Students can find motivation from various sources, be they internal (values, interests and competence beliefs such as self-efficacy) or external (affordances provided in the design of ALT features, context of the learning environment, and learner-centered instruction), and levels of motivation can fluctuate over the course of a learning task (Ainley & Ainley, 2019). Despite its importance, research shows that students face several challenges in monitoring and regulating their motivation during learning, problem solving and so forth (see Renninger & Hidi, 2019). Wolters (2003, p. 190) defines motivation regulation as "the activities through which individuals purposefully act to initiate, maintain, or supplement their willingness to start, to provide work toward, or to complete a particular activity or goal". Students particularly need to accurately monitor and regulate their motivation when experiencing setbacks or after making mistakes which affect their motivation levels and engagement with a learning task (Wolters, 2003; Reindl et al., 2020). However, unlike contemporary research using trace methods to measure cognitive and metacognitive SRL processes, most of the motivation research has focused on self-reports and therefore not captured the temporally unfolding dynamics of motivational processes during learning, problem solving and reasoning with ALTs.

In the context of SRL, the most important component of affect is arguably emotion regulation, though we must not neglect its links with mood, feelings and attitudes (Efklides et al., 2018; McRae & Gross, 2020; Taub et al., 2019). Students with strong emotion-regulation skills may use these skills to develop positive affect and engage with learning more deeply than their peers. They develop adaptive emotions towards learning (e.g., curiosity and enjoyment), they experience less intense negative and potentially detrimental emotions and they are more inclined to use more metacognitive strategies such as monitoring their progress toward meeting their goals (Lajoie et al., 2020; Price et al., 2018). By contrast, students with negative affect (e.g., anxiety, frustration) tend to display lower levels of motivation, do not persevere as much, and engage in superficial learning (Gabriel et al., 2020; Pekrun et al., 2011; Pekrun, 2013).

6.2 Measuring Motivational and Affective Processes Using Multimodal Data

Motivation and affect can be conceptualized as either static traits or dynamic states (Pekrun, 2006; Ainley & Ainley, 2019). Traits can be described as a general orientation or disposition, whereas states are experienced on task at a specific point in time and tend to fluctuate (Pekrun, 2006). The main difference between traits and states is the temporal element. Traits tend to be more stable, consistent and predictable relative to states, which are temporary and influenced by individual and situational variables (e.g., context, task, prior knowledge; Pekrun, 2020; Robinson & Clore, 2002; Winne, 2020). Before the advent of learning analytics, researchers were limited in their ability to measure dynamic states, and studies have understandably relied on traditional measures that captured the static aspect of motivation and affect. It has been reported that up to 90% of these studies rely upon self-reported instruments (Klassen & Usher, 2010), which are famously problematic in that students do not and cannot always report their motivational and affective strategies accurately for a variety of reasons (Panadero et al., 2016; Zimmerman, 2008). For example, students may inaccurately recall their motivational and emotional strategies as the intensity of their emotions and motivation changes over time depending on the learning context; they may misunderstand questions; or they simply may not have sufficient declarative knowledge to correctly label the strategies they used (Karabenick et al., 2007; Rovers et al., 2019; Veenman, 2011). These issues challenge the validity of such measures in accurately detecting, measuring, tracking, modeling, and supporting students' learning with ALTs. By only measuring static traits that do not fluctuate during learning, we end up missing key information that might provide insight into the temporal and dynamic nature of these constructs.

Our capacity to capture and analyze multimodal data during learning activities has grown exponentially over the last decade (D'Mello, 2017; Ochoa, 2017). This has caused a noteworthy shift in how researchers define and measure motivation and affect, with models such as D'Mello and Graesser's Model of Affective Dynamics (2012) becoming widely accepted and techniques such as eye-tracking, facial monitoring, learner-system interactions and psycho-physiological indicators becoming commonplace to measure dynamic states (Azevedo & Gašević, 2019; Lodge et al., 2019). The next section will focus on novel techniques (e.g., biometric sensors, trace data) aimed at detecting, tracking, modeling, and fostering students' motivational (e.g., interest, task value, self-efficacy) and affective (e.g., emotion regulation) processes during learning, problem solving, and reasoning with ALTs.

6.2.1 How Can These New Techniques Be Used to Detect, Track, Model, and Foster Students' Motivation and Affect?

The rise of novel ALTs, such as immersive virtual environments, and computer power has provided an ubiquitous platform for capturing and analyzing multimodal data across spatial and temporal dimensions (i.e., multimodal learning analytics) to study learning with emerging technologies (Azevedo & Gašević, 2019; Dindar et al., 2019; Järvenoja et al., 2020; Lajoie et al., 2020; Mu et al., 2020; Noroozi et al., 2020). The majority of studies have used multimodal learning analytics to study cognitive and metacognitive aspects of the learning process (Azevedo &

Gašević, 2019; Sharma & Giannakos, 2020). This focus likely results from the zeitgeist of the late 1980s, which had deep roots in understanding learning using information-processing theory, especially in regard to cognitive sciences and its contributions to advancing our understanding of human cognition and metacognition using computers and artificial intelligence such as ACT-R (Anderson & Fincham, 2014) and SOAR architectures (Laird, 2012). This movement ultimately provided a foundation for building intelligent systems capable of capturing cognitive and metacognitive processes as they occurred over time such as students' initiated metacognitive monitoring using an SRL palette designed into MetaTutor (Azevedo et al., 2018; Cloude et al., 2020b; Mangaroska et al., 2020). However, SRL does not only involve cognition and metacognition and would not exist without motivational and affective processes (Schunk & Greene, 2018; Winne & Azevedo, 2022). Few studies have leveraged new techniques to capture motivational and affective processes during learning activities with ALTs that go beyond self-report methodologies (i.e., studying affect and motivation as traits; Ainley & Ainley, 2019). We argue this gap in literature misses key information that could inform effective instructional and educational practices. Further, since contemporary theoretical frameworks describing motivation (e.g., metamotivation; Miele et al., 2020) and affect (e.g., emotion regulation; McRae & Gross, 2020) explain that several key factors influence motivational and affective processes, such as the context and individual variables (Murayama & Elliot, 2011; Pintrich, 2000, 2003), it is essential to capture multiple streams of multimodal data during learning sessions to define these constructs as dynamical states. These data provide insight into the complex and dynamic nature of motivation and affect to understand their role in developing SRL, task, and domain-specific knowledge and skills with ALTs.

Other methodologies have been developed with aims to reduce the reliability/validity concerns associated with using self-report data, such as the microanalytic approach for studying SRL (e.g., Cleary et al., 2019; Follmer & Sperling, 2019; Zimmerman, 2008). This approach uses both self-report items as well as semi-structured interview questions that are administered at specific points in the learning session to gauge contextual and task-specific information (e.g., students' metacognition related to a specific problem or topic) rather than more general construct information (e.g., overall metacognitive awareness). For example, to study metacognition using the microanalytic approach during learning about complex biology topics with ALTs might involve administering both open- and close-ended questions during certain time points (e.g., moving to a new content page) across the learning session. The system might prompt the learner during a reading task to assess their feelings about how well they understand the content using a Likert scale ranging from 1 to 6 (e.g., 1 = *I feel I strongly do not understand* to 6 = *I feel strongly I understand*) and then administer a follow-up question to gauge why they felt the way they did and whether it relates to their learning goals. This method generates both qualitative and quantitative data before, during, and after specific learning activities to measure task- and context-specific information as it relates to metacognition based on what the learner is doing. While this is a notable step toward measuring SRL as both a dynamic and fluctuating states that accounts for the context and task the learner is engaging in, the approach is not impartial to reliability and validity concerns due to the very nature of self-reporting that relies on subjective data measuring the students' perception of their metacognitive awareness (Ainley & Ainley, 2019; Cloude et al., 2018).

In this chapter, we argue that supplementing multimodal (objective) data with other methodologies, such as the microanalytic approach, might provide a grounding framework to assist researchers in mitigating the challenges plaguing the field to augment our understanding of the role of motivation and affect on SRL across spatial and temporal dimensions. A systematic review conducted by Noroozi et al. (2020) assessed studies using multimodal data to capture motivational and affective processes during learning with emerging technologies. While they found that the majority of studies still used survey methods, they also reported a significant growth in the proportion of studies leveraging other modalities such as eye movements, concurrent verbalization, and trace data to capture motivational and affective processes during learning activities. For instance, a longitudinal study by Wong and colleagues (2016) captured data on motivation using surveys, interviews, and reflective essays, while another study by Shahrokni and Talaeizadeh (2013) used trace data, interviews, and interactions with peers via forums, messages, and/or chats to capture motivation and affect during language learning. These studies demonstrate that triangulating multiple data channels such as capturing and triangulating both objective (i.e., trace data) and subjective (i.e., interviews, surveys) data points over time and at critical points during learning (e.g., administering self-report items measuring motivation and affect when students change their strategies during learning) has the potential to capture motivational and affective processes during learning activities.

A more recent study conducted by Emerson et al. (2020) used continuous data to generate multimodal analytics and assess its relation to self-reported interest, a motivation construct, by comparing different machine-learning models to assess the accuracy rate of either unimodal and multimodal data channels in their ability to predict interest after learning with a game-based learning environment. They used a combination of the following modalities to predict interest: (1) trace data of in-game actions, (2) eye movements, and (3) facial expressions of emotions. After comparing different models and their parameters, the results showed the most accurate model in predicting high, medium, or low interest was using both eye movements and trace data during learning, achieving an accuracy rate of 0.59. While the accuracy was moderate (i.e., 40% error rate), this study demonstrated significant progress in implementing a novel approach that used multiple streams of data captured during learning activities to examine whether it was predictive of interest with emerging technologies (Emerson et al., 2020). A similar study by Taub et al. (2020) examined the impact of agency during game-based learning on problem solving, scientific reasoning, and interest using multimodal data. They used the following data streams in their analyses: (1) trace data on in-game actions, (2) facial expressions of emotions, and (3) performance on content assessments. Results showed that facial expressions of emotions and trace data during game-based learning across the agency conditions were related to self-reported interest (Taub et al., 2020). These studies illustrate a start in the right direction, as relationships revealed between self-reported motivation and multimodal data might provide insight into how to capture motivation in real-time during learning activities. For example, what might facial expressions of emotions and eye tracking reveal about motivation and how it changes over time? Do fixations associated with particular interface features reveal students' motivation as it relates to achieving specific goals like learning about the topic versus indicating no interest in learning about the topic? Further, could self-reported perceptions of motivation offer a 'grounding' technique for what the learner perceives their motivation to be and then assess the extent to which multimodal data generated before, during, and after reporting motivation might reveal relationships between eye gaze and facial expressions of emotions as indicators of motivation in real-time?

A study by Järvenoja et al. (2020) used a similar approach and examined motivation and emotion regulation during collaborative-learning activities at the individual- and group-level using video and physiological instruments. Specifically, the video recordings provided data on individual students' voice, facial expressions of emotions, and interactions among other group members, while the physiology instrument measured electrodermal activity and heart rate which have been previously correlated with intensive, temporal information on motivation, affect, and cognitive constructs during task execution (Efklides et al., 2018). Other studies have used this novel approach as well, highlighting its capacity to capture motivational and affective fluctuations across learning activities (Bakhtiar et al., 2018; Dindar et al., 2019; Goetz et al., 2016; Ketonen et al., 2018). However, major challenges continue to exist as most studies using multimodal data to examine fluctuations in motivation and affective processes continue to use traditional modeling techniques that cannot handle the nonlinear and complex nature of motivation and affective constructs, presenting more issues on the current state of challenges plaguing the field.

6.2.2 Modeling Motivational and Affective Processes

Common tools used to analyze multimodal data are traditional linear statistics (e.g., regression, clustering, correlation, analysis of variance, etc.), which test relationships between outcome and predictor variables. These tools require adhering to three primary assumptions to reveal meaningful results: observation independence, normality of frequency distribution, and equal variance. While linear techniques account for unsystematic measurement variation, they do not account for systematic sources of variation (e.g., dynamical states such as intrinsic fluctuations like emotion regulation during learning; Amon & Holden, 2019) versus unsystematic variation (e.g., measurement errors). Since contemporary theoretical models emphasize the dynamic, interdependent, and adaptive nature of motivation and affect, it draws us to question the appropriateness of using linear statistics to analyze data capturing motivational and affective processes during learning with ATLs. Do linear techniques align with contemporary theories and their underlying assumptions? Since variation in motivational and affective processes often results from countless and intertwined intrinsic interactions sensitive to contextual circumstances, it is imperative that we implement modeling techniques built to handle the non-linear nature of motivational and affective constructs in order to partition the potential measurement errors that may exist in multimodal datasets and study individual differences in its variation.

Other limitations involve the very nature of the multimodal data we analyze, which more often than not fail to meet statistical assumptions of linear modeling. Linear techniques are based on the notion that averaging across data points reveals the 'ground truth' about the phenomena being studied (Laplace, 1812). But we are drawn to question this assumption since studies also reveal that the fluctuating dynamics of motivational and affective processes impact learning processes and academic achievement (Bakhtiar et al., 2018; Cloude et al., 2020a; Goetz et al., 2016; Moeller et al., 2018). We argue that in order to model the temporal and spatial dimensions related to motivational and affective processes, future studies need to implement non-linear techniques, such as fractals or recurrence quantification analysis, which are built to handle the continuous (or reoccurring) changes in motivation and affective processes rather than averaging across those dynamics (e.g., Amon etal., 2018).

Few studies have begun using non-linear techniques to examine a range of topics such as biological, psychological, and physiological processes (Favela, 2020; Amon et al., 2018). For instance, Amon and others (2018) modeled cognitive control using fractal equations. Their results illustrated that this scaling technique could handle changes over time as well as the interplay between the learner and context in which they were learning. Other studies have found similar oscillatory fluctuations in cognition during reaction-time tasks using fractal analysis, where their findings emphasized that changes in cognition were not due to randomness (e.g., measurement errors) but rather individual differences (e.g., prior knowledge; Gilden, 2001). These findings highlight that using non-linear techniques may provide insight into motivational and affective processes as they occur in the real world to study how they relate to learning and achievement. However, in order to leverage non-linear techniques, we recommend future studies reconsider motivational and affective processes as complex systems. Referring to complexity science may offer novel frameworks and tools that could provide benefit to studying motivational and affective processes with ALTs by incorporating concepts, methods, and theories of complexity science (e.g., embodied cognition). If motivation and affect contemporary frameworks represent distinctive features that characterize complex systems (i.e., self-organization, emergence, and interaction dominance), they ought to be investigated via complexity science (Favela, 2020). In the next section, we describe the implications for multimodal motivation and affect SRL data for researchers, students, and educators.

6.3 Implications of Measuring Multimodal Motivational and Affective Processes for Researchers, Students, and Educators

6.3.1 Implications for Researchers

A major concern for researchers examining multimodal data is how to use, modify, and/or develop new techniques to capture the dynamics of motivational and affective processes during learning, reasoning, and problem solving with ALTs. While much work has been done in the area of affect and learning analytics in terms of using facial expressions of emotions and physiological sensors (e.g., D'Mello, 2017), the question remains whether the same methods can be used to capture motivational states. For example, how can researchers validly and reliably measure (detect), track and model motivational states during learning with ALTs? What do a student's fluctuations in their interest, task value, self-efficacy, etc. look like, and can they be captured using facial expressions and physiological data? If not, then can they be combined with other methods such as eye-movements and verbalizations which have predominantly been used to measure cognitive and metacognitive processes? Or, do we have to develop new methods or use the ALTs themselves as the research and instructional tool such as taking advantage of the affordances of immersive technologies? For example, imagine using traditional self-reports measures (e.g., sub-scales of the MSLQ, AEQ) in a stealthy manner by having artificial agents embedded in an immersive virtual environment administer the questions included in these questionnaires (e.g., "*I think the instructional material in this VR system is useful for me to learn. Do you feel they are useful? If not, then tell me why? I'm really interested in understanding the basic concepts of the cardiovascular system in this VR system. Do you feel the same levels of interest? If not, then what should we do to generate more interest?*") as part of their behavioral repertoire. This would serve as combining the microanalytic approach with objective measures that are triggered by embedded artificial agents to gauge motivation and affect questions during a natural exchange between them and the students (during learning) using natural language processing (NLP) and be strategically timed based on (1) theoretical assumptions and research questions testing specific hypotheses (e.g., reactivity to explicitly asking about motivational and affective states induces the learner to reveal their motivational and affective states, beliefs, etc.), (2) time thresholds (e.g., ask about task value after 10 minutes of learning and 10 minutes before a learning session), and (3) intelligently based on learning analytics and other real-time analyses of multimodal data (and may include explicit or implicit modeling of specific motivational and affective states by the artificial agents). Using embedded artificial agents as research tools in immersive virtual environments can significantly advance the field by allowing research to systematically design and test theoretical assumptions related to the complexity of capturing the temporal dynamics of motivation and affective processes using multimodal data. The advances will allow for fine-grained specification of time-scale of the phenomena (from milliseconds to hours and days), temporal fluctuations based on internal and external constraints, sequencing, occurrence, re-occurrence, monitoring and regulating, and transfer of knowledge and skills to other ALTs and non-ALT learning contexts and tasks. Lastly, issues of ethics and privacy will continue to be a major concern as researchers strive to collect significantly more multimodal data from students, including negative emotions which have been shown to be influential in enhancing learning. In addition, these issues remain a constant concern for researchers especially when negative personal or socially "undesirable" motivational (e.g., prolonged low situation interest) and affective (e.g., extended episodes of confusion which under certain conditions may be beneficial to learning) processes may have negative repercussions on students' well-being and impact their educational opportunities, academic achievement and success.

6.3.2 *Implications for Students*

A major concern for students is their general lack of ability to dynamically and accurately monitor and regulate their motivational and affective states during learning with ALTs. Recent advances in AI, learning analytics, and educational data mining have enabled new opportunities for visualizing data including open learner models (OLMs) to foster students' understanding of their cognitive, affective, metacognitive, and motivational strategies and skill development while using ALTs (Bull, 2020; Bull & Kay, 2016). OLMs are a visualization of the system's internal beliefs about the student's current knowledge, skills, and abilities. They externalize the system's internal model of the student to aid their monitoring of learning processes. OLMs are a means of visualizing the current knowledge or skill levels of students in various ways, such as helping students independently track, reflect on, and pace their learning processes so they can learn more effectively (Bull, 2020). We propose that OLMs will be ideal for supporting and fostering the awareness needed to monitor, regulate, and reflect on motivational and affective processes (e.g., students can inspect, edit, and/or negotiate their motivational and affective states presented in OLMs). However, a neglected aspect of OLMs has been motivation and affect, which is critical for successful learning, problem solving, and reasoning with ALTs. Imagine a learner inspecting an OLM and it illustrates the system's beliefs about their current motivational and affective states (e.g., low task value, low situational interest). The learner could use such data to (1) inspect the OLM and raise their awareness that they have not been monitoring their confusion, thus allowing them to identify the source of the confusion and potentially resolving their impasse; (2) reflect on their levels of interest and task value and edit their OLM beliefs about their interest and task value because the system beliefs are not accurate and this gives the learner autonomy to correct the systems' beliefs; and (3) negotiate with the system as the learner justifies why they believe the OLM does not accurately represent their current motivational and affective states. The design of the OLMs can include data analytics from the various data and the major challenge will be the sensing of the changes in motivational and affective states—e.g., when do we know the ALT detected change in student's self-efficacy? What was or were the sources of data? How reliable and valid are they? How does the sensing of this motivational state translate to a quantifiable index (and qualitative value) that can be represented in the students' OLM? When, how, what does the OLM illustrate regarding the learner's self-efficacy? Is it the current state, how the current state differs from previous state (and the time between states), the history of self-efficacy states during learning (and the contextual information as to when the changes occurred), does it project future self-efficacy states based on beliefs about the learner as well as other learning and multimodal data? These are open questions that can advance the current state of measuring motivational and affective states and their implication to improve students' self-regulation.

6.3.3 Implications for Educators

The motivational and affective data presented in OLMs can come from several sources, including self-report measures, learner-system interactions with artificial pedagogical agents (see previous sections), learning analytics from other instructional resources (e.g., log-files from learning with MOOCs), and other external regulating agents (e.g., peers, teachers, parents, etc.). Similar to the scenarios and questions raised in the previous section (*Implications for students*), these same data can be accessed by educators to understand the fundamental motivational and affective processes of students by representing the same data using learning analytics onto teachers dashboards (Klerkx et al., 2017; Molenaar & Knoop-van Campen, 2019; Wiedbusch et al., 2021). Teacher dashboards based on learning analytics of students' multimodal motivational and affective (and other SRL) data can be objectively presented to teachers to enhance their instructional decision-making. By focusing on motivational and affective states, we can significantly enhance the current states of teacher dashboards and learning analytics that tend to focus on behavioral indices of instruction (e.g., number of problems completed, time spent on specific tasks) while ignoring the complexities of motivational and affective states. By providing teachers with dashboards that include these data, teachers may have opportunities to more effectively isolate specific areas of self-regulation that need attention in the classroom (e.g., emotional dysregulation). The area of teacher dashboards opens new opportunities to enhance instructional practices. For example, imagine a teacher inspecting the edits a student has made to their OLM that are inaccurate and, based on this data, the teacher "pushes-out" a message that explains to the student that their edits to their motivational states on their OLM are inaccurate and provides an explanation. In addition, instead of the system's beliefs about the students being presented on their OLMs, the teacher overrides the system and may "manipulate" the OLM states' values to encourage all students to raise their awareness and subsequently engage in more accurate monitoring and regulating of motivation and affect during learning with ALTs. In summary, implications of measuring multimodal motivational and affective processes align with current national and international reforms aimed at modernizing methods of assessing student learning with ALTs using multimodal data and learning analytics (NASEM, 2018; OECD, 2020).

References

Ainley, M., & Ainley, J. (2019). Motivation and learning: Measures and methods. In K. A. Renninger & S. E. Hidi (Eds.), *The Cambridge handbook of motivation and learning* (pp. 665–688). Cambridge University Press.

Amon, M. J., & Holden, J. G. (2019). The mismatch of intrinsic fluctuations and the static assumptions of linear statistics. *Review of Philosophy and Psychology, 12*, 1–25. https://doi.org/10.1007/s13164-018-0428-x

Amon, M. J., Pavlov, O. C., & Holden, J. G. (2018). Synchronization and fractal scaling as foundations for cognitive control. *Cognitive Systems Research, 50*, 155–179.

Anderson, J. R., & Fincham, J. M. (2014). Extending problem-solving procedures through reflection. *Cognitive Psychology, 74*, 1–34.

Azevedo, R., & Gašević, D. (2019). Analyzing multimodal multichannel data about self-regulated learning with advanced learning technologies: Issues and challenges. *Computers in Human Behavior, 96*, 207–210.

Azevedo, R., Taub, M., & Mudrick, N. V. (2018). Using multi-channel trace data to infer and foster self-regulated learning between humans and advanced learning technologies. In D. Schunk & J. A. Greene (Eds.), *Handbook of self-regulation of learning and performance* (2nd ed., pp. 254–270). Routledge.

Azevedo, R., Mudrick, N. V., Taub, M., & Bradbury, A. (2019). Self-regulation in computer-assisted learning systems. In J. Dunlosky & K. Rawson (Eds.), *Handbook of cognition and education* (pp. 587–618). Cambridge University Press.

Bakhtiar, A., Webster, E. A., & Hadwin, A. F. (2018). Regulation and socio-emotional interactions in a positive and a negative group climate. *Metacognition and Learning, 13*(1), 57–90. https://doi.org/10.1007/s11409-017-9178-x

Bull, S. (2020). There are open learner models about! *IEEE Transactions on Learning Technologies, 13*(2), 425–448.

Bull, S., & Kay, J. (2016). SMILI☺: A framework for interfaces to learning data in open learner models, learning analytics and related fields. *International Journal of Artificial Intelligence in Education, 26*(1), 293–331.

Cleary, T. J., Konopasky, A., La Rochelle, J. S., Neubauer, B. E., Durning, S. J., & Artino, A. R. (2019). First-year medical students' calibration bias and accuracy across clinical reasoning activities. *Advances in Health Sciences Education, 24*(4), 767–781.

Cloude, E. B., Taub, M., & Azevedo, R. (2018). Investigating the role of goal orientation: Metacognitive and cognitive strategy use and learning with intelligent tutoring systems. In R. Nkambou, R. Azevedo, & J. Vassileva (Eds.), *ITS'18: Proceedings from the international conference of intelligent tutoring systems* (pp. 44–53). Springer. https://doi.org/10.1007/978-3-319-91464-0_5

Cloude, E. B., Dever, D. A., Wiedbusch, M. D., & Azevedo, R. (2020a). Quantifying scientific thinking using multichannel data with Crystal Island: Implications for individualized game-learning analytics. *Frontiers in Education, 5*, 572546. https://doi.org/10.3389/feduc

Cloude, E. B., Wortha, F., Dever, D., & Azevedo, R. (2020b). How do emotions change during learning with an intelligent tutoring system? Metacognitive monitoring and performance with MetaTutor. In S. Denison, M. Mack, Y. Xu, & B. C. Armstrong (Eds.), *Proceedings of the 42nd annual conference of the cognitive science society* (pp. 423–429). Cognitive Science Society.

D'Mello, S. (2017). Emotional learning analytics. In C. Lang, G. Siemens, A. Wise, & D. Gašević (Eds.), *The handbook of learning analytics* (pp. 115–127). Society for Learning Analytics Research.

D'Mello, S., & Graesser, A. (2012). Dynamics of affective states during complex learning. *Learning and Instruction, 22*(2), 145–157.

Dindar, M., Malmberg, J., Järvelä, S., Haataja, E., & Kirschner, P. A. (2019). Matching self-reports with electrodermal activity data: Investigating temporal changes in self-regulated learning. *Education and Information Technologies, 25*, 1785–1802. https://doi.org/10.1007/s10639-019-10059-5

Efklides, A. (2011). Interactions of metacognition with motivation and affect in self-regulated learning: The MASRL model. *Educational psychologist, 46*(1), 6–25.

Efklides, A., Schwartz, B. L., & Brown, V. (2018). Motivation and affect in self-regulated learning: Does metacognition play a role? In D. H. Schunk & J. A. Greene (Eds.), *Educational psychology handbook series. Handbook of self-regulation of learning and performance* (pp. 64–82). Routledge/Taylor & Francis Group.

Emara, M., Hutchins, N., Grover, S., Snyder, C., & Biswas, G. (2021). Examining student regulation of collaborative, computational, problem-solving processes in open-ended learning environments. *Journal of Learning Analytics, 8*, 49–74.

Emerson, A., Cloude, E. B., Azevedo, R., & Lester, J. (2020). Multimodal learning analytics for game-based learning. *British Journal of Educational Technology, 51*(5), 1505–1526.

Favela, L. H. (2020). Cognitive science as complexity science. In *Wiley interdisciplinary reviews: Cognitive science* (p. e1525). Wiley-Blackwell.

Follmer, D. J., & Sperling, R. A. (2019). Examining the role of self-regulated learning microanalysis in the assessment of learners' regulation. *The Journal of Experimental Education, 87*(2), 269–287.

Gabriel, F., Buckley, S., & Barthakur, A. (2020). The impact of mathematics anxiety on self-regulated learning and mathematical literacy. *Australian Journal of Education, 64*(3), 227–242. https://doi.org/10.1177/0004944120947881

Gilden, D. L. (2001). Cognitive emissions of 1/f noise. *Psychological Review, 108*(1), 33–56. https://doi.org/10.1037/0033-295X.108.1.33

Goetz, T., Sticca, F., Pekrun, R., Murayama, K., & Elliot, A. J. (2016). Intraindividual relations between achievement goals and discrete achievement emotions: An experience sampling approach. *Learning and Instruction, 41*, 115–125.

Järvelä, S., Gašević, D., Seppänen, T., Pechenizkiy, M., & Kirchener, P. (2020). Bridging learning sciences, machine learning, and affective computing for understanding cognition and affect in collaborative learning. *British Journal of Educational Technology, 51*, 2391–2400.

Järvenoja, H., Malmberg, J., Törmänen, T., Mänty, K., Haataja, E., Ahola, S., & Järvelä, S. (2020). A collaborative learning design for promoting and analyzing adaptive motivation and emotion regulation in the science classroom. *Frontiers in Education, 5*, 111. https://www.frontiersin.org/article/10.3389/feduc.2020.00111

Karabenick, S. A., Woolley, M. E., Friedel, J. M., Ammon, B. V., Blazevski, J., Bonney, C. R., … Kelly, K. L. (2007). Cognitive processing of self-report items in educational research: Do they think what we mean? *Educational Psychologist, 42(3)*, 139–151.

Ketonen, E. E., Dietrich, J., Moeller, J., Salmela-Aro, K., & Lonka, K. (2018). The role of daily autonomous and controlled educational goals in students' academic emotion states: An experience sampling method approach. *Learning and Instruction, 53*, 10–20.

Klassen, R., & Usher, E. (2010). Self-efficacy in educational settings: Recent research and emerging directions. In T. Urdan & S. Karabenick (Eds.), *The decade ahead: Theoretical perspectives on motivation and achievement* (pp. 1–33). Emerald Group Publishing Limited.

Klerkx, J., Verbert, K., & Duval, R. (2017). Learning analytics dashboards. In C. Lang, G. Siemens, A. Wise, & D. Gašević (Eds.), *The handbook of learning analytics* (pp. 143–150). Society for Learning Analytics Research.

Laird, J. E. (2012). *The SOAR architecture*. MIT Press.

Lajoie, S. P., Pekrun, R., Azevedo, R., & Leighton, J. P. (2020). Understanding and measuring emotions in technology-rich learning environments. *Learning and Instruction, 70*, 101272.

Lang, C., Siemens, G., Wise, A., & Gasevic, D. (Eds.). (2017). *Handbook of learning analytics* (pp. 115–127). Solar, Society for Learning Analytics Research.

Laplace, P. S. marquis de. (1812). *Théorie analytique des probabilités*. V. Courcier.

Lodge, J. M., Panadero, E., Broadbent, J., & De Barba, P. G. (2019). Supporting self-regulated learning with learning analytics. In *Learning analytics in the classroom: Translating learning analytics research for teachers* (pp. 45–55). Routledge.

Mangaroska, K., Sharma, K., Gasevic, D., & Giannakos, M. (2020). Multimodal learning analytics to inform learning design: Lessons learned from computing education. *Journal of Learning Analytics, 7*(3), 79–97. https://doi.org/10.18608/jla.2020.73.7

McRae, & Gross, J. J. (2020). Emotion regulation. *Emotion, 20*, 1–9.

Miele, D. B., Scholer, A. A., & Fujita, K. (2020). Metamotivation: Emerging research on the regulation of motivational states. In A. J. Elliot (Ed.), *Advances in motivation science* (pp. 1–42). Elsevier.

Moeller, J., Ivcevic, Z., Brackett, M. A., & White, A. E. (2018). Mixed emotions: Network analyses of intra-individual co-occurrences within and across situations. *Emotion, 18*(8), 1106.

Molenaar, I., & Knoop-van Campen, C. (2019). How teachers make dashboard information actionable. *IEEE Transactions on Learning Technologies, 12*, 347–355.

Mu, S., Cui, M., & Huang, X. (2020). Multimodal data fusion in learning analytics: A systematic review. *Sensors, 20*(23), 6856. https://doi.org/10.3390/s20236856

Murayama, K., & Elliot, A. J. (2011). Achievement motivation and memory: Achievement goals differentially influence immediate and delayed remember–know recognition memory. *Personality and Social Psychology Bulletin, 37*(10), 1339–1348.

National Academies of Sciences, Engineering, and Medicine (NASME). (2018). *How people learn II: Learners, contexts, and cultures*. NASME.

Noroozi, O., Pijeira-Díaz, H. J., Sobocinski, M., Dindar, M., Järvelä, S., & Kirschner, P. A. (2020). Multimodal data indicators for capturing cognitive, motivational, and emotional learning processes: A systematic literature review. *Education and Information Technologies, 25*, 5499–5547.

Ochoa, X. (2017). Multimodal learning analytics. In C. Lang, G. Siemens, A. Wise, & D. Gašević (Eds.), *The handbook of learning analytics* (pp. 129–141). Society for Learning Analytics Research.

Organisation for Economic Co-operation and Development (OECD). (2020). *PISA 2025 assessment and analytical framework*. OECD.

Panadero, E., Klug, J., & Järvelä, S. (2016). Third wave of measurement in the self-regulated learning field: When measurement and intervention come hand in hand. *Scandinavian Journal of Educational Research, 60*(6), 723–735.

Pekrun, R. (2006). The control-value theory of achievement emotions: Assumptions, corollaries, and implications for educational research and practice. *Educational Psychology Review, 18*(4), 315–341.

Pekrun, R. (2013). Emotion, motivation, and self-regulation: Common principles and future directions. In *Emotion, motivation, and self-regulation: A handbook for teachers* (pp. 167–188). Emerald.

Pekrun, R. (2020). Self-report is indispensable to assess students' learning. *Frontline Learning Research, 8*, 185–193.

Pekrun, R., Goetz, T., Frenzel, A. C., Barchfeld, P., & Perry, R. P. (2011). Measuring emotions in students' learning and performance: The Achievement Emotions Questionnaire (AEQ). *Contemporary Educational Psychology, 36*(1), 36–48.

Pintrich, P. R. (2000). An achievement goal theory perspective on issues in motivation terminology, theory, and research. *Contemporary Educational Psychology, 25*(1), 92–104.

Pintrich, P. R. (2003). A motivational science perspective on the role of student motivation in learning and teaching contexts. *Journal of Educational Psychology, 95*(4), 667–686. https://doi.org/10.1037/0022-0663.95.4.667

Price, M. J., Mudrick, N. V., Taub, M., & Azevedo, R. (2018). The role of negative emotions and emotion regulation on self-regulated learning with metatutor. In *International conference on intelligent tutoring systems* (pp. 170–179). Springer.

Reindl, M., Tulis, M., & Dresel, M. (2020). Profiles of emotional and motivational self-regulation following errors: Associations with learning. *Learning and Individual Differences, 77*, 101806.

Renninger, K. A., & Hidi, S. E. (Eds.). (2019). *The Cambridge handbook of motivation and learning*. Cambridge University Press.

Robinson, M. D., & Clore, G. L. (2002). Belief and feeling: Evidence for an accessibility model of emotional self-report. *Psychological Bulletin, 128*(6), 934.

Rovers, S. F. E., Clarebout, G., Savelberg, H. H. C. M., et al. (2019). Granularity matters: Comparing different ways of measuring self-regulated learning. *Metacognition Learning, 14*, 1–19. https://doi.org/10.1007/s11409-019-09188-6

Schumacher, C., & Ifenthaler, D. (2018). The importance of students' motivational dispositions for designing learning analytics. *Journal of Computing in Higher Education, 30*(3), 599–619.

Schunk, D., & DiBenedetto, M. (2020). Motivation and social cognitive theory. *Contemporary Educational Psychology, 60*, 101832.

Schunk, D., & Greene, J. A. (Eds.). (2018). Historical, contemporary, and future perspectives on self-regulated learning and performance. In D. Schunk & J. A. Greene (Eds.), *Handbook of self-regulation of learning and performance* (2nd ed., pp. 254–270). Routledge.

Schunk, D. H., Meece, J. R., & Pintrich, P. R. (2014). Attribution theory. In *Motivation in education: Theory, research and affiliation* (pp. 91–138). Pearson.

Shahrokni, S. A., & Talaeizadeh, A. (2013). Learning processes in blended language learning: A mixed-methods approach. *TESL-EJ, 17*(3), 3.

Sharma, K., & Giannakos, M. (2020). Multimodal data capabilities for learning: What can multimodal data tell us about learning?. *British Journal of Educational Technology, 51*(5), 1450–1484.

Taub, M., Azevedo, R., Rajendran, R., Cloude, E. B., Biswas, G., & Price, M. J. (2019). How are students' emotions related to the accuracy of cognitive and metacognitive processes during learning with an intelligent tutoring system? *Learning and Instruction, 72*, 101200. https://doi.org/10.1016/j.learninstruc.2019.04.001

Taub, M., Sawyer, R., Smith, A., Rowe, J., Azevedo, R., & Lester, J. (2020). The agency effect: The impact of student agency on learning, emotions, and problem-solving behaviors in a game-based learning environment. *Computers & Education, 147*, 103781.

Veenman, M. V. J. (2011). Alternative assessment of strategy use with self-report instruments: A discussion. *Metacognition and Learning, 6*(2), 205–211.

Wiedbusch, M. D., Kite, V., Yang, X., Park, S., Chi, M., Taub, M., & Azevedo, R. (2021). A theoretical and evidence-based conceptual design of metadash: An intelligent teacher dashboard to support teachers' decision making and students' self-regulated learning. In *Frontiers in education* (Vol. 6, p. 570229). Frontiers Media SA.

Winne, P. H. (2018). Cognition and metacognition within self-regulated learning. In D. Schunk & J. A. Greene (Eds.), *Handbook of self-regulation of learning and performance* (2nd ed., pp. 36–48). Routledge.

Winne, P. H. (2020). A proposed remedy for grievances about self-report methodologies. *Frontline Learning Research, 8*, 164–173.

Winne, P. H., & Azevedo, R. (2022). Metacognition. In R. K. Sawyer (Ed.), The *Cambridge handbook of the learning sciences* (3rd ed.). Cambridge University Press.

Wolters, C. A. (2003). Understanding procrastination from a self-regulated learning perspective. *Journal of Educational Psychology, 95*(1), 179–187. https://doi.org/10.1037/0022-0663.95.1.179

Wong, E., Leslie, J. J., Soon, J. A., & Norman, W. V. (2016). Measuring interprofessional competencies and attitudes among health professional students creating family planning virtual patient cases. *BMC Medical Education, 16*(1), 1–9.

Zimmerman, B. J. (2008). Investigating self-regulation and motivation: Historical background, methodological developments, and future prospects. *American Educational Research Journal, 45*(1), 166–183.

Chapter 7
SR-WMS: A Typology of Self-Regulation in Writing from Multiple Sources

Mladen Raković and Philip H. Winne

Abstract When writers mine information from multiple sources to develop an essay, they reinterpret and reorganize their knowledge as they pursue and, possibly, reshape goals for rhetorical structure. Such writing tasks are popular across age levels and domains. It is assumed cognitive processes engaged in this kind of task provide practice that improves writing skills and deepens engagement with content. However, writing grounded in multiple and typically diverse sources is a demanding task. Successfully synthesizing information across multiple sources calls on multiple and interwoven cognitive and metacognitive processes as authors balance work in rhetorical, content and metacognitive spaces. To successfully traverse this complex and evolving cognitive landscape shaped by multidimensional goals, writers need procedural knowledge that operationalizes skills plus broad conditional knowledge to guide using those skills. For these reasons, success in multi-source writing tasks requires extensive and productive self-regulation. To advance research on these issues and give direction to engineering writing analytics to support productive self-regulation in multi-source writing, we synthesized research accessing and synthesizing content across multiple sources (Cho et al., Strategic processing in accessing, comprehending, and using multiple sources online. In: Handbook of multiple source use. Routledge, pp 133–150, 2018; Perfetti et al., Toward a theory of documents representation. In: The construction of mental representations during reading. Psychology Press, p 88108, 1999; Rouet et al., Educ Psychol 52(3):200–215, 2017; Rouet and Britt, Relevance processes in multiple document comprehension. In: Text relevance and learning from text. Information Age Publishing, Inc., pp 19–52, 2011), writing processes (Bereiter and Scardamalia, The psychology of

M. Raković (✉)
Centre for Learning Analytics, Monash University, Clayton, VIC, Australia
e-mail: mladen.rakovic@monash.edu

P. H. Winne
Faculty of Education, Simon Fraser University, Burnaby, BC, Canada
e-mail: winne@sfu.ca

Y. E. Wang et al. (eds.), *Social and Emotional Learning and Complex Skills Assessment*, Advances in Analytics for Learning and Teaching,
https://doi.org/10.1007/978-3-031-06333-6_7

written composition. Hillsdale, 1987) and self-regulated learning (SRL; Winne, Cognition and metacognition within self-regulated learning. In: Schunk D, Greene J (eds) Handbook of self-regulation of learning and performance, 2nd edn. Routledge, pp 36–48, 2018; Winne and Hadwin, Studying as self-regulated learning. In: Hacker DJ, Dunlosky J, Graesser A (eds) Metacognition in educational theory and practice, Erlbaum, pp 277–304, 1998). The result is a two-dimensional typology of cognitive and metacognitive processes in self-regulated writing using multiple sources (SR-WMS) spanning two problem spaces in writing tasks, rhetorical and content.

Keywords Self-regulation · Multi-source writing · Multiple source comprehension · nStudy · Learning analytics

7.1 Introduction

Students in post-secondary education are commonly assigned essays that require searching and synthesizing information across multiple sources. It is widely believed these kinds of assignments create opportunities for students to develop writing skills, deepen engagement with course content and practice composing within disciplinary genres. For instance, a chemistry student may be assigned a lab report in which evidence is extracted from theoretical papers to explain experimental findings. A literature student may be assigned an argumentative essay dissecting and resolving a controversial interpretation of an author's corpus. An economics student may need to synthesize diverse accounts of factors affecting economic growth in a multinational market. And a graduate student reviewing literature on a dissertation topic may weigh approaches to researching a phenomenon.

7.1.1 Challenges in Multi-source Composition

Students engaging in multi-source writing assignments are theorized to benefit in at least two ways. First, they develop skills in producing genre-appropriate written compositions (e.g., an evidence-based lab report, a persuasive argumentative essay). Second, they extend knowledge in the domain by carefully parsing and synthesizing with source materials they mine for this kind of assignment (e.g., Graham et al., 2012; Klein & Boscolo, 2016).

Multi-source writing is, however, a demanding task that unfolds in two problem spaces: rhetorical and content domain (Bereiter & Scardamalia, 1987). In the rhetorical space, writers work out rhetorical problems of the composition (e.g., presenting a claim, crafting a rebuttal). Simultaneously, in the content space, writers tackle challenges in identifying, contrasting and synthesizing knowledge in a domain. Successful writers strategically engage multiple cognitive and metacognitive operations to select and transform information in sources, and blend propositions and larger rhetorical structures into a coherent written product that meets goals for the assigned composition. These processes can broadly be classified as setting goals, comprehending information presented in multiple texts and other media, producing text, and metacognitively monitoring and controlling work and work flow. In this chapter, we assume cognitive and metacognitive operations are applied consciously and guided by goals. This aligns with Winne's (2018) model of self-regulated learning and Cho et al.'s (2018) definition of "…strategies [as] deliberate applications of one's processes…" (p. 135). We use the terms process and strategy interchangeably.

Multi-source writers set rhetorical goals in a context guided by task and genre requirements, e.g., "Convince the reader of a claim that plastic materials harm people's health." This overarching goal is translated into multiple subgoals in the content space, e.g., find then synthesize information in sources that support and subvert this claim. To approach these subgoals, writers work to comprehend source texts by integrating propositions, identifying textual and conceptual relations, reflecting on how they process sources (reading paths; Cho et al., 2018), sorting and selecting propositions, transforming bundles of information to assemble new (for them) content knowledge, and (re)presenting these in a clear and coherent manner (Bereiter & Scardamalia, 1987). At the same time, writers metacognitively monitor the evolving draft for fit to rhetorical and other goals such as format and length. Given results of these monitoring operations, writers selectively engage metacognitive control processes (e.g., re-searching, editing, revising) to modify their developing product and, perhaps, reshape goals in relation to revised perceptions of the assigned task (Winne, 2018; Winne & Hadwin, 1998). To navigate productively within the multivariable rhetorical and content problem spaces in a multi-source writing task, writers need more than sufficient command of text comprehension strategies and text production processes. They also need to know standards used to metacognitively monitor when to invoke and how to manage those multiple processes. In other words, to be effective, writers need to self-regulate (Greene et al., 2018). Self-regulation in multi-source writing may, however, encounter several major challenges.

Comprehending information within individual sources is just the first step in a multi-source writing assignment. Sampling and coordinating information across multiple sources adds demands to this work. The writer needs to keep text representations distinct while assembling inter-text relationships (Britt et al., 2018). It is not uncommon to encounter multiple yet differentially patterned representations of the same situation or phenomena (i.e., situational incoherence), as well as genuine informational discrepancies and conflicts (Braasch et al., 2012; Bråten et al., 2018). These challenges are further compounded by divergent lexical and semantic phrasing often present across sources, particularly when source documents were written for purposes differing from those of the assigned writing task. Limitations of working memory pose another challenge. Consequently, understanding textual material often evolves over successive cycles involving external supports, e.g., notes and highlighted selections (Braasch et al., 2012).

At the same time, challenges to generating draft text compete for scarce cognitive and metacognitive resources as authors strive to address rhetorical goals for a composition (Aull, 2015; Aull & Lancaster, 2014). Common among these challenges are under-developed composing skills, ignorance or misconceptions about genre-appropriate standards, and difficulties in establishing text coherence. These are compounded by demands arising from needs to transform and assemble source content in a composition ranging from mere paraphrasing to constructing a deeper account of a source information.

Adding to all these demands, many students lack self-regulatory skills to strategically engage in goal-oriented processing critical to successfully navigating multi-source writing tasks (Hyytinen et al., 2017; Klein & Boscolo, 2016). For instance, goals writers construct for the writing task as they perceive it often do not go beyond simply specifying the genre. This leaves vague or may omit particular standards of the genre (Klein & Boscolo, 2016). As well, many writers tend to underuse metacognitive monitoring (e.g., for fit of source material) and control (e.g., text revisions) as they draft and revise the composition. Taken together, all these demands thwart many students from productively engaging in multi-source writing. The result may well preclude reaching main goals for which multi-source compositions are assigned: developing composing skills and expanding domain knowledge.

7.1.2 Foundations for Modelling Multi-source Composition

Prior research has generated seminal theoretical frameworks, e.g., the Document Model Framework (DMF; Britt et al., 1999; Perfetti et al., 1999), the MD-TRACE model (Rouet & Britt, 2011), the RESOLV model (Britt et al., 2018; Rouet et al., 2017), and accounts of cognitive and metacognitive processes in writing (e.g., Bereiter & Scardamalia, 1987). Models of SRL highlight key roles for cognitive and metacognitive processes (e.g., Winne, 2018; Winne & Hadwin, 1998). Moreover, researchers have recently integrated research on SRL and writing processes (Graham et al., 2012; Klein & Boscolo, 2016), and SRL and multiple source use (Greene et al., 2018). Klein and Boscolo (2016) noted that, from a cognitive perspective, writing grounded in multiple sources requires writers to strategically cycle between reviewing source documents and composing can usefully be examined as a self-regulatory process. As Greene et al. (2018) pointed out "…expanding the SRL connections to multiple source use research, and vice versa, reveals many promising directions and implications for future theory, research, and practice" (p. 321).

To advance research on multi-source writing processes, we integrated research on multiple source use, writing processes and SRL to assemble the SR-WMS (Self-Regulation in Writing from Multiple Sources) typology of interweaving cognitive, metacognitive and self-regulatory processes in multi-source writing. The SR-WMS model provides educational researchers and practitioners with a new theoretical lens for studying writers' strategic engagement during multi-source writing which, in turn, may spark developing new instructional interventions to promote students' success in this complex task.

The remainder of the chapter is organized in three sections. In Sect. 7.2 we review relevant theoretical literature on multiple source comprehension, writing processes and self-regulated learning. In Sect. 7.3 we present the SR-WMS typology. In Sect. 7.4 we discuss opportunities to use the nStudy software system (Winne et al., 2019) to collect trace data for engineering learning analytics about processes represented in the SR-WMS typology.

7.2 Literature Review

7.2.1 Multiple Source Comprehension

Researchers have proposed several theoretical models to represent comprehension of information spread across multiple sources. Perfetti et al. (1999) modelled multi-source comprehension as an interaction between the Intertext and Situations component (submodels of a more general Documents Model). The Intertext component contains metadata about documents (e.g., author information, document type, document context, inferred audience, language style, rhetorical goals, content summary) and functional relationships (e.g., *supports*, *opposes*) among the documents comprising the document space. The Situations model includes real and reader-constructed situations, and also captures interrelated situations in the document space. In this way, the Documents Model accounts for how readers move beyond comprehension of a single document to integrate information across documents to create an overall understanding of a situation, e.g., a controversial issue (Bråten et al., 2018). As one way to remedy conflicting information found across documents, Britt et al. (1999) suggested readers consider each document's metadata (e.g., the authority ascribed to a source, publication date), relationships between author profiles and document content, and relationships among the documents themselves (e.g., common citations, succession from an earlier document to a later one).

Rouet and Britt (2011) modelled comprehension of multiple documents as an iterative cycle unfolding over five core processing steps in their MD-TRACE model: constructing a model of the task, assessing information needs, processing of each document, creating a product and assessing product quality. In the first step, the reader creates representations of the comprehension task and the associated writing task by surveying task instructions, recalling prior domain knowledge and experience(s) with a similar task, and indexing the availability of relevant resources. The reader's mental representation of the task includes goals given in the task description as a basis for a plan and associated standards for monitoring work on the task, e.g., monitoring available documents for relevance. Next, the reader estimates their information needs, essentially the gap between current knowledge and information needed to complete the task. This underlies decisions about whether and how to search for external resources if recalled domain knowledge is judged insufficient to satisfy goals. In step 3, the reader accesses source documents, assesses them for their topical and task relevance, selects documents judged relevant and analyses them to identify and integrate content information. This step forms a Documents Model (Perfetti et al., 1999). In step 4, the reader harnesses source information to begin crafting a response to the writing task. This step may involve various degrees and forms of knowledge transformation as dictated by task requirements (Wiley & Voss, 1999), e.g., to summarize relevant papers versus to fashion an argument supported by empirical evidence. In the final step, the reader compares a current draft to task goals and exits the task or, if the product is monitored as misaligned to task goals, recycles to previous steps. This triggers reassessing content in source documents and potentially revising the product.

Rouet et al. (2017) extended this line of research to include reader's motivation, interest and values in their REading as Problem SOLVing (RESOLV) framework. The RESOLV framework introduces two additional mental models for Context and Task. A Context Model is the reader's representation of the task and conditions circumscribing the reading situation. It incorporates information about task instructions, attributes of the authority who set the writing assignment (e.g., course instructor), forecasts about characteristics of the intended audience for the essay, available external resources, and personal characteristics of the reader, namely, knowledge, skills, interest and perceived task value (Britt et al., 2018). A Task Model represents subgoals, plans, strategies and values (e.g., importance of a particular strategy) associated with the student's understanding of the assignment. Importantly, the RESOLV model is based on several assumptions about cognition. First, reading is an adaptive and goal-directed activity. Second, processing resources are constrained. Third, readers intrinsically monitor feelings-of-knowing to estimate whether information can be retrieved from memory. And, fourth, readers analyse costs and benefits of pursuing goal-directed actions, and tune their decision thresholds as the task unfolds.

Parallel to the complex RESOLV model, Cho et al. (2018) argued: "Sophisticated readers use diverse strategies, monitoring the function of each, and deciding upon alternative choices of strategies if progress is halted." (p. 144). Accessing, comprehending and using multiple sources is thus complemented by the student's strategic processing. Cho et al. (2018) proposed three layers of strategic processing. Constructive-integrative processing involves information search and identification of important ideas and knowledge building. Critical-analytical processing require evaluating sources' metainformation and content. And metacognitive-reflective processing consists of monitoring one's knowledge and beliefs, reading paths, and properties of meaning constructed by work on the task.

These empirically supported models mark the reader's prior knowledge as a critical feature in multi-source comprehension. Readers use their knowledge to construct and enhance meaning from source texts (Stein, 1989) through a generic process of elaboration. According to Spivey (1990), three cognitive operations are central to elaboration. One is organizing propositions to reorder or recombine multi-source material and to group propositions to form chunks. The second is selecting relevant content per goals for the task, The third process is connecting elements to form relationships new to content in relation to prior knowledge.

Generally, these representative models of multiple source comprehension converge on the view that cyclically constructing a global representation of source documents is critical when students develop provisional solutions to task requirements. Beyond this accomplishment, however, students must coordinate these processes and integrate the products they generate in the content space with processes that produce text aligned to standards in the rhetorical space (Bereiter & Scardamalia, 1987). This entails retrieving, activating and monitoring composing strategies to approach rhetorical goals. We review these in the next section.

7.2.2 Processes in Writing

According to Magliano et al. (2018), writers not only create a task model (Rouet & Britt, 2011) and documents model (Perfetti et al., 1999), they also construct a product model. It is a mental representation of what has been written. The product model includes representations of two important kinds of differences: how much and in what ways the draft differs from sources, and how much and in what ways the draft deviates from standards for the writing assignment.

In this context, Stein (1990) identified four major groups of cognitive processes that occur in reading to comprehend and writing from multiple sources: planning, monitoring, elaborating, and structuring. Planning and monitoring processes concern procedural aspects of cognition in a problem-based learning environment. They are commonly considered metacognitive processes that set standards for metacognitive monitoring. Stein's (1990) processes of elaborating and structuring represent actions students take after metacognitive monitoring, i.e., metacognitive control (Son & Schwartz, 2002). In this section we discuss planning, monitoring and structuring processes with respect to their significance for writing production.

In the process of planning, the writer sets preliminary goals for the composition. Alternative plans are considered as means to build from available content knowledge and accounting for rhetorical constraints the writing task imposes on discourse. Flower et al. (1989) distinguished schema-driven, knowledge-driven and constructive planning strategies in writing. The standard for schema-driven planning is a framework explicit in or implied by the writing task. An example is this task instruction: *Develop an argumentative essay that conveys a main claim, and provides three supporting sub-claims and one counterclaim, all backed up by empirical evidence.* A writer who employs schema-driven planning "could concentrate on filling in the slots with appropriate information" (Flower et al., 1989, p. 4). Knowledge-driven planning strategies, in contrast, draw on the writer's personal representation of content knowledge. Writers adopting this kind of plan decide to organize the essay based on the structure of their integrated knowledge. Constructive-planning strategies subsume schema- and knowledge-driven planning. These plans include generating interconnected goals, organizing goals hierarchically, monitoring progress toward goals, instantiating goals in the written product, and identifying and resolving conflicts between goals.

While processing content spanning multiple sources and drafting fragments of an essay, authors monitor whether and how well the meaning of their evolving text agrees with goals they set for the composition (Hacker et al., 2009). For example, a skilled author might reflect: "After reviewing my goals for the essay and re-reading this paragraph, I am not sure if it effectively supports my main claim." Metacognitive monitoring can also be focused on schema-driven goals. For example, the author monitors the state of a draft relative to perceived task requirements, e.g., "According to the professor's instructions, I think I have not yet provided a counterargument." Or "Is the evidence I borrowed from sources empirical or rational?" According to Hacker et al. (2009), four primary goal-oriented monitoring strategies are used to meet goals: reading, re-reading, reflecting and reviewing.

To judge whether they have operated validly on content information while drafting an essay, writers also monitor their comprehension (Stein, 1990) relative to sources. For example, "Have I correctly paraphrased this paragraph from the source article?" Another potential topic for metacognitive monitoring concerns the extent to which information embedded in the draft meets rhetorical goals for the composition, e.g., "Is this really a compelling example that supports my argument?" Metacognitive monitoring can lead writers to re-evaluate and modify their written product and/or goals they previously set for the composition. This is one way the writer engages in metacognitive control to improve text quality (Hayes, 2000). According to Hacker et al. (2009), writers can draw on six metacognitive control strategies in writing: editing, drafting, generating ideas, producing words, translating information into new expressions, and revising.

As writers monitor and polish the structure of a draft, they manipulate multiple connected (or partially connected) propositions mined from sources. In this process, writers strive to ensure included content is appropriately shaped in the composition. Examples of structuring activities include subsuming source information under superordinate categories, organizing text into high- and low-level propositions, and illuminating relations between ideas in the text that might have been otherwise overlooked (Stein, 1990).

It is important to note that, during multi-source writing, metacognitive monitoring and control strategies interweave in and across the content and rhetorical spaces. For example, as the writer reads source texts and generates potentially appropriate ideas, those ideas are evaluated relative to goals for writing and adjudicated as to whether they already have been imported into the essay. Monitoring continues as other ideas are being shaped into appropriate rhetorical form in the essay draft. If meaning conveyed by the draft is a poor match to the author's goals, one option is to rewrite to conform better to writing goals. The writer is an agent in this endeavour. In these activities, the writer's self-regulatory tactics and strategies contribute to attaining goals and the overall quality of written products (Graham et al., 2012) and learning gains (Berthold et al., 2007; Klein et al., 2007). In the next section we discuss powerful processes of self-regulated learning (SRL) and cast light on their connections to multi-source use and writing processes.

7.2.3 *Self-Regulated Learning*

SRL is a complex approach learners take to monitoring, managing and manipulating cognitive operations, motivations and emotions to optimize learning as a process and the products of those learning processes (Winne, 2018; Winne & Marzouk, 2019). To represent this process, several models have been proposed (e.g., Boekaerts, 1991; Efklides, 2011; Hadwin et al., 2011; Pintrich, 2000; Winne & Hadwin, 1998; Zimmerman, 1989). For an overview of most influential SRL models and an empirical evidence supporting them, see Panadero (2017).

We use Winne and Hadwin's (1998); Winne & Marzouk, 2019) model of SRL to guide the typology we propose. Their model highlights the roles of metacognitive monitoring and control in ways that closely align to how these operations are modelled when students ground compositions in multiple sources.

Winne and Hadwin theorize SRL as unfolds over four recursive and loosely ordered stages: defining features of the task, setting goals and forging plans to meet them, enacting tactics and strategies, and, optionally, making large-scale adaptations to the current task or for future ones. Five dimensions of a task can be observed in each stage: conditions in which current work is situated, operations the learner performs on information, products of those operations, and evaluations of operations (e.g., pace, difficulty) and their products relative to standards. A shorthand for this model is COPES (Winne, 2018).

In the task definition stage, students survey task requirements and internal (e.g., prior knowledge, skills, interests, preferences) and external (e.g., available source articles) factors they perceive have bearing on an assigned task. This product of this stage of SRL is a perception about the task resembling the context model component in Britt et al.'s (2018) framework of reading as problem solving. In the next stage of SRL, students set goals and produce plans to approach those goals. Goals are influenced by standards perceived for the task and by each student's personal standards, including motivational features such as self-efficacy and incentives perceived about the task (Winne & Marzouk, 2019). Plans include sets of tactics and strategies to attain the goal. Stein (1990) and Flower et al. (1989) considered goals and plans critical for success in writing assignments. Moreover, goals and associated plans for organizing tactics and strategies are factors embedded in task models of multi-source reading comprehension and writing production described by Rouet and Britt (2011) and Magliano et al. (2018). In the stage 3, learners enact tactics and strategies (e.g., compare, contrast, integrate across multiple sources; Bråten et al., 2014) that they planned in the previous stage. Learners create successive and interrelated products as work proceeds, e.g., drafts of opening sentences and entire sections, and monitor each against the product model (Magliano et al., 2018). At points of their choosing, writers internally evaluate whether what has been written so far accords with perceptions of the task, goals and plans. As judged worthwhile, students may exercise metacognitive control to modify products "just in time" created in stages 1–3. This may include revising goals, activating new strategies and adapting foci and a schedule for monitoring the evolution of the essay draft. The adaptation stage takes place after the main task is completed. At this stage, learners reflect on the whole of their work on the task (metacognitive monitoring) and may develop long-term adaptations for similar tasks in the future (metacognitive control). The result is forward-reaching transfer (Salomon & Perkins, 1989). Greene et al. (2018) point out, "There has been very little research on the processes learners use after reading or completing a multiple source use learning task (p. 327)" despite their theorized significance.

We enhance existing models of multi-source use in our SR-WMS typology to include various adaptive processes that occur after the task is complete. Equally important, we bring forward how the Winne-Hadwin model embraces key aspects of SRL relating to motivation and self-efficacy (Bråten et al., 2014; Zimmerman, 2013), affect and emotions (Bowler, 2010), and epistemic beliefs and cognition (Greene et al., 2015).

7.3 SR-WMS – A Typology of Self-Regulation in Writing from Multiple Sources

Drawing on sources reviewed in the preceding sections, the SR-WMS typology of self-regulation in writing synthesizes research on using multiple sources in self-regulated researching, reading and writing (Table 7.1). As an organizer, we used the four stages of Winne and Hadwin's (1998) model of SRL because it conceptually organizes multiple cognitive, metacognitive and motivational processes previous researchers described about multiple source writing. In particular, the Winne-Hadwin framework emphasizes adaptation and the cyclical nature of engagement characteristic of multi-source writing tasks. We identify self-regulatory constructs salient to each stage and project them onto the rhetorical and content spaces of multi-source writing tasks (Bereiter & Scardamalia, 1987). In the rhetorical space, writers work on solving problems of text production and in the content space they work on solving problems attending comprehending and coherently integrating information provided in multiple sources.

In stage 1 the writer makes efforts to understand the task. They strive to develop familiarity with rhetorical constraints the task imposes, e.g., number of claims to present and validate in the essay, the minimum/maximum number of source texts to be cited. They survey resources available for support, including: their own composing skills (self-efficacy), knowledge of the assigned genre, motivation and interest in task and content, and time constraints. In the content space, the writer surveys their own information problem solving skills for searching, filtering, selecting relevant documents and harnessing meta-textual information (e.g., linguistic cues,

Table 7.1 The SR-WMS typology of self-regulation in writing from multiple sources; constructs are organized over the four recursive and loosely ordered stages of SRL (Winne & Hadwin, 1998) and projected onto rhetorical and content space of written composition

		Rhetorical space		Content space	
SRL stage	Definition	Construct	Example	Construct	Example
Stage 1: Task understanding	Surveying task requirements and internal and external resources available for writing task	Composing skills	Word production, idea generation, drafting, editing, translating, revising	Information problem solving skills	Searching, filtering and collecting (1) relevant and trustworthy source texts and (2) relevant information within texts collected
		Knowledge of genre	Argument essay vs expository text vs lab report…	Knowledge of text comprehension tactics	Knowledge of reading, re-reading, highlighting…
		Task constraints	Number of claims that should be presented and defended, number of source texts required, citation style…	Prior knowledge of content	(Feeling of) knowing terms, understanding major concepts and misconceptions…
		Motivation and interest in task	Getting targeted grade, convincing wider audience of important argument	Interest in topic	Curious to further explore the topic…
		Self-efficacy	Confidence to perform well	Present emotions	Excitement for the content
		Cognition	Logical reasoning		
		Epistemic beliefs	Knowing how to compose an argument is necessary for every modern professional	Epistemic beliefs	Knowledge is provided by authorities
		Time constraints	Time available to create the first draft, time available to revise…	Time constrains	Time available to study source texts

(continued)

Table 7.1 (continued)

SRL stage	Definition	Rhetorical space		Content space	
		Construct	Example	Construct	Example
Stage 2: Goal settings and planning	Goals are designing and composing a written product using evidential information from source texts Plans include sets of strategies to attain the goal	Schema-driven	A framework implied by the writing task	Rhetorical goals are translated into subgoals in content space	Locate and integrate information in the source articles that support claims in the rhetorical space
		Knowledge-driven	Writer's initial representation of content knowledge or of knowledge in sources		
		Constructive planning	Generating interconnected goals, organizing goals hierarchically and identifying/resolving conflicts between goals		
Stage 3: Enactment	Applying strategies to complete writing task while monitoring for products and processes, against goals	Object-level strategies	Word production, idea generation, drafting, editing, translating, revising	Object-level strategies	Reading, re-reading, highlighting, tagging, note taking, reviewing, reflecting, pausing, questioning, elaborating
		Process-level strategies	Properties and usefulness of composing strategies	Process-level strategies	Properties and usefulness of strategies for text comprehension
Stage 4: Adaptation	Evaluation of prior task activities and creation of forward-reaching plans for future tasks	Forward reaching adaptation for similar writing episodes in the future	Adaptations to domain knowledge, interests, goals, plans and strategies	Forward reaching adaptation for similar writing episodes in the future	Adaptations to domain knowledge, interests, goals, plans and strategies

rhetorical devices, headings; Greene et al., 2018). These skills are applied to locate relevant and trustworthy source texts, if sources are not already provided, and to locate relevant information within source texts. The writer also brings to the task prior knowledge about the content, knowledge of text comprehension tactics (e.g., highlighting, note taking, summarizing), and interest in and emotional stance about the topic. As well, the writer's epistemic beliefs may play a critical role in the writer's perception of the task and standards selected for engaging with it (e.g., opting only for sources written by authorities).

In stage 2, the writer sets goals and creates plans. Rhetorical goals relate to designing and composing a written product using evidential information from multiple source texts. These can be created by strategies guided by schemas, knowledge or constructions generated uniquely for the task (Flower et al., 1989). Content goals, on the other hand, deal with processing information selectively sampled from source texts to fill slots in the schema governing the genre's rhetorical space, e.g., warrants for evidence supporting claims. Klein (1999) explains that "…writers set rhetorical goals … [t]hen set sub goals in content space that subserve these rhetorical goals" (p. 244). For example, if a writer is composing an argumentative essay arguing against using plastic bags, they may set a goal in the rhetorical space to convince the reader of a claim that plastic materials harm people's health. This goal then is translated into a subgoal in the content space, e.g., find and synthesize information in source articles that support this claim taking care (metacognitively monitoring) to note potential counterclaims and evidence supporting them.

In stage 3, the writer enacts tactics and strategies planned in the previous stage. Given the nature of multi-source writing tasks, the writer needs to deftly interweave text production and text comprehension strategies. We categorize these as object level (vs. metacognitive) strategies as they represent operations on information comprising the essay itself. Examples of object level processes the writer enacts in the rhetorical space include word production, idea generation, drafting, editing, translating, revising (Hacker et al., 2009). Examples of object level processes in content space include reading, re-reading, highlighting, tagging, note taking (e.g., defining, summarizing, debating, comparing and contrasting), reviewing and reflecting.

The writer also engages metacognitive strategies to monitor emerging properties of the draft and the utility of selected object-level composing strategies in the rhetorical space as well as text comprehension strategies in the content space. This includes monitoring the alignment of products and properties of processes (e.g., ease, efficiency) relative to goals and related subgoals in the content space to assure meaning represented in the draft composition agree with goals they set (Hacker et al., 2009). As well, the writer monitors whether they validly represented content information (Stein, 1990). To address any discrepancies revealed by metacognitive monitoring, the writer engages in metacognitive control to enact what are judged to be corrective object level strategies for text production, e.g., translating, editing (Hacker et al., 2009); and for reading comprehension, e.g., modifying a reading path, revisiting, note taking, pausing, questioning (Cho et al., 2018; Goldman et al., 2012; Greene et al., 2015; Greene & Azevedo, 2007). Metacognitive monitoring and control can also be enacted regarding products of previous stages, potentially modifying perceptions about the task, revising goals and reforming plans.

Once the task is complete, at the adaptation stage (stage 4), the writer may evaluate the entire scope of activities and products across previous stages. One aim may be to profile how they engaged in the task, e.g., to identify text comprehension strategies that did versus those that did not work well. When this happens, the writer may develop forward-reaching plans to improve performance in future similar tasks (Winne & Azevedo, 2022).

It is important to note the four stages of self-regulated learning may not unfold in this linear way. Writers exercising agency can retreat and jump over stages. For example, upon becoming aware of task requirements and recalling prior content knowledge (stage 1), the writer may immediately begin drafting (stage 3) without carefully setting goals (stage 2), then survey (monitor) their internal and external resources to check for contributions and obstacles present in the task (stage 1).

7.4 Learning Analytics About Processes in Multi-source Writing

7.4.1 nStudy – Software to Support Multi-source Writing

Digital data about student's writing activities in technology-enhanced learning environments can be harnessed to map and support student progress in writing from multiple sources. Winne et al. (2019) designed and developed a software suite called nStudy. nStudy has two main components. One is an extension to the Chrome web browser. It provides a wide array of structured cognitive tools learners can use to operate on information they survey and learn (Fig. 7.1). The other component is a pair of backend databases and systems. These record and assemble logged events as a time-stamped array of learner activity and as multi-part artifacts that amalgamate those events in coordinated structures. Among a variety of functionalities, nStudy's features support students to research, index by tags, annotate and cross-link information in source materials. Other features support drafting and revising compositions that incorporate artifacts learners construct by mining and organizing information situated in multiple sources. For example, writers can annotate an entire bookmark as well as selections of information within a website. Notes are web forms that can be configured by an instructor to nudge cognition and provide standards for metacognition. Note forms can be configured to operationally define facets of goals, schemas for analysing and evaluating information according to discipline-specific genres or almost any structure of information or procedure, such as monitoring a task's conditions in preparation for framing goals. A Goal note can be configured to document constructive-planning strategies (Flower et al., 1989). It can include fields to guide the writer to organize rhetorical goals hierarchically, identify interconnected goals and resolve potential conflicts between goals. As

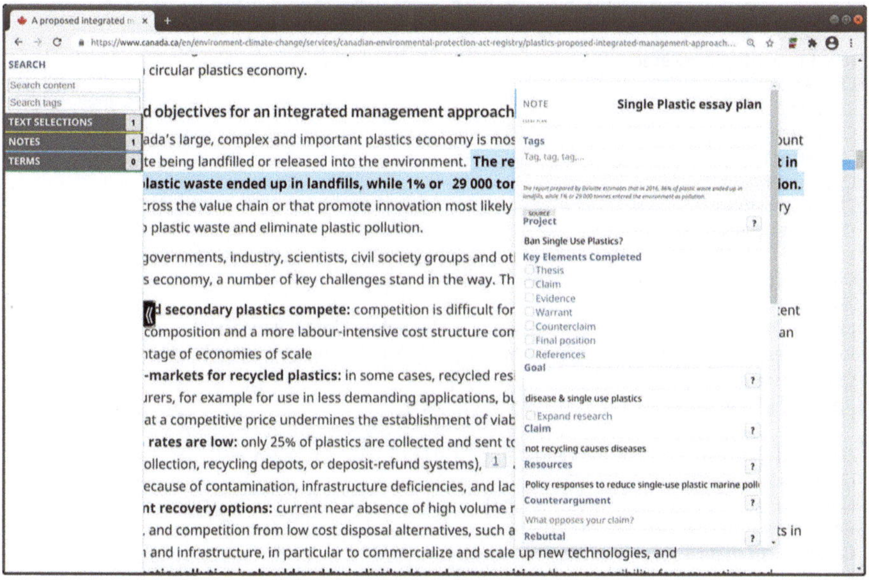

Fig. 7.1 Study view in nStudy

another example, a Debate note can be designed so the writer can enter information about main claims, evidence and warrants from the source articles. The writer can also link multiple Debate notes to create a larger argument structure. Students can search for and organize artifacts in nStudy's Library view and index them using task specific tags, e.g., "Reliable?" or "Needs Research." Every engagement a student has with information is timestamped and stored to represent which operations a student applied to particular information. When these structured data have strong links to theoretical constructs, they form trace data describing how learners work and what they work with (Winne, 2020). Trace data provide rich material for developing learning analytics describing and encouraging productive self-regulation in multi-source writing.

7.4.2 *Example*

Ann is an undergraduate student studying environmental science. She has been assigned a term paper – an essay to argue why using single-use plastics (bags, take out containers, etc.) should be abandoned in modern societies. Ann has previously written several essays in this genre but she is aware there are several new, specific requirements for this particular assignment. She interprets her instructor's oral description of the essay as requiring her to provide at least two arguments supported with evidence, at least one counterargument supported with evidence, and a rebuttal to the counterargument. Ann recalls from in-class discussions several propositions

Essay Plan (note form)

Key Elements Completed
☐ Thesis
☐ Claim
☐ Evidence
☐ Warrant
☐ Counterclaim
☐ Final position
☐ References

Claim 1 [plastics recyclers need reliable markets]
☒ Expand research
Notes [China has a large recycling market; see UN Single Use whitepaper]
Resources [Canadian Gov't Management Plan, UN Single Use whitepaper]
Counterargument [what opposes this claim?]
Rebuttal [what invalidates the counterargument?]

Claim 2 [enter a claim; click + to add another] [+]
☐ Expand research
Notes [what's key to address this claim?]
Resources [drop bookmarks here]
Counterargument [what opposes this claim?]
Rebuttal [what invalidates the counterargument?]

Notes to self
1 [check out Canadian recycling marketers like Return-It]
2 [what should you not forget?] [+]

Fig. 7.2 Essay Plan note form in nStudy. (Green shading indicates material that can be cloned to create a blank copy of the field array. Gray text is replacement text, meant to guide the learner, disappears as the learner enters text in the text field)

forwarded by peers, including that plastic materials (a) may harm human health and (b) may present significant challenges to being recycled even if consumers deliver used plastics to recycling depots. She decides she will research multiple sources to extract and organize information to satisfy these requirements. Accordingly, she sets two rhetorical goals using the Essay Plan note form in nStudy (Fig. 7.2).

In the Topic Goals note form (Fig. 7.3), Ann creates the subgoals in the content space and links them to corresponding rhetorical goals in the Essay Plan. For example, Ann's subgoals linked to her first rhetorical goal could be to find examples of diseases caused by plastic materials, rank those diseases by severity and elaborate on the most severe diseases to create a persuasive case for her rhetorical argument. She considers, however, that plastic bags may have one advantage over the paper bags – plastic bags are more durable and that can save money. Ann decides she will include this claim as a counterargument and sets the corresponding goal in the Essay Plan note template. In this way, Ann's goal setting and planning is guided by the task schema (e.g., she dedicated two template slots for arguments and one slot for counterargument) and content knowledge (e.g., she recalled information from in-class discussion). As well, Ann engaged in productive constructive planning, e.g., by hierarchically organizing and interconnecting her goals and subgoals.

Topic Goals (note form)

Claim [what is your target claim?]

Topic 1 [challenges to consumer recycling]
Targets [details to flesh out]
Resources [drop bookmarks here]

Topic 2 [diseases caused by plastics]
Targets [details to flesh out]
Resources [drop bookmarks here]

Topic 3 [describe topic] [+]
Targets [details to flesh out]
Resources [drop bookmarks here]

Fig. 7.3 Topic Goals note form in nStudy

Debate (note form)

Claim [state your claim]
Evidence [summarize evidence for your claim] [+]
Resource(s) for evidence [drop artifacts here to link]
Warrant [Why is this evidence relevant to the claim?]
How sturdy is this reasoning chain?
soft |----------O---------| solid

See also [drop artifacts to link]

Fig. 7.4 Debate Note form in nStudy

Ann searches for relevant source documents using her content subgoals as standards for monitoring information she finds in the Internet. She wants to source only peer-reviewed academic articles, as she deems those documents are trustworthy. This reflects her epistemic beliefs. She creates a bookmark for each article she judges potentially relevant based on reading its abstract. In the Library view, Ann revisits the collected bookmarks and reads each, highlighting relevant information. She completes multiple Debate note forms (Fig. 7.4), jotting down claims, examples and warrants identified in these sources that she judges are relevant to her goals. Ann reorganizes her notes in the Library view, e.g., by collecting notes with information about diseases caused by plastic materials into a subfolder. In this way, she further elaborates source information while, again, monitoring the fit of these artifacts to her goals. Ann noticed that some of the collected documents did not contain goal-relevant information and removed them from the nStudy Library.

Ann begins drafting her argument in nStudy's Report tool. First, she generates a list of ideas by dragging relevant notes into the tool. The contents of those notes becomes available for editing. She works out rhetorical connections among these ideas and develops the skeleton for her first draft. In this process, Ann navigates between the rhetorical and content spaces, often re-reading parts of source texts using the link nStudy automatically established between a note and information selected in an article. She monitors her comprehension of information in sources as she re-evaluates how that information fits rhetorical roles in her evolving draft. Ann compares draft text to goals in her Essay Plan note form about evidence for causes of diseases (monitoring for goals). She wants to enrich the essay with a few details and switches to re-reading her selections highlighted in source articles (metacognitive monitoring and control of tactics). As she assembles information, she realizes she did not bookmark articles that advocate for use of plastic bags, the concept she originally wanted to use as a counterargument. She, however, remembers a few websites that documented how plastic bags were useful as a kitchen aid. So, Ann returns to her Essay Plan note and modifies her rhetorical goal for the counterargument, an instance of metacognitive control. Once she has drafted what she judges (monitors) forms a the full set of arguments, counterargument and rebuttal, Ann reviews and makes minor edits before submitting her essay to the instructor.

Now that her task is complete, Ann reflects on her activities throughout this project. She judges she learned a lot about the topic. Tactics and strategies she used to select information to highlight and using nStudy's note forms helped her comprehend source materials. She observes that reviewing her Essay Plan note and, occasionally, updating it to match information available in source texts was productive. She plans to reuse these tactics in the next essay assigned. She also plans a modification to her just-used strategy. She will skip the idea generation activity. Instead, she will start drafting her essay by reviewing highlighted selections and notes in the nStudy Library because that approach seems more efficient.

7.4.3 Implications

Analysing fine-grained trace data about Ann's writing activities in terms of the SR-WMS typology expands opportunities to develop learning analytics to address challenges many writers face in activating productive self-regulation when developing essays sourced from multiple documents. For example, notes Ann created in nStudy when she dealt with problems in the content space could be analysed using natural language processing methods to detect inter-textual relationships she may have initially missed (Britt et al., 1999, 2018). Such a semantic similarity- and/or ontology-based analysis (Foltz et al., 2010; Harispe et al., 2015) of source documents Ann consulted may shed light on representations of the essay topic across sources and help Ann identify similarities and discrepancies between sources and her essay (Braasch et al., 2012; Bråten et al., 2017). Also, characterizing semantic overlap between the task requirements/goals Ann set and her draft may help her identify and develop productive revisions. An analysis of text cohesion (Graesser et al., 2004) may signal need for other revisions in both the content and rhetorical spaces.

New learning analytics also can be generated from nStudy's trace data as Ann worked on her research and writing activities. A "playback" of her work, e.g., "This is how you set your goals, searched and filtered sources … ." may support Ann's work at phase 4 of the Winne-Hadwin SRL, charting forward-reaching adaptations for her next assignment. Analytics like these are designed to promote metacognitive monitoring and control critical to productive self-regulation (Winne, 2018). With practice across multiple assignments, these analytics can help Ann and her classmates construct effective genre- and task-appropriate compositions as they expand domain knowledge.

Analytics like these also can help instructors provide Ann and her classmates personalised support to improve writing. This is often a challenge for instructors given the massive and diverse enrolments in contemporary higher education. Finally, the SR-WMS framework may be used to explore a range of research questions. Examples include: To what extent do information-problem solving skills and prior knowledge about the content predict learning gains? Can engagement in constructive planning and goal setting predict essay quality? How does students' self-regulation of writing change within single assignments and across assignments? Which analytics are particularly critical for supporting students' self-regulation in writing?

References

Aull, L. (2015). *First-year university writing: A corpus-based study with implications for pedagogy*. Springer.

Aull, L. L., & Lancaster, Z. (2014). Linguistic markers of stance in early and advanced academic writing: A corpus-based comparison. *Written Communication, 31*(2), 151–183. https://doi.org/10.1177/0741088314527055

Bereiter, C., & Scardamalia, M. (1987). *The psychology of written composition*. Hillsdale.

Berthold, K., Nückles, M., & Renkl, A. (2007). Do learning protocols support learning strategies and outcomes? The role of cognitive and metacognitive prompts. *Learning and Instruction, 17*(5), 564–577.

Boekaerts, M. (1991). Subjective competence, appraisals and self-assessment. *Learning and Instruction, 1*, 1–17. https://doi.org/10.1016/0959-4752(91)90016-2

Bowler, L. (2010). Talk as a metacognitive strategy during the information search process of adolescents. *Information Research, 15*(4), 1–17.

Braasch, J. L., Rouet, J. F., Vibert, N., & Britt, M. A. (2012). Readers' use of source information in text comprehension. *Memory & Cognition, 40*(3), 450–465.

Bråten, I., Anmarkrud, Ø., Brandmo, C., & Strømsø, H. I. (2014). Developing and testing a model of direct and indirect relationships between individual differences, processing, and multiple-text comprehension. *Learning and Instruction, 30*, 9–24.
Bråten, I., Stadtler, M., & Salmerón, L. (2018). The role of sourcing in discourse comprehension.
Britt, M. A., Perfetti, C. A., Sandak, R., & Rouet, J. F. (1999). Content integration and source separation in learning from multiple texts. In *Narrative comprehension, causality, and coherence: Essays in honor of Tom Trabasso* (pp. 209–233). Routledge.
Britt, M. A., Rouet, J. F., & Durik, A. (2018). Representations and processes in multiple source use. In *Handbook of multiple source use* (pp. 17–33). Routledge.
Cho, B. Y., Afflerbach, P., & Han, H. (2018). Strategic processing in accessing, comprehending, and using multiple sources online. In *Handbook of multiple source use* (pp. 133–150). Routledge.
Efklides, A. (2011). Interactions of metacognition with motivation and affect in self-regulated learning: The MASRL model. *Educational Psychologist, 46*, 6–25. https://doi.org/10.1080/00461520.2011.538645
Flower, L. R., Schriver, K. A., Haas, C., Carey, L., & Hayes, J. (1989). *Planning in writing: The cognition of a constructive process*. Defense Technical Information Center. https://doi.org/10.21236/ADA210434
Foltz, P. W., Gilliam, S., & Kendall, S. (2010). Supporting content-based feedback in on-line writing evaluation with LSA. *Interactive Learning Environments, 8*, 111–127. https://doi.org/10.1076/1049-4820(200008)8:2;1-B;FT111
Goldman, S. R., Braasch, J. L. G., Wiley, J., Graesser, A. C., & Brodowinska, K. (2012). Comprehending and learning from internet sources: Processing patterns of better and poorer learners. *Reading Research Quarterly, 47*(4), 356–381.
Graesser, A. C., McNamara, D. S., Louwerse, M. M., & Cai, Z. (2004). Coh-Metrix: Analysis of text on cohesion and language. *Behavior Research Methods, Instruments, & Computers, 36*(2), 193–202.
Graham, S., McKeown, D., Kiuhara, S., & Harris, K. R. (2012). A meta-analysis of writing instruction for students in the elementary grades. *Journal of Educational Psychology, 104*(4), 879.
Greene, J. A., & Azevedo, R. (2007). A theoretical review of Winne and Hadwin's model of self-regulated learning: New perspectives and directions. *Review of Educational Research, 77*(3), 334–372.
Greene, J. A., Bolick, C. M., Jackson, W. P., Caprino, A. M., Oswald, C., & McVea, M. (2015). Domain-specificity of self-regulated learning processing in science and history digital libraries. *Contemporary Educational Psychology, 42*, 111–128.
Greene, J. A., Copeland, D. Z., Deekens, V. M., & Freed, R. (2018). Self-regulated learning processes and multiple source use in and out of school. In *Handbook of multiple source use* (pp. 320–338). Routledge.
Hacker, D. J., Keener, M. C., & Kircher, J. C. (2009). Writing is applied metacognition. In *Handbook of metacognition in education* (pp. 154–172). Routledge.
Hadwin, A. F., Järvelä, S., & Miller, M. (2011). Self-regulated, co-regulated, and socially shared regulation of learning. In B. J. Zimmerman & D. H. Schunk (Eds.), *Handbook of Self-regulation of learning and performance* (pp. 65–84). Routledge.
Harispe, S., Ranwez, S., Janaqi, S., & Montmain, J. (2015). Semantic similarity from natural language and ontology analysis. *Synthesis Lectures on Human Language Technologies, 8*(1), 1–254.
Hayes, J. R. (2000). A new framework for understanding cognition. In *Perspectives on writing: Research, theory, and practice* (p. 6). International reading association (IRA).
Hyytinen, H., Löfström, E., & Lindblom-Ylänne, S. (2017). Challenges in argumentation and paraphrasing among beginning students in educational sciences. *Scandinavian Journal of Educational Research, 61*, 411–429. https://doi.org/10.1080/00313831.2016.1147072
Klein, P. D. (1999). Reopening inquiry into cognitive processes in writing-to-learn. *Educational Psychology Review, 11*(3), 203–270.

Klein, P. D., & Boscolo, P. (2016). Trends in research on writing as a learning activity. *Journal of Writing Research, 7*(3), 311–350. https://doi.org/10.17239/jowr-2016.07.03.01

Klein, P. D., Boman, J. S., & Prince, M. P. (2007). Developmental trends in a writing to learn task. In *Writing and cognition* (pp. 201–217). Brill.

Magliano, J. P., Hastings, P., Kopp, K., Blaum, D., & Hughes, S. (2018). Computer-based assessment of essays based on multiple documents. In *Handbook of multiple source use* (pp. 502–526). https://doi.org/10.4324/9781315627496-28

Panadero, E. (2017). A review of self-regulated learning: Six models and four directions for research. *Frontiers in Psychology, 8*, 422.

Perfetti, C. A., Rouet, J. F., & Britt, M. A. (1999). Toward a theory of documents representation. In *The construction of mental representations during reading* (p. 88108). Psychology Press.

Pintrich, P. R. (2000). The role of goal orientation in self-regulated learning. In M. Boekaerts, P. R. Pintrich, & M. Zeidner (Eds.), *Handbook of self-regulation* (pp. 452–502). Academic.

Rouet, J. F., & Britt, M. A. (2011). Relevance processes in multiple document comprehension. In *Text relevance and learning from text* (pp. 19–52). Information Age Publishing, Inc.

Rouet, J. F., Britt, M. A., & Durik, A. M. (2017). RESOLV: Readers' representation of reading contexts and tasks. *Educational Psychologist, 52*(3), 200–215.

Salomon, G., & Perkins, D. N. (1989). Rocky roads to transfer: Rethinking mechanisms of a neglected phenomenon. *Educational Psychologist, 24*, 113–142.

Son, L. K., & Schwartz, B. L. (2002). The relation between metacognitive monitoring and control. In T. J. Perfect & B. L. Schwartz (Eds.), *Applied metacognition* (pp. 15–38). Cambridge University Press. https://doi.org/10.1017/CBO9780511489976.003

Spivey, N. N. (1990). Transforming texts: Constructive processes in reading and writing. *Written Communication, 7*, 256–287. https://doi.org/10.1177/0741088390007002004

Stein, V. (1989). *Elaboration: Using what you know* ((Reading-to-write report no. 6). Technical Report No. 25). Center for the Study of Writing.

Stein, V. (1990). Exploring the cognition of reading-to-write. In L. Flower, V. Stein, J. Ackerman, M. Kantz, K. McCormick, & W. Peck (Eds.), *Reading to write: Exploring a cognitive and social process* (pp. 119–143). Oxford University Press.

Wiley, J., & Voss, J. F. (1999). Constructing arguments from multiple sources: Tasks that promote understanding and not just memory for text. *Journal of Educational Psychology, 91*(2), 301.

Winne, P. H. (2018). Cognition and metacognition within self-regulated learning. In D. Schunk & J. Greene (Eds.), *Handbook of self-regulation of learning and performance* (2nd ed., pp. 36–48). Routledge.

Winne, P. H. (2020). Construct and consequential validity for learning analytics based on trace data. *Computers in Human Behavior, 112*, 106457.

Winne, P., & Azevedo, R. (2022). Metacognition and self-regulated learning. In R. Sawyer (Ed.), *The Cambridge handbook of the learning sciences*. (Cambridge Handbooks in Psychology, pp. 93–113). Cambridge: Cambridge University Press. https://doi:10.1017/9781108888295.007

Winne, P. H., & Hadwin, A. F. (1998). Studying as self-regulated learning. In D. J. Hacker, J. Dunlosky, & A. Graesser (Eds.), *Metacognition in educational theory and practice* (pp. 277–304). Erlbaum.

Winne, P. H., & Marzouk, Z. (2019). Learning strategies and self-regulated learning. In J. Dunlosky & K. Rawson (Eds.), *Cambridge handbook of cognition and education* (pp. 696–715). Cambridge University Press.

Winne, P. H., Teng, K., Chang, D., Lin, M. P. C., Marzouk, Z., Nesbit, J. C., … Vytasek, J. (2019). nStudy: Software for learning analytics about processes for self-regulated learning. *Journal of Learning Analytics, 6*(2), 95–106.

Zimmerman, B. J. (1989). A social cognitive view of self-regulated academic learning. *Journal of Education & Psychology, 81*, 329–339. https://doi.org/10.1037/0022-0663.81.3.329

Zimmerman, B. J. (2013). From cognitive modeling to self-regulation: A social cognitive career path. *Educational Psychologist, 48*(3), 135–147.

Chapter 8
Identifying Tertiary Level Educators' Needs and Understanding of the Collaboration Process Analytics

Mutlu Cukurova, Carmel Kent, and Abayomi Akanji

Abstract There is little doubt about the significant role the educators play in supporting the collaboration process through monitoring and supporting effective interactions. However, little work explores the educators' needs and understandings of the analytics generated to measure the process of collaboration in online learning settings. In this chapter, we first explain a new method of measuring the process of collaboration (CLaP) by drawing upon the collaborative cognitive load theory and utilising social network analysis. Then, we report the results of two educator workshops and a survey that investigated the educators' understanding of the collaboration process visualisations compared to more commonly used participation measures such as the number of posts and the number of views. Our results show that although educators can indeed gain more insights into the collaboration process with CLaP visualisations, these are still considered limited and too complex to be easily adopted in practice. Moreover, currently, many educators are not evaluating the collaboration process in online settings at all, or when they do, they only rely on participation measures. We conclude the chapter with a discussion on the findings and their future implications.

Keywords Collaboration analytics · Teacher evaluations · Learning analytics

M. Cukurova (✉) · A. Akanji
University College London, London, UK
e-mail: m.cukurova@ucl.ac.uk; abayomi.akanji.19@alumni.ucl.ac.uk

C. Kent
EDUCATE Ventures, London, UK
e-mail: carmel@educateventures.com

Y. E. Wang et al. (eds.), *Social and Emotional Learning and Complex Skills Assessment*, Advances in Analytics for Learning and Teaching,
https://doi.org/10.1007/978-3-031-06333-6_8

8.1 Introduction

Education is going through unprecedented changes across the globe. During the year 2020, 165 countries have entirely closed their primary, secondary, and higher education institutes in an attempt to stop the spread of the ongoing coronavirus pandemic (UNESCO, 2021). Even though schools were closed, many countries' educational systems made significant efforts to provide for continuity of learning through distance education and online teaching. However, as over half of the students never worked together during the school closures (Parkin et al., 2020), collaborative learning opportunities were far from reaching their potential.

Learning analytics (LA), as a field, has a significant role in facilitating collaborative learning in the classroom, whether remote or not. LA can be used to inform educators, administrators, parents, students and other educational stakeholders with actionable insights about the collaborative learning processes of students. LA has been significantly evolving as a research field, contributing to our understanding of how collaboration occurs and can be supported in digital learning environments. However, real-world adoption and impact of learning analytics research are scarce, and far from their actual potential (Ferguson & Clow, 2017; Dawson et al., 2019; Alwahaby et al., 2021). In part, this is due to the limited amount of research focusing on the adoption and use of collaboration analytics solutions by key stakeholders in real-world settings (i.e., Zhou et al., 2021a, b).

This chapter presents the findings of our investigations on educators' expectations and their perceptions of collaboration analytics generated from student interactions in an online collaborative learning platform. More specifically, we first present a new method of evaluating the process of collaboration from students' online interaction data with the help of collaborative cognitive load theory and social network analysis (Kent & Cukurova, 2020). Then, we present the results of our fieldwork, investigating educators' requirements, insights, and iterative suggestions to the visualisations of these analytics. Specifically, two research questions are of interest;

1. To what extent are tertiary-level educators evaluating the collaborative processes in digital learning environments and the value of descriptive metrics to do so?
2. What is the added value of collaboration process analytics provided by CLaP compared to more traditional participation metrics for educators?

8.2 Background and Previous Work

For educators to provide adequate support to their collaborating students, they need to understand students' patterns of behaviour within the collaboration process (Van Leeuwen & Rummel, 2020). Unlike the outcome of collaboration where the impact can be measured through pre and post-test analyses, an understanding of *the process of collaboration* is not as direct. It involves the consideration of both "cognitive and social (interaction) aspects of the collaborative process" (Kaendler et al., 2015; Greiffenhagen, 2012). Over time, some orchestration dashboards have been devised to gather, analyse and interpret students' digital traces in a bid to understand their collaborative learning (Van Leeuwen et al., 2019; Van Leeuwen & Rummel, 2020). These dashboards serve as visual representations, informing educators about their learners' work, to help them track progress (Verbert et al., 2014). They also offer useful insights for teachers to track and stimulate different communication mechanisms among learners that are contributory to learning (Van Leeuwen et al., 2019), and recognise students and/or groups who need particular support (Molenaar & Knoop-van Campen, 2017).

As presented by Van Leeuwen et al. (2019) collaborative learning dashboards can be categorised into mirroring, alerting and advising dashboards. Mirroring dashboards offer information about learners to support monitoring of collaborative activity but leave all subsequent detection and interpretation of relevant information to the teacher (Van Leeuwen & Rummel, 2020). On the other hand, *alerting* dashboards provide alerts about classified groups that need the teachers' support, and the *advising* dashboards also provide an interpretation and advice on top of the information provided to the teacher (Van Leeuwen et al., 2019). Recent LAK and CSCL publications have good examples of all three categories of collaborative learning dashboards (i.e., Schwarz et al. (2018); Voyiatzaki and Avouris (2014); Casamayor et al. (2009); Martinez-Maldonado et al. (2015); Gerard and Linn (2016); Segal et al. (2017)).

Although LA dashboards have the potential to enable educators to reflect and gain insights on their students' collaboration, Van Leeuwen et al. (2017) revealed that the method of how educators identify and interpret the information presented on these dashboards remains predominantly uninvestigated. McCoy and Shih (2016) report that one of the contributing factors to this difficulty is the perception of teachers as mere users of LA technologies rather than considering them as co-creators of the data and visualisations. Additionally, some educators are not well equipped with the necessary data literacy skills to make sense of LA and their visualisations (McCoy & Shih, 2016). Moreover, it is important to note that most collaborative learning dashboards fulfil the mirroring function. That is, the interactions are visualized only in a descriptive manner without any meaningful interpretation of what it might mean for collaborative learning as a whole or what the teacher should do next. When teachers use such a dashboard, they have to take an interpretative stance by themselves and make the connections between observed events to the pedagogical aims (Van Es & Sherin, 2002). Since taking an interpretative approach is not what teachers instinctively do or are routinely trained to do (Van Es & Sherin, 2008), there is an urgent need to investigate educators' understanding of collaborative learning visualisations and analytics, to (i) help them adapt their pedagogy to the observed collaborative learning analytics; (ii) better adapt the design of LA to the needs and requirements of educators.

8.2.1 Teacher Evaluations of Collaboration Analytics

In the design of effective collaborative LA, most available research highlights the significance of robust technical approaches (Rosé et al., 2019) and the use of learning sciences principles (Luckin & Cukurova, 2019). There is a range of other factors that are frequently overlooked, such as teachers' preferences, the reason and usage of the collaboration analytics, or the social context in which the analytics will be used. Understanding the perceptions of educators, their needs in collaborative learning support and their perceptions of the collaborative learning visualisations are crucial for the successful adoption and the wider impact of learning analytics. Recently, there has been scrutiny of the limitations of modern LA systems (Prestigiacomo et al., 2020). This investigation is a result of the challenges that students (Jivet et al., 2018; Matcha et al., 2019) and teachers (Mangaroska & Giannakos, 2018) experience in understanding and acting upon data to enhance learning. This examination is important because the effectiveness of collaboration analytics visualisations is highly dependent on the application of insights to achieve the desired goal. Yet, very little work has been done to address teachers' needs and understanding of collaboration analytics in real-world teaching contexts.

According to Gibson and Martinez-Maldonado (2017), teachers frequently extract "irrelevant" interpretations from collaboration analytics. As a result, they find it challenging to apply insights from the visualisation to improve learning. This research indicates that stakeholders (such as students and teachers) should be involved in the design process of collaboration analytics to support their needs. If teachers are excluded from the design process, likely, the generated analytics will not fulfil their needs and understanding. Thus far, few studies specifically focus on engaging teachers in the design of LA or undertaking significant teacher evaluation studies in collaborative learning contexts. Chen and Zhu (2019), Holstein et al. (2017), and Holstein et al. (2018) are some relevant emerging examples, yet not specifically focusing on collaboration analytics.

For instance, Prestigiacomo et al. (2020) suggested a human-centred design strategy via the concept of social translucence that can be used to design effective learning analytics. According to Prestigiacomo et al. (2020), the *"visibility"* principle advocates for the need to make relevant information available about a specific task. The *"awareness"* concept aims, above the *visibility* principle, to enable an interpretation of a noticeable situation to facilitate evidence-based decision-making. Finally, the concept of *"accountability"* is the dimension for social regulation to keep individuals accountable for the data they share with others. To put this into practice, Prestigiacomo et al. (2020) involved six high school teachers in implementing the three principles of Social Translucence. Their analysis showed that, under visibility, teachers want the following information to be available: tracking students' (reading and writing) behaviour, collaboration, affect, engagement, orchestration, learning modalities, feedback and assessment. More specifically on collaboration, the teachers wanted to know how well the students work collectively and have an understanding of the individuals that participate in the group work.

More specifically in the context of collaborative learning, Martinez-Maldonado (2019) studied teachers' preferences in collaborative LA visualisations. Focusing on the perspective of user experience, this study confirms that teachers prefer graphical explanations in the form of tracking visualizations, to text-based explanations. On the other hand, the teachers found that text-based explanations were easier for students to follow when they are linking ideas. In essence, the choice of analytics to use could be influenced by the specific pedagogy implemented. Moreover, Martinez-Maldonado's (2019) findings show that teachers want the flexibility to configure the data the collaborative learning visualisations display. Teachers are generally under pressure to keep up to date with all activities and have to continuously decide which group or student receives their attention at any given moment (Greiffenhagen, 2012). Given the dynamic nature of collaborative learning, such flexibility offers good chances for the adoption of collaboration analytics visualisations. Similar points were raised by Holstein et al. (Holstein et al., 2018) in individual learning settings. The authors report that the value of teacher's visualisations may depend on the extent to which they are involved in their design decisions. In collaborative learning contexts, Van Leeuwen et al. (2014) also affirm that the teacher's beliefs of what is accountable for effective collaboration can significantly affect their use of the collaboration analytics visualisations.

Swidan et al. (2019) investigated how teachers comprehended the progression of multiple groups through collaborative visualisations. The authors found that incorrect solutions, explanations or challenges, technical problems, confusion, off-topic discourse, idleness and correct solutions are some of the situations that teachers can detect using their collaboration analytics visualisations. It's interesting to note how teachers' experience can also affect how they interpret the visualisations of collaborative learning. Teachers with more years of expertise tend to respond based on their preferences of the situation regardless of what is presented with the visualisations. However, novice teachers respond to the learners in a sequential pattern, as the dashboard informs them. However, as argued by Van Leeuwen et al. (2019), the mechanisms by which educators discover and perceive important data on collaboration visualisations remain understudied. The authors explain that the pattern or sequence in which teachers navigate through the visualisations affects their understanding of the data, and consequently, the decision they make to support the students learning in groups (Van Leeuwen et al., 2019). Therefore, they argue that in addition to having a data-rich visualisation, teachers must also have an inbuilt guide to enhance their use of collaboration analytics visualisations. In their meta-analyses of 26 papers on collaboration orchestration tools, Van Leeuwen and Rummel (2019) emphasised the need to investigate further how teachers engage and interact with collaboration analytics and their visualisations.

To be able to create collaboration analytics and visualisations that would be meaningful for the educators' practice, it is expedient to (a) "understand the teachers' needs;" (b) "understand the particular context of usage;" and (c) "understand how the design of the analytics can be aligned with their pedagogical intentions," (Martinez-Maldonado, 2019). Before exploring teachers' needs and understandings with regards to collaboration process analytics, we present in the next section the specific analytics built and evaluated in this study.

8.3 Specific Collaboration Analytics Investigated by the Study

In this study, we focused on collaboration analytics that are inspired by the collaborative cognitive load theory -CCLT- (Kirschner et al., 2018). In our previous work, we suggested a new method of measuring the process of collaboration using social network analysis to evaluate the balance between interactivity gains and coordination costs of learner communities (Kent & Cukurova, 2020). The following section presents a slight overview of this approach in support of the chapter's interpretations. For more detailed explanations of the approach, as well as the detailed explanations of the connection between CCLT and the social network analysis metrics we used in the analysis please refer to Kent and Cukurova (2020).

Kirschner et al. (2018) argue in CCLT that the limitations of working memory (WM) result mainly from a high cognitive load, thus necessitating the need to combine multiple WMs to work collectively on an assignment. The combination of multiple working memories makes it easier for students to perform tasks in groups. In this case, the working memory capacity may increase without necessarily increasing the cognitive load of the tasks. So rather than an individual focusing their limited working memory to solve a problem, more than one learner's working memory can be combined to solve the same problem. In such a scenario, collaboration becomes useful because it reduces extraneous cognitive load. Therefore, the goal of the collaboration is, to a certain extent, to provide just enough collective WM to overcome cognitive overload. On the other hand, when the collective WM is significantly higher than the cognitive load, the task loses its complexity; therefore, participants become less engaged since their cognitive resources become redundant. Moreover, groups in collaborative settings require significant effort to be coordinated and organised. This also negatively impacts the outcome of the collaborative process since the learners would have to put in more effort to get things going. If the learners have prior experience of working together, the opposite may be the case. Thus, collaborative learning becomes an act of striking the right balance between the WM gains from interactions (interactivity gains) and the costs of coordinating the challenges associated with operating with (and in the context of) others (coordination costs).

8.3.1 *Analytics of Collaboration as a Process (CLaP)*

Leveraging CCLT, in our recent work, we suggested a new method of measuring the process of collaboration irrespective of its particular indirect outcome evaluations (i.e., grades, group project outcomes etc.) (Kent & Cukurova, 2020). In this work, we use social network analysis to examine the relationship between the interactivity gains (IG) and coordination costs (CC) in group learning to make sense of learners' *collaboration process*.

Table 8.1 Top-level breakdown of the collaboration

Data	Community 1	Community 2
Total number of students	42	32
Total number of interactions	7600	15,515
Total number of interactions per student	7600/42 = 181.0	15,515/32 = 484.8
Total number of posts	253	408
Total number of posts per student	253/42 = 6.0	408/32 = 12.8
Total number of cross-reference	24	222
Total number of cross-references per student	24/42 = 0.57	222/32 = 6.94

Interactivity Gains (IGs) IGs are the cognitive wealth and benefit resulting from interactions with co-learners. Interactions, using social influence, are known to boost collaborative learning and collective performance (Bernstein et al., 2018).

Coordination Costs (CCs) CCs are cognitive resources needed by participants to participate in the collaboration process effectively, manage their interdependencies and ideas, and complete the task collaboratively. The CCs affect collaborative learning; the more resources that are needed to coordinate the collaboration, the lower the effectiveness of collaboration (Nokes-Malach et al., 2012).

To examine the process of collaboration, we analysed the collaboration process of two communities using the CC and IG. The CC is proxied as *variance in degree*, and the IG is proxied as *reciprocity*. The students were drawn from two different postgraduate cohorts, pseudonymised Community 1 and Community 2, with 42 and 32 students, respectively.

An online discussion tool was used for data collection, and the collaboration activities were recorded for 7 weeks. The tasks for the two communities were the same and involved building a collective concept map via online discussions. At the end of the 7 weeks, Community 1, which has 42 students generated 7600 total number of interactions, 253 total number of posts and 24 total number of cross-references. On the other hand, Community 2 with 32 participants aggregated 15,515 total number of interactions, 408 total number of posts and 22 total number of cross-references among their posts. Table 8.1 shows the breakdown.

Collaboration as a Process (CLaP) Analytics The interactions among the students are categorised into three dimensions, namely: *Contribution Interactions.* These are interactions that are related to the creations, updates and deletion of posts. *Consumption Interactions.* These are interactions that are related to viewing posts, viewing a map of posts, viewing attachments, searching and refreshing sub-posts (Kent & Rechavi, 2020). *Organisational Interactions.* These involve the connection of non-connected posts, voting and "un-voting" posts, following and unfollowing posts/learners.

Figure 8.1 shows that the two communities exhibit different patterns of growing interactions. Community 1 started with fewer interactions compared with

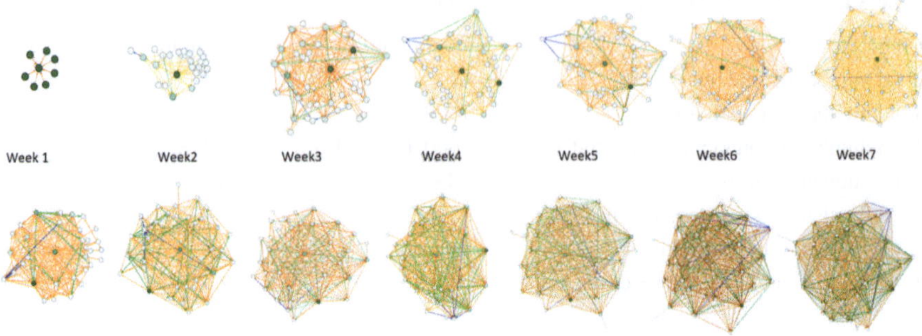

Fig. 8.1 Evolving interaction networks of Community 1 (top) and Community 2 (bottom). Consumption type interactions are orange, contributions are blue, and organisational interactions are green

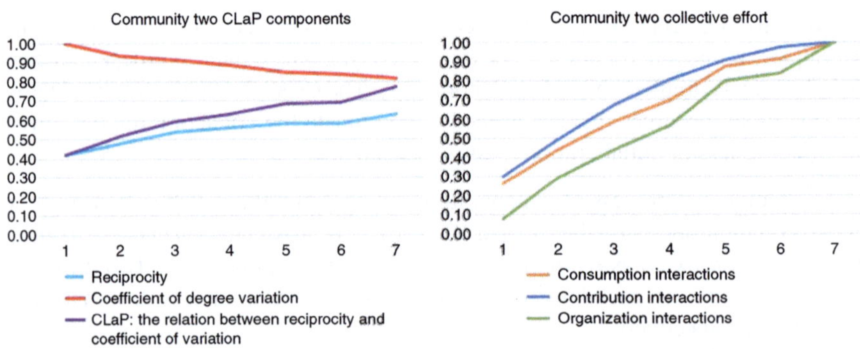

Fig. 8.2 (Left) Community 2's CLaP components, normalised to a 0–1 scale; (right) Community 2's number of interactions, normalised to the size of the community and a 0–1 scale.

Community 2, which started on a good note and continues to grow throughout the 7 weeks.

As can be seen in Fig. 8.2 above, in community 2, the number of contributions for all three types of interactions show increase from week 1 until the end of week 7. On the other hand, reciprocity (proxy for IG) and degree of variation (proxy for CC) values present different degree and directions of change. For instance, it seems that the higher the CC, the lower the IG and vice versa. Despite the sharp increase in the IG from weeks 6 to 7, for Community 1, the CCs did not experience a sharp decrease. This experience can be attributed to the fact that the tasks introduced were grade related, which by default could compel all students to participate in the collaboration. The impact on learning design decisions, including assessment, on the observed learning analytics, has been well established in the literature (i.e., Zhou et al., 2021a). Based on this insight, the CLaP analysis was assumed to also be used by instructors to understand possible ways to manipulate the learning activities to enhance collaboration. In this study, we intend to examine the value, or not, of CLaP visualisation for educators and test such assumptions.

8.4 Methodology

To choose educators from different backgrounds and experiences to investigate their needs and understanding with regards to the visualisations of collaboration process analytics, we asked faculty members at five different institutes to respond to a survey and invited them to two workshops for introducing the ClaP analytics for focus group discussions. The details of the workshops and survey are discussed below, and the questions used can be found in the Appendices.

8.4.1 Participants

In total, 19 participants were recruited based on convenience sampling. The sample included both experienced online teachers and those with fewer or no years of experience. The diversity of participants was ensuring, as we were able to craft a good image of both novices, as well as experienced instructors. The participants were all based in the departments of psychology, education, and computer science and were teaching a variety of subjects related to these fields. All participants had some research interests in the area of educational technology and/or learning analytics. However, they are not actively engaged in this area. As we highlight below, most of them did not also have any particular experience in teaching in online settings. Although 19 participants are not a large sample size, compared to previous research (i.e., Prestigiacomo et al. 2020), it was a sufficient sample size for in-depth teacher evaluations.

8.4.2 Data Collection Phases

Workshop Part 1 In the first phase, we organised a workshop for the participants to introduce them to the CLaP analysis and the visualisations of IG and CC. The workshop was interactive, as the attendees were encouraged to react/respond to the subject of discussion in real-time. The discussion included the theoretical considerations of CLaP analytics (i.e., how the working memory and long-term memory relates to the collaboration process). We also explained the IGs and CCs in relation to the proxies that we derived from the social network analysis. Finally, we displayed some of the visualisations to the participants to have a better understanding of the metrics (posts, views, reciprocity, and the variance of note degree) we are using to analyse the process of collaboration.

Survey After the first phase, we asked participants to fill out the survey to express their understanding of student collaboration in online asynchronous learning settings as well as their needs in interpreting those visualisations. We grouped the survey questions into two categories as summarised below. The survey involved multiple-choice questions, Likert scales, open and closed-ended questions, multiple selection questions. All survey questions used in this study can be found in the link provided in Appendix 8.1.[1]

1. Previous Experience in Online Collaboration. – Survey Section 1, all seven questions
2. Collaboration Visualisations
    2.1. Charts showing students' participation. – Survey section 2, questions 1, 2, and 3
    2.2. Charts showing students' interaction and coordination. – Survey section 2, questions 4, 5, 6, and 7
    2.3. Comparing the participation charts to interaction and coordination charts. – Section 2, question 8
    2.4. Views about online collaboration charts – Survey Section 2, questions 9, 10, and 11

Workshop Part 2. Open discussion After the participants completed the surveys, in the second half of the workshop, we invited participants for an open-ended discussion on the collaboration visualisations. This phase aimed to generate in-depth probes on the participants' views of the collaborations charts as well as any potential aspects the survey didn't cover but the participants considered as significant. The workshop was run in online settings with all participants attending together. The discussion was facilitated by an experienced researcher inviting contributions to pre-set questions in Appendix 8.2 (as derived from the survey items), and questions were followed up with explorative probes inviting explanations to answers. The visualisations were introduced with a presentation by the expert, each phase of the workshop took 90 min, and was recorded for qualitative data analysis of the discussion transcript.

8.4.3 Data Analysis

The survey data was analysed using SPSS 24.0 software. For the analysis, all data were inputted into the software and descriptive statistics were applied. The qualitative data analysis was conducted using Braun and Clarke's six phases of thematic analysis (Braun & Clarke, 2006). Transcribing workshop recordings was the first step. An independent researcher developed the first thematic codes of critical moments. Afterwards, codes and data were shared with other researchers to be discussed and revised to ensure that emerging themes and quotes covered all of the collected data and that they can be audited.

[1] https://forms.gle/qknjnTkpQ16JwXf8A

8.5 Results

8.5.1 *Participants Experiences in Online Teaching and Their Confidence in Reading Basic Visualisations*

Based on previous research (Swidan et al., 2019; Martinez-Maldonado, 2019) and our assumption that participants' knowledge in collaborative learning and their confidence in reading basic visualisations both might be moderating factors on their understanding of the process visualisations we studied, next we summarise their answers to these two questions.

Figures 8.3a, b summarise the experience of the participants in online teaching as well as their confidence in reading visualisations. Six of the participants did not have any experience in online teaching, while 5 of them had 1–3 years of experience in online teaching. Only 1 participant had 3–5 years of experience in online teaching, while 7 of them had over 5 years of experience. Mean = 2.9 years and SD = 2.5 years. To compliment that, 7 out of the 19 participants had no experience in designing online collaboration, while 8 of them had 1–3 years of experience in designing online collaboration. No one had 3–5 years of experience, while only 4 of them had over 5 years of experience. Mean = 2.1 years and SD = 2.2 years. On a scale of 1 (Not at all) to 7 (I am an expert), 5 participants indicated having level 5 of knowledge about CL (Collaborative Learning), while only two indicated having the same confidence level (5) of reading basic visualisations. Although none indicated having a level 7 of CL knowledge, five participants indicated having level 7 of confidence in interpreting basic visualisations. Out of the five participants who reported having a level 5 of CL knowledge, only two reported on level 5 confidence level of reading basic visualisation. Although only one participant reported having level 1 of CL knowledge, this single participant is among the four participants who declares having level 6 confidence in reading basic visualisation.

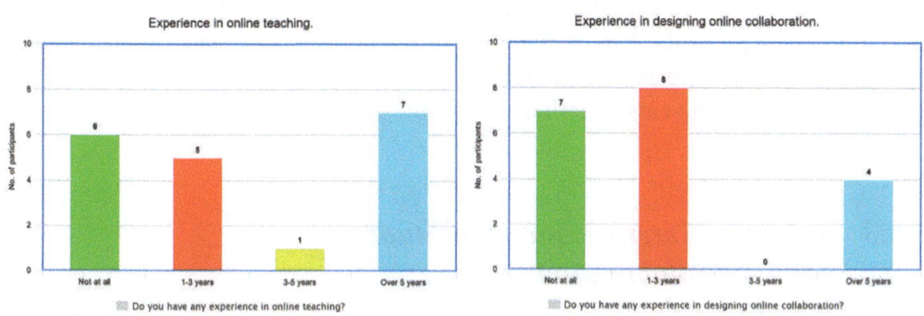

Fig. 8.3 (**a**) Participants' years of experience in online teaching and designing online collaboration. (**b**) Participants' self-declared knowledge of collaborative learning and confidence in reading basic visualisations

8.5.2 *Evaluation of Collaboration in Online Classes*

The survey reports showed that 12 out of the 19 participants do not evaluate online collaboration as part of their teaching practice. Only one participant evaluates the collaboration competence of every learner individually. Three participants evaluate collaboration competence as a group while another three participants argued that they assess both the collaboration of individuals and groups. In general, we had four participants who evaluated the collaboration competence of individuals and six who preferred to evaluate the collective collaboration competence of groups. In general, even though most of the participants used online collaboration to engage their students in active learning, most of them did not evaluate collaboration at all.

Participants' Pedagogical Purposes for Using Online Collaboration (Table 8.2) 15 out of the 19 participants use online collaboration to engage their students in active learning. Out of these fifteen participants, ten use online collaboration for knowledge building, nine use it to build their students' collaborative skills and another eight for communication. Only two participants said they use all six pedagogical purposes included as options in the question, and only three participants (P6, P7, and P8) listed other purposes for using online collaboration. P6 said that they use online collaboration to support *"master's projects remotely." and* P7 uses online collaboration for *"Community building – developing a sense of community among the students as well as partially developing a sense of ownership."*

Criteria Participants Use to Evaluate Online Collaboration (Table 8.3) Out of all the criteria teachers might use to evaluate online collaborations, "Quality of each student's posts evaluated from a subject domain perspective" received the highest response; 11 of the 19 participants chose this one. Out of these 11 participants, eight also evaluate the collaboration by checking the "quality of each student's posts evaluated from a dialogic perspective". Interestingly, three of the seven participants with experience in teaching online also evaluate the online collaboration based on the quality of each student's posts from the perspective of the subject domain. Notably, only P14 said they evaluate collaboration based on the "Number of posts for the whole group". Out of the five participants (P3, P13, P14, P17, and P2) who evaluated the "Number of posts each student replies to others," four (P13, P14, P17, and P2) have more than 5 years experience in online teaching.

8.5.3 *Educator Interpretations of Students' Online Collaboration*

Interpretations of Student Collaboration Based on the Descriptive Statistics of Student Interactions (Table 8.4) First, participants are shown basic descriptive metrics and asked to interpret and determine the collaboration process between two

Table 8.2 Participants' pedagogical purpose of using online collaboration

Purposes	P1	P2	P3	P4	P5	P6	P7	P8	P9	P10	P11	P12	P13	P14	P15	P16	P17	P18	P19	%
Engagement	■	■		■				■	■		■		■	■			■			47
Active learning	■	■	■		■		■	■	■	■	■	■	■		■	■	■		■	78
Assessment			■				■	■	■		■							■		31
Collaborative skills	■	■			■			■		■	■		■	■			■	■	■	57
Knowledge building	■		■		■	■	■	■		■	■					■	■	■	■	63
Communication	■				■			■		■	■	■	■			■				42

Table 8.3 Survey results of criteria the practitioners use to evaluate online collaboration

Criteria	P1	P2	P3	P4	P5	P6	P7	P8	P9	P10	P11	P12	P13	P14	P15	P16	P17	P18	P19	%
I do not evaluate collaboration in online settings.															◘				◘	10
Number of posts each student contributes	◘						◘			◘	◘		◘	◘		◘	◘			42
Number of posts each student replies to others		◘	◘										◘	◘			◘			26
Number of comments each student makes			◘									◘	◘	◘			◘			26
Number of posts for the whole group														◘						5
Number of views on posts for each student				◘									◘							10
Number of views each student takes on other students' posts			◘										◘							10
Number of views of the whole group														◘						5

The amount of time each student spends in the discussion			■										■	■			■			21
Quality of each student's posts evaluated from the subject domain perspective	■	■	■		■	■		■			■	■	■			■		■		57
Quality of each student's posts evaluated from the dialogic perspective		■	■		■	■					■	■				■		■		42
Quality of each student's posts evaluated from the learning perspective	■		■			■	■		■		■					■		■		42

Table 8.4 High-level descriptive metrics of the two cohorts

Class name	Term	Total number of interactions	Total number of posts	Number of students	Total number of cross-references
Community 1	Autumn 2015	7600	233 (246 after deletions)	42	24
Community 2	Spring 2017	15,515	408 (399 after deletions)	32	222

cohorts of students. The descriptive metrics included the total number of interactions, the total number of posts, the number of students, and the total number of cross-references.

Although 17 out of the 19 participants expressed that Community 2 was more collaborative, four of them were sceptical about their decisions. P3 said that *"the number of interactions and cross referencing indicated Community 2 is more collaborative than Community 1"*. This participant also suggested that more data such as the "*quality of interaction*" could confirm the decision to go for Community 1. Although P9 preferred to have more data to determine that Community 2 was more collaborative than Community 1, they used the "*total number of cross references*" as the yardstick for the choice of the more collaborative community. Among the 15 participants who argued that Community 2 was more collaborative, five (P6, P7, P12, P19, and P1) made their decision based only on the total number of interactions and cross-references.

In addition to the two factors mentioned above, P6 argued that familiarity among the students and the pedagogy/course context can impact their online collaboration. More specifically, the participant said that *"I think maybe students knew better each other in spring, maybe there is something different in the context"*. In addition, P7 asserted that having access to the quality of interactions instead of just the total number of interactions and cross-references can give more insight into the community with better collaboration. P11 who based the choice between Community 1 and Community 2 on the "total number of interactions" had a similar opinion: *"The number of posts might not be an accurate metric of collaborative learning, because there are other, more accurate metrics (quality over quantity)"*. The four participants (P5, P13, and P14, P2) who have more than 5 years of experience designing online collaboration had slightly different reasons for their choice of Community 2 over Community 1. P5 only mentioned that *"Community 2 looks better from the outside"*. P13 and P2 said Community 2 was more interactive because they had more interactions, posts, and cross-references. For P14, the choice of Community 2 was determined only by the number of cross-referencing.

Besides basing their decisions on metrics, P15, P17, P19, P1 and P2 also based their decisions on the ratio of students to posts/interactions/cross-references. Although Community 2 had fewer students than Community 1, the posts/interactions/cross-references were higher than those of community 1.

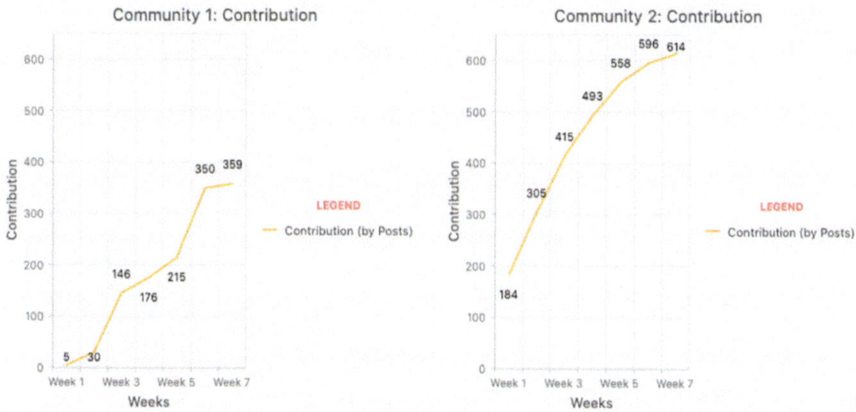

Fig. 8.4 Line graphs of the number of contributions (left – Community 1; right – Community 2)

Next, Fig. 8.4, showing the number of posts per week for each community, were shown to the participants. Participants were required to read the visualisation to determine which community is more collaborative and provide reasons to support their choice. Additionally, they were asked if their choice remains the same throughout the 7 weeks.

Out of the 19 participants who responded to this question, 15 asserted that Community 2 is more collaborative than Community 1. P17 observed that there is an increase in Community 1's number of contributions during Week 5. On the other hand, P12 said that Community 1 is more active in Week 3. Though most participants came up with their decision by looking at the number of posts, P9 took it one step further by looking at the ratio of students to posts. Two participants (P3 and P5) said that they could not make sense of which community is more collaborative - using the available data (Contribution by post).

According to P14 and P19, Community 2 is preferred because it yields higher contributions, which increase consistently over the weeks. P19 noted some ambiguities in certain weeks: *"Overall contributions in community 2 are higher and consistently increasing. However, different periods reveal different trends. Weeks 2 and 5 appear to show a much greater increase in collaboration in community 1. Rate of increase per week would be a useful metric."*

On the other hand, P2, a participant with over 5 years of experience in online collaboration expressed that "*Community 2 is more collaborative because there are more posts every week than community 1. I'm not sure of the trend since the rise of community 1 is much bigger than 2.*"

Following was Fig. 8.5 below, showing the number of posts (yellow line) and views (purple line) for each community based on the number of posts and views per

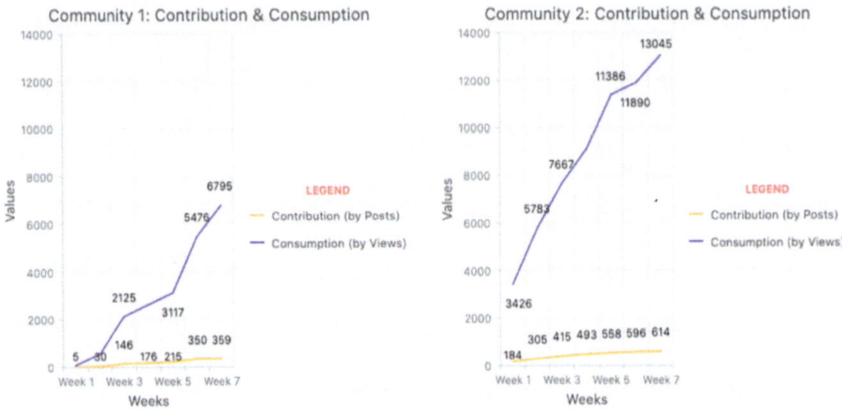

Fig. 8.5 Line graphs of contributions and views (left – Community 1; right – Community 2)

week. The participants were invited to view the visualisations to determine which community was more collaborative. Further, they were encouraged to provide reasons for their choice as well as indicate whether their choice remained the same for all 7 weeks.

In this case, only 12 participants (P5, P7, P8, P9, P12, P13, P15, P16, P18, P19, P1 and P2) chose Community 2 as more collaborative than Community 1. Only nine of these 12 participants (P7, P9, P12, P15, P16, P17, P18, P19, and P1) explained the reason behind their choice of Community 2. It is worth taking note that P3, P14 and P17 who have a minimum of 1–3 years of experience in designing online collaboration found it quite challenging to decide which of the communities is more collaborative. According to P3, the data (the number of posts and views) were insufficient to make conclusions while P14 and P17 expressed that they aren't sure which community is more collaborative.

According to P7, the choice of Community 2 was heavily predicated on the steady increase in views and postings over the weeks. Further, the participant noticed a sharp increase for Community 1 in Week 6. The participant explained that this might be because it is toward the end of the course, and everyone is trying to address everything they might have missed. On the other hand, the participant observed that the interaction in Community 2 was less during this week (6) and said that this may be because they have a steady interaction before that week and they would not have so many backlogs to address. *"They may be more interested in self-learning or self-review of course-related stuff,"* the participant added. According to P12, although Community 2 had more posts and views than Community 1, the Communities can be said to have almost the same level of collaboration if the ratio of views per post is considered. On account of this, the participant asserted that the number of posts is not enough metric to measure students' collaboration. According

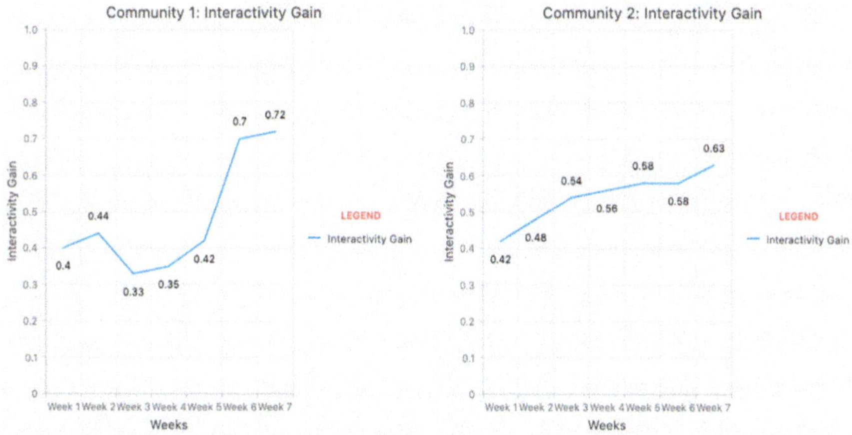

Fig. 8.6 Line graphs of Interactivity Gain (left – Community 1; right – Community 2)

to P19, *"The overall number of posts is much higher for community 2 so I would say they are more collaborative. However, both week 2 and 5 show community 1 collaborations increasing more rapidly."*

Interpretations of Student Collaboration Based on the Collaboration Process Analytics In this section, we first presented the graphs in Fig. 8.6 showing the reciprocity of learners' interactions per week for each community. The reciprocity of the learners' social network (i.e., to what extent they respond to each other) was measured to represent their interactivity gains (IG). The participants were asked which community is more collaborative and why. Additionally, they were asked if their choice would remain the same for all 7 weeks.

Ten participants (P3, P5, P8, P11, P12, P13, P16, P18, P19 and P2) chose Community 2 to be more collaborative than Community 1 based on the IG. Overall, they chose Community 2 because in most weeks the normalised value of the IG was higher than in Community 1. Additionally, P5, P7, and P8 explained that the steadiness and consistency of the IG of Community 2 indicated that there is better collaboration. For P18, Community 2 looked more collaborative in Weeks 2, 3 and 4. However, there were no meaningful differences in the other weeks. P2, on the other hand, used the average IG over the weeks to decide the most collaborative community. Here is what the participant observed: *"Community 2 is more collaborative because the average gain of community 1 is around 0.48 while the average of community 2 is around 0.54. I am not sure of the future situation according to the bigger rise of community 1."*

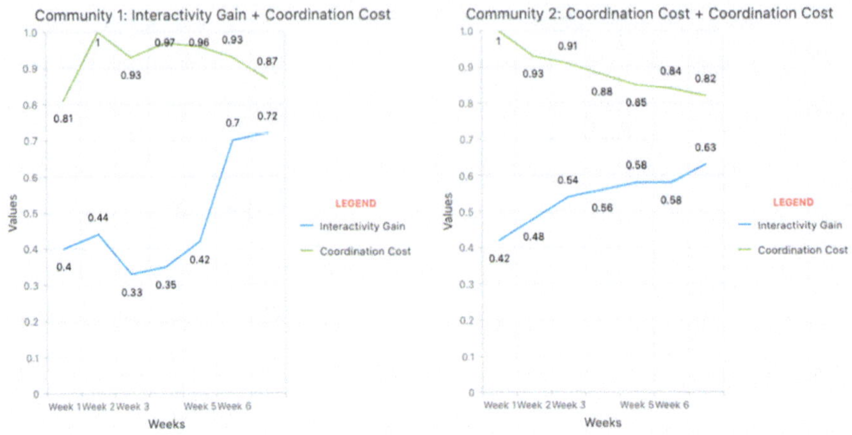

Fig. 8.7 Line graphs of Interactivity Gains and Coordination Cost (left – Community 1; right – Community 2)

According to P11, the choice of Community 2 was also based on the decrease in the IG of Community 1 in the middle of the course (Week 3 and 4). Although P12, P13, P16, and P19 chose Community 2, they explained that the experience was not the same throughout the entire period. P16 stated that Community 1 was more collaborative in Week 6 & 7 and for P19, it was more collaborative in Weeks 5 to 6. *"Overall community 2 is higher, but not for all of the weeks,"* said P12. For P13, Community 1 showed a higher level of collaboration only during Week 5.

Although not being specific about which is the more collaborative community, P7 expressed that Community 2 had a steady IG, while Community 1 still has a higher IG despite the drop in weeks 2 and 3. P15 was similarly not clear about which is the more collaborative community but explained that while Community 1 had the highest value (presumably Week 7) of IG, Community 2 had a steady value. P17 had a hard time interpreting the visualisation in a short amount of time.

We next presented Fig. 8.7 depicting the mutuality of learners' interactions (blue line) and the cognitive cost for learners to coordinate their actions (green line) per week for each community. This value is applied by examining the degree of heterogeneity within learners' social networks to represent their CCs. The participants observed the visualisation to examine which group, Community 1 or Community 2, is more collaborative, and why. Similar to previous questions, they were also asked whether their choice remained the same for all 7 weeks.

P19 offered further insight into the graphs: *"The results for community 2 seem to indicate that as mutual interactivity increases, cognitive cost decreases. This indicates a greater ability to collaborate. Community 1 does not follow this pattern in week 1 and 2, where mutual interactivity also correlates with higher cognitive costs and vice versa".*

P16 argued that *"Community 2 was more collaborative", and explained*: *"due to an increasing trend in interactivity gain and a decrease trend in coordination cost throughout"*. It is quite interesting to note that this participant neither has a long experience in online teaching nor online collaboration. However, they seemed to be comfortable interpreting the graphs. However, P13 noted that they do not understand how CC is calculated despite their extensive experience in both online teaching and online collaboration. P5 (also with the same experience in online collaboration) chose Community 2 and expressed that the *"Community 2 process grows more steadily, which I believe is better than trying to catch up in the last week."*

Similar to their previous responses to the descriptive metric graphs, P17 also emphasise the difficulty of interpreting the visualisation in a short amount of time (within the workshop presentation period). P6, who did not give a definite conclusion, suggested that the two communities differ between the IG and the CC in Week 7. According to P18, Community 1 was more collaborative because the interactions were more heterogeneous. Other comments were from P1 *"Community 2 as they were consistent in their collaboration"* and from P2 – *"Community 2 is more collaborative because its ratio of gain and cost is higher than community 1. I think it will remain."*

Following were the graphs in Fig. 8.8 showing the difference between the learners' gains per week for each community and their CCs. The participants were again posed with the same questions above, to decide which community is more collaborative and also provide reason(s) for their decision.

Eight participants (P3, P5, P6, P8, P10, P11, P16 and P2) chose Community 2 to be more collaborative. P3, P6, P11 and P16 provided explanations that relate to the highlighted differences. These participants indicated that Community 2 is more collaborative since the difference between the CC and IG was lower in Community 2.

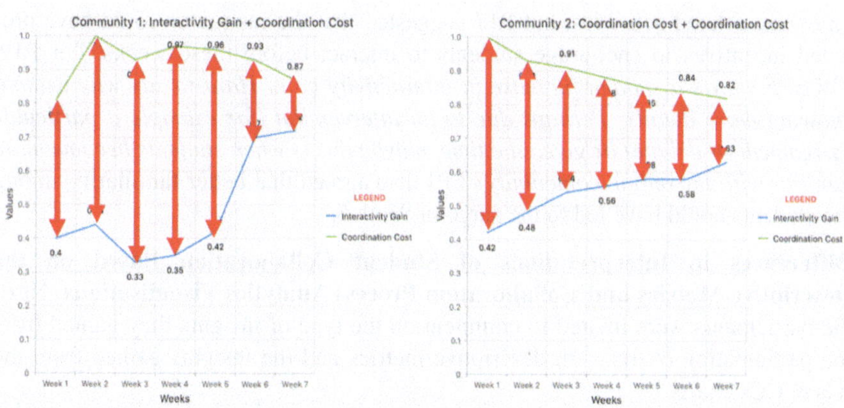

Fig. 8.8 Line graphs of Interactivity Gain, Coordination Cost and the highlighted difference between the two measures (left – Community 1; right – Community 2)

The participants explained that the closer the lines (CC & IG) are to each other, the greater the chance of Collaboration. Notably, only P6 and P16 out of the four participants observed the difference between Week 6 and Week 7. According to P6, Community 1 was more efficient than Community 2 from Week 6 through Week 7. P16, however, provided a different conclusion. The participant suggested that the two communities had the same level of collaboration in Week 6 and 7.

P2 concluded that Community 2 was more collaborative because "...*its average ratio of gain to cost is smaller than community 1, and the ratio becomes smaller and smaller*". P4 and P15 found this visualisation intriguing and argued that they serve as a better "*representation of the collaboration*" than the previous charts with descriptive metrics. P19 was not able to make sense of these graphs. The participant expressed "*I am not convinced of how this 'difference' is useful in determining which community is more collaborative*." Finally, P18 noted the changes across the weeks. The participant argued that "*At weeks 1, 2 and 6, community 2 looks more collaborative, other weeks community 1 looks more collaborative.*"

When the participants probed if they spot a change in Community 1's behaviours in different weeks, 17 participants (P3, P5, P6, P7, P8, P9, P10, P11, P12, P13, P15, P16, P17, P18, P19, P1 & P2) observed a change in Community 1's behaviour after Week 5. The explanations given by P5, P6, P7, P8, P9 and P11 concerned exam, end-of-term, or grade-related activities. They were all aware of the peculiarities of students' activities near assignment periods. P7 speculated that the changes in students' collaboration could be because they needed to complete a final project. Therefore, they needed to make up for the time they may have lost before that time. Additionally, P6 indicated that since the students are likely to have known each other better and shared a common purpose, that could also influence their interaction rate during those 2 weeks.

According to P13, P16, and P2, the changes might stem from the introduction of a new intervention/instructional design. Additionally, P16 mentioned the possibility that assessments could have been assigned at those points. Most participants focused on course-related activities, but P17 suggested that the instructor could have provided incentives to encourage students to interact before these weeks. For P19, "*Week 5 shows a sudden increase in interactivity gain. Hard to say why without knowing more details- Perhaps due to an intervention (for example a particular assignment or debate) or given the time, maybe this is when students become more familiar with the online community?*" P1 also agreed that better familiarity among the students could have led to the surge in Week 5.

Differences in Interpretations of Student Collaboration Based on the Descriptive Metrics and Collaboration Process Analytics Visualisations Next, the participants were invited to comment on the type of insights they gained from the participation charts with descriptive metrics and the insights gained from the IGs & CCs charts.

The reports showed that half of the participants (for example, P13, who has over 5 years' experience in online teaching and online collaboration) had difficulty making sense of the CC and the IG charts. On account of that, the participants could not interpret collaboration based on these graphs. On the other hand, the other half indicated that the process analytics visualisations provided better metrics (than the number of posts and views) to determine the quality of collaboration. For instance, P19 expressed that the process graphs showed how collaboration relates to cognition: *"I believe the second two charts show more granular information to understand the collaborative process more deeply and its effects (or gains) on cognition as a consequence."*

Similarly, P3, P5, P8, P11 and P17 indicated that the process graphs provided a better perspective about the collaboration process. Unlike the descriptive metrics, they give an understanding of whether the students are only posting randomly, or they also respond to one another (Reciprocity as IG). Even though P17 acknowledged that the IG and CC charts provided more information about the collaboration, the participant requested that examples/training should be provided to educators to minimise the difficulty in explaining the concepts to avoid giving an explanation that is out of context. P10 was satisfied with the descriptive metrics to make necessary inferences about collaboration, whereas P7 and P1 were unsure whether the process charts offer better insights than others. P1 explained that regardless, the familiarity among the students would influence their rate of collaboration so data from other contextual information should be provided in addition to descriptive graphs and/or process graphs.

Finally, we asked the participants what other metrics educators would appreciate seeing to interpret students' collaboration process. In addition to the number of social interactions among the students, the participants reported they would like to have metrics about the quality of the (interaction) contents. While the number of social interactions among the students could also have been counted as an indication of collaboration, that might not be an accurate indicator of the quality of the collaboration. In essence, many participants wanted the quality of the content to be taken into consideration as well as the amount of interaction. As P11 argued, *"From my point of view, the number of social interaction might not be an accurate indicator as students might replies to each other with a short sentence like 'that is a great idea' especially if they have been given marks according to their interaction, therefore content quality should be associated with the number of social interaction."*

In addition to the quality metrics, the participants wanted to know the level of agreement or disagreement amongst the participants as well as the number of arguments initiated. Similarly, the level of off-task discussions and the specific tasks that lead to collaboration were considered equally important measures to interpret collaboration. P15 said that it would be interesting if the data graphs were linked to specific tasks to know the nature of the tasks. P19 wanted to know the exact type of collaboration activity students engages in. Another metric considered as important was the number of contributions in a specific discussion topic thread rather than accumulated measures. Finally, the participants wanted to have data about the specific individual students who participated in the collaboration including their existing knowledge of the topic discussed, their previous experience in collaboration and their familiarity with each other.

To gain a general sense of how useful the graphs discussed above would be to the participants in their practice, we asked whether they would be open to seeing any graphs from the lessons they teach online. Eleven participants (P7, P9, P10, P11, P12, P13, P14, P15, P16, P17, & P1) were interested in descriptive metrics graphs because they deemed them useful for preparing feedback to students. Interestingly, eight (P7, P9, P10, P11, P12, P13, P15, & P16) were also interested in having access to the same graphs to improve the online teaching design, and seven (P9, P10, P11, P12, P13, P16, & P1) were also interested for evaluating the collaboration process.

Out of these 11 participants, six (P7, P11, P12, P14, P15, & P16) were interested in the process analytics graphs with IGs and CCs for preparing feedback on the collaboration process. Moreover, five (P11, P12, P15, 16, and 17) of the 11 participants were interested in them for improving online teaching design, and ten were interested in accessing them to measure the students' collaborative efforts. Most practitioners thought that the graphs can help them provide *feedback* and *assess* the process of collaboration. In general, only three (P11, P12, & P16) of all the participants wanted to have access to *all the graphs* for providing feedback, assessment and improvement of the collaboration process. Rarely, P4 was the only participant who did not find any of the graphs useful.

Finally, inspired by the accountability dimension of the social transparency principles (Prestigiacomo et al., 2020) we investigated to what extent practitioners agreed on who should have access to the graphs. Most participants (16) agreed that the individual learner and the course management team should have access to the graph. A significant number of participants (9) also stated that individuals should have access to other groups' graphs. Except for P10, all participants did not consider it necessary to share the data with the public. P17 indicated that learners and instructors should have access to the data alongside training and explanations of concepts.

8.6 Discussion

This section is an examination of the research questions based on the participants' interpretations of the various collaboration analytics graphs presented above.

8.6.1 The Extent Tertiary-Level Educators Evaluate Their Students' Online Collaboration and the Value of Descriptive Metrics

The results above show that most of our tertiary-level educators (12 out of 19 participants) do not evaluate their students' online collaboration activities as part of their teaching practice at all. Those who do, prefer to evaluate the students collectively as a group rather than assessing individual student's collaborative actions. This is quite interesting to observe since all of the participants said that they use collaborative activities in their teaching. One of the workshop participants said: *"I speak for myself, and probably for a lot of teachers, these kinds of things are very implicitly measured, and when they do, they usually follow heuristics. I mean a very implicit approach, like... okay ... are students posting messages here? Or is the forum empty? Do they share messages and materials? To what extent are they getting on the task? To what extent they are dividing the task – you know – like you do this, you do this, and I do that, and we just assemble the different pieces. For most of us, usually, the assessment is very implicit and very heuristic, even when it is online."*

While most of them use collaborative learning to engage their students in active learning, some use it for knowledge building, few others use it for skill development. Those who said that they evaluate students' online collaboration, said that they look at the quality of students' posts from the domain knowledge and dialogic perspectives. Although most Learning Management Platforms do not provide a lot more than merely the number of posts and views to evaluate online discussions, none of the educators emphasized in their free responses that the number of posts and views are very valuable sources to gauge how much their students are collaborating. However, the educators' interpretations of the graphs of the total number of posts and views suggested that they are conversant with the metrics. Using the total number of posts and views for the two communities, educators can make some sense of what is going on in their students' online collaboration activities. These metrics give the teachers some understanding about participation levels, but less on the interactive and dialogic levels. Asides from the total number of posts and views, the educators indicate the consistent and steady increment of the metrics as exhibited by Community 2 as a reason for their decisions. Although educators were familiar and comfortable interpreting these descriptive metrics, they were not considered these very insightful to interpret students' online collaboration.

We found that participants' experiences with online teaching, online collaborations, and their general knowledge of graphs had little influence on their interpretation of collaboration from descriptive metrics. However, this is likely to be due to the participants' high level of digital and data literacy in general, as the sample of our participants were selected from university-level educators. Although both novice and experienced teachers made sense of the descriptive metrics and their graphs, they expressed that the number of posts and views in online discussions is not a good indicator of collaboration. On the contrary to what is commonly found from "human-computer interface research" (Dix et al., 2004), participants did not state they needed some time to get familiar with these kinds of charts, which might be considered an advantage.

8.6.2 The Added Value of CLaP Analysis (IGs and CCs)

Our results show that the educators are neither aware of nor familiar with the theoretical considerations of collaborative cognitive load theory: namely interactivity costs and coordination costs. Furthermore, the number of participants who interpret the data from CLaP analysis accurately is lower in comparison to previous descriptive metrics. However, participants extract more detailed information about the process of collaboration from the graphs of CCs and IGs. The graphs of CCs and IGs provided practitioners with a better understanding of the learners' cognitive process during collaboration. One of our workshop participants said: *"From my own teaching experience, I used to evaluate my students upon their posts number and the quality of the posts themselves. But now after reading the paper and being involved in this interaction, I have changed my mind as my eyes have been opened... I think the posts number is not enough by itself, and we need to collect a lot of metrics together to evaluate the collaboration process... for me, as a teacher, to see this graph is really useful as opposed to the basic analytics as just posts number."*

Educators have appreciated the perception of the cognitive costs that the students have invested and the benefits they were able to get during this process. It is interesting to note that they "decoded" that the higher the CCs, the lower the IG scores the learners would be able to attain, and vice versa. Only in weeks 6 and 7 did the CC not decrease substantially; the IGs line, however, showed a steep increment. This peculiarity was due to the introduction of exam-related activities during the six-week period in our dataset. The participants have encountered situations like this before. They were, therefore, able to explain that the teacher must have introduced some exam or grade related activities to motivate the students and enhance their collaboration. The educators specifically reflected based on the temporal dimension of the analysis: *"Looking at these two graphs, I get the impression that both groups ended up more or less at the same level, but the process is very different. I mean community two was collaborating in a more steady fashion, while community 1 is trying to catch up more on the last day. In my experience when they try to catch up at the last day, the product may be good, but given that the process is not good, learning tends to be not so good."*

These kinds of interpretations and opportunities to reflect on the collaborative patterns of behaviours of their students can potentially be valuable for teachers to improve their practice. Learning analytics graphs typically capture and visualise traces of learning events, to facilitate understanding and contemplation (Verbert et al., 2013). That is, beyond the recognition of what happened to learners during the collaboration process, the graphs should effectively initiate reflection in the mind of the instructors to make sense of how different activities or interventions they introduced influence the students' collaboration. Although the participants didn't have any prior knowledge of the CLaP analysis, the results indicated that it had the potential to assist educators in understanding online collaboration. Such a theory-based understanding of collaboration may help instructors to connect to learning theories to carry out various manipulations in a bid to improve the students' interactions. While some of these strategies might yield positive outcomes, others might have a negative/null effect on the process. For example, when reflecting on being able to follow the CC angle of learning one of the workshop participants said: *"If my goal for students was to develop collaboration skills (rather than domain knowledge), I might appreciate coordination costs because students can then reflect on how they collaborated and what was hindering/helping"*. Another participant tried to connect CC to the pedagogical aim: *"I am struck by these two graphs which actually shows that the cost is much higher in both communities, and I ask myself why are students continuing to interact… if my goal is for students to learn from the interaction – that is striking."*

It is worth noting that participants with higher years of experience sometimes ignore metrics that they are unfamiliar with. This is similar to what Swidan et al. (2019) observed when educators engage with the dashboard. Less experienced participants chose to follow the instructions on the dashboard, while more experienced participants followed their preferences. This is perhaps an example of "Illusion of Validity" (Kahneman, 2011) – where they trust their judgment more than what the dashboard indicates. Alternatively, it might be due to experts' tendency to 'automate' their decision making, and not deliberate on details. These results further highlight the need for human-centred approaches (Buckingham Shum et al., 2019) to address the needs of educators.

To our surprise, most practitioners found it difficult to understand the changes in Community 1 during the last 2 weeks when we asked them to observe the changes alongside other interpretations. While this was not the case when the interpretation was made separately. This is most likely due to the density of information on the collaboration process graphs (Lim et al., 2019). Indeed, previous research indicates that some learning analytics visualisations in collaborative learning settings might increase teachers' cognitive load (van Leeuwen, 2015), and may be perceived as "extra workload" (Chounta & Avouris, 2016). Therefore, we noted the importance of simplifying the graphs to enhance the educators' cognitive ease when interpreting the results. Recent research into data storytelling approaches in learning analytics contexts suggests potential ways to accomplish this (Martinez-Maldonado et al., 2020). We argue that collaboration analytics visualisations may be more beneficial if they focus on fewer points/features for educators to interpret at a time (Echeverria et al., 2018). In ClaP visualisations, this might potentially be achieved through clear explanations of indexes used to calculate IG and CC values to teachers, IG and CC values' presentation in separate graphs, and weekly progress of the CLaP values for each cohort rather than the presentation of the graph based on accumulated values. However, these assumptions should be studied experimentally, and potential solutions to supporting educators' interpretations of analytics visualisations should come from co-design and participatory design sessions with educators.

8.7 Conclusions

The process of collaboration is not as straightforward to evaluate and support as its learning outcomes (i.e., students' academic grades, group project outcomes, etc.). In contrast to the latter, where pre and post-test results can be examined to comprehend the impact of collaboration, cognitive processes must be explored in the former; not just individually, but also collectively. For more details on the operationalisation of the collaborative cognitive load theory with social network analysis and metrics, readers are referred to Kent and Cukurova (2020). However, the purpose of this chapter was to study educators' needs concerning such analytics, as well as their interpretations of them.

The visualisations of cognitive processes must be easily understood by instructors and learners so that they can reflect on the collaboration process and glean insight from it and intervene accordingly. Here, we investigated the tertiary-level educators' needs and interpretations of the online collaboration process. More specifically, we looked at to what extent they evaluate students' online collaboration in their teaching, how much they appreciate descriptive metrics on students' contributions to and views in online discussions, and to what extent the visualisations of IGs and CCs can help them get better insights into the *process of collaboration*.

Overall, the participants found the descriptive metrics of the total number of posts and views to be useful for "broad interpretations" and a "superficial understanding". However, the visualisations of CCs and IGs appeared to strengthen their understanding of the collaborative and cognitive processes of communities. Such an understanding was also connected to more detailed and timely interventions for learning which would not be possible only from the number of posts and views. On the other hand, all graphs studied here were still considered limited by educators in various ways to assess online collaboration and were also considered as too complex to be quickly adopted in practice.

8.7.1 Limitations and Future Research

All our participants were recruited from departments that are somehow associated with learning analytics including educational psychology, learning sciences, and computer science. Although participants didn't have direct experience in using learning analytics in their teaching frequently, it is important to acknowledge that they all had some research interest in this space and wanted to explore more in the future. This background of the participants is likely to skew some of the results presented here. Besides, knowing that this research was conducted during the height of the Covid19 pandemic, it would be interesting to know if the findings remained the same pre or post-COVID. Since most academic activities have migrated online, educators needed to have a good understanding of their learners' online activities. It was surprising to find out that although all of them use collaborative activities in their teaching, most educators do not evaluate their students' online collaboration at all, which might affect their ability to provide helpful feedback. Furthermore, we recognized that the participants not having detailed information about specific activities learners undertook to generate the metrics they are presented with put them in a disadvantageous position. On account of this, we consider it important to conduct similar research with participants who have a good understanding of the tasks the learner completed, and investigate their interpretations of the CLaP analytics.

Acknowledgement We would like to thank all our workshop participants for granting permission to collect data for this study. We would like to particularly thank Marc Lafuente Martinez from École polytechnique fédérale de Lausanne - EPFL as well as Arif Altun and Gokhan Akcapinar from Hacettepe University for their help with the organisation of two workshops.

Appendices

Appendix 8.1: Link to the Participant Survey

https://forms.gle/qknjnTkpQ16JwXf8A

Appendix 8.2: Below Are the Survey Questions We Used to Probe Discussions in the Second Half of the Workshop

1. Using the HLDs (total number of posts, views and cross-reference), which community is more collaborative and why?
2. Using the graphs of the number of posts per week for each community. Which community is more collaborative and why? Does your choice remain the same for all 7 weeks?
3. Using the number of posts and views per week for each community, which community is more collaborative and why? Does your choice remain the same for all 7 weeks?
4. Using the graphs of the mutuality of learners' interactions per week for each community. This is measured through the reciprocity of learners' social network (that is, to what extent they respond to each other) to represent their IGs. Which community is more collaborative and why? Does your choice remain the same for all 7 weeks?
5. Using the graphs showing the mutuality of learners' interactions and the cognitive cost for learners to coordinate their actions per week for each community. This value is measured through the degree of heterogeneity in learners' social network to represent their CC. Which community is more collaborative and why? Does your choice remain the same for all 7 weeks?

6. Using the graphs showing the difference between the learners' gains from interactions and their CCs per week for each community. Which community is more collaborative, and why? Does your choice remain the same for all 7 weeks?
7. Can you spot a change in Community 1's behaviours after week 5? What would you guess might have caused it?
8. When comparing the type of insights you gain from the participation charts to the insights gained from the IGs & CCc charts, how (if at all) would you say the last two add to what you might observe in online collaboration?
9. What additional information would you like to see in online collaboration graphs? What metrics would be valuable to see?
10. Would you like to see any of the graphs from the lessons you teach online? If so, how might you use them?
11. Who should have access to these graphs generated from online learning environments?

References

Alwahaby, A., Cukurova, M., Papamitsiou, C., & Giannakos, M. (2021). The evidence of impact and ethical considerations of multimodal learning analytics: A systematic literature review. In Giannakos et al. (Eds.), *The handbook of multimodal learning analytics*. Springer.

Bernstein, E., Shore, J., & Lazer, D. (2018). How intermittent breaks in interaction improve collective intelligence. *PNAS: Proceedings of the National Academy of Sciences of the United States of America, 115*(35), 8734–8739. https://doi.org/10.1073/pnas.1802407115

Braun, V., & Clarke, V. (2006). Using thematic analysis in psychology. *Qualitative Research in Psychology, 3*(2), 77–101.

Buckingham Shum, S., Ferguson, R., & Martinez-Maldonado, R. (2019). Human-centred learning analytics. *Journal of Learning Analytics, 6*(2), 1–9.

Casamayor, A., Amandi, A., & Campo, M. (2009). Intelligent assistance for teachers in collaborative e-learning environments. *Computers & Education, 53*(4), 1147–1154.

Chen, B., & Zhu, H. (2019). Towards value-sensitive learning analytics design. In *Proceedings of learning analytics and knowledge conference, LAK'19* (pp. 343–352).

Chounta, I. A., & Avouris, N. (2016). Towards the real-time evaluation of collaborative activities: Integration of an automatic rater of collaboration quality in the classroom from the teacher's perspective. *Education and Information Technologies, 21*(4), 815–835.

Dawson, S., Joksimovic, S., Poquet, O., & Siemens, G. (2019). Increasing the impact of learning analytics. In *Proceedings of the 9th international conference on Learning Analytics & Knowledge (LAK19)* (pp. 446–455). https://doi.org/10.1145/3303772.3303784

Dix, A., Finlay, J., Abowd, G. D., & Beale, R. (2004). *Human-computer interaction* (3rd ed.). Pearson.

Echeverria, V., Martinez-Maldonado, R., Granda, R., Chiluiza, K., Conati, C., & Shum, S. B. (2018). Driving data storytelling from learning design. In *Proceedings of the 8th international conference on learning analytics and knowledge* (pp. 131–140).

Ferguson, R., & Clow, D. (2017). Where is the evidence? A call to action for learning analytics. In *LAK'17 proceedings of the seventh international learning analytics & knowledge conference* (pp. 56–65). https://doi.org/10.1145/3027385.3027396

Gerard, L. F., & Linn, M. C. (2016). Using automated scores of student essays to support teacher guidance in classroom inquiry. *Journal of Science Teacher Education, 27*(1), 111–129.

Gibson, A., & Martinez-Maldonado, R. (2017). That dashboard looks nice, but what does it mean?: Towards making meaning explicit in learning analytics design. In *Proceedings of the 29th Australian conference on computer-human interaction (OZCHI'17)* (pp. 528–532). https://doi.org/10.1145/3152771.3156171

Greiffenhagen, C. (2012). Making rounds: The routine work of the teacher during collaborative learning with computers. *International Journal of Computer-Supported Collaborative Learning, 7*(1), 11–42. https://doi.org/10.1007/s11412-011-9134-8

Holstein, K., McLaren, B. M., & Aleven, V. (2017). Intelligent tutors as teachers' aides: Exploring teacher needs for real-time analytics in blended classrooms. In *Proceedings of learning analytics and knowledge conference, LAK'17* (pp. 257–266).

Holstein, K., Hong, G., Tegene, M., McLaren, B. M., & Aleven, V. (2018). The classroom as a dashboard: Co-designing wearable cognitive augmentation for K-12 teachers. In *Proceedings of the 8th international conference on learning analytics and knowledge (LAK'18)* (pp. 79–88). Association for Computing Machinery. https://doi.org/10.1145/3170358.3170377

Jivet, I., Scheffel, M., Specht, M., & Drachsler, H. (2018). License to evaluate: Preparing learning analytics dashboards for educational practice. In *Proceedings of learning analytics and knowledge conference, LAK'18* (pp. 31–40).

Kaendler, C., Wiedmann, M., Rummel, N., & Spada, H. (2015). Teacher competencies for the implementation of collaborative learning in the classroom: A framework and research review. *Educational Psychology Review, 27*(3), 505–536.

Kahneman, D. (2011). *Thinking, fast and slow* (pp. 209–221). Farrar, Straus and Giroux.

Kent, C., & Cukurova, M. (2020). Investigating collaboration as a process with theory-driven learning analytics. *Journal of Learning Analytics, 7*(1), 59–71. https://doi.org/10.18608/jla.2020.71.5

Kent, C., & Rechavi, A. (2020). Deconstructing online social learning: Network analysis of the creation, consumption and organization types of interactions. *International Journal of Research & Method in Education, 43*(1), 16–37. https://doi.org/10.1080/1743727X.2018.1524867

Kirschner, P., Sweller, J., Kirschner, F., & Zambrano, R. J. (2018). From cognitive load theory to collaborative cognitive load theory. *International Journal of Computer-Supported Collaborative Learning, 13*, 213–233. https://doi.org/10.1007/s11412-018-9277-y

Lim, L., Dawson, S., Joksimovic, S., & Gasevic, D. (2019). Exploring students' sensemaking of learning analytics dashboards: Does frame of reference make a difference? *ACM International Conference Proceeding Series*, 250–259. https://doi.org/10.1145/3303772.3303804

Luckin, R., & Cukurova, M. (2019). Designing educational technologies in the age of AI: A learning sciences-driven approach. *British Journal of Educational Technology, 50*, 2824–2838. https://doi.org/10.1111/bjet.12861

Mangaroska, K., & Giannakos, M. N. (2018). Learning analytics for learning design: A systematic literature review of analytics-driven design to enhance learning. *IEEE Transactions on Learning Technologies*, 1–19.

Martinez-Maldonado, R. (2019). A handheld classroom dashboard: Teachers' perspectives on the use of real-time collaborative learning analytics. *International Journal of Computer-Supported Collaborative Learning, 14*, 1–29. https://doi.org/10.1007/s11412-019-09308-z

Martinez-Maldonado, R., Clayphan, A., Yacef, K., & Kay, J. (2015). MTFeedback: Providing notifications to enhance teacher awareness of small group work in the classroom. *IEEE Transactions on Learning, 8*(2), 187–200.

Martinez-Maldonado, R., Echeverria, V., Nieto, G. F., & Buckingham, S. S. (2020). From data to insights: A layered storytelling approach for multimodal learning analytics. In *Proceedings of the 2020 CHI conference on human factors in computing systems (CHI'20)* (pp. 1–15). https://doi.org/10.1145/3313831.3376148

Matcha, W., Gasevic, D., & Pardo, A. (2019). A systematic review of empirical studies on learning analytics dashboards: A self-regulated learning perspective. *IEEE Transactions on Learning Technologies, 13*, 226–245.

McCoy, C., & Shih, P. C. (2016). Teachers as producers of data analytics: A case study of a teacher-focused educational data science program. *Journal of Learning Analytics, 3*(3), 193–214. https://doi.org/10.18608/jla.2016.33.10

Molenaar, I., & Knoop-van Campen, C. (2017). Teacher dashboards in practice: Usage and impact. In *Proceedings of European conference on technology enhanced learning* (pp. 125–138). Springer.

Nokes-Malach, T. J., Meade, M. L., & Morrow, D. G. (2012). The effect of expertise on collaborative problem solving. *Thinking and Reasoning, 18*(1), 32–58.

Parkin, T., Caunite-Bluma, D., Ozolins, K., & Jenavs, E. (2020). Report 3: Technology use in schools during Covid-19. *Findings from the Edurio Covid-19 impact review.*

Prestigiacomo, R., Hadgraft, R., Hunter, J., Locker, L., Knight, S., Hoven, E. V., & Maldonado, R. M. (2020). Learning-centred translucence: An approach to understand how teachers talk about classroom data. In *Proceedings of the tenth international conference on learning analytics & knowledge (LAK'20)* (pp. 100–105). https://doi.org/10.1145/3375462.3375475

Rosé, C. P., McLaughlin, E. A., Liu, R., & Koedinger, K. R. (2019). Explanatory learner models: Why machine learning (alone) is not the answer. *British Journal of Educational Technology, 50*, 2943–2958. https://doi.org/10.1111/bjet.12858

Schwarz, B. B., Prusak, N., Swidan, O., Livny, A., Gal, K., & Segal, A. (2018). Orchestrating the emergence of conceptual learning: A case study in a geometry class. *International Journal of Computer-Supported Collaborative Learning, 12*(3), 1–23.

Segal, A., Hindi, S., Prusak, N., Swidan, O., Livni, A., Schwarz, B., & Gal, K. (2017). Keeping the teacher in the loop: Technologies for Monitoring Group learning in real-time. In *Proceedings of artificial intelligence in education* (pp. 64–76).

Swidan, O., Prusak, N., Livny, A., Palatnik, A., & Schwarz, B. B. (2019). Fostering teachers' online understanding of progression of multiple groups towards the orchestration of conceptual learning. *Unterrichtswissenschaft, 47*, 159–176.

UNESCO. (2021). *COVID-19 impact on education*. Retrieved from https://en.unesco.org/covid19/educationresponse

Van Es, E. A., & Sherin, M. G. (2002). Learning to notice: Scaffolding new teachers' interpretations of classroom interactions. *Journal of Technology and Teacher Education, 10*(4), 571–596.

Van Es, E. A., & Sherin, M. G. (2008). Mathematics teachers' learning to notice in the context of a video club. *Teaching and Teacher Education, 24*(2), 244–276.

Van Leeuwen, A. (2015). Learning analytics to support teachers during synchronous CSCL: Balancing between overview and overload. *Journal of Learning Analytics, 2*(2), 138–162.

Van Leeuwen, A., & Rummel, N. (2019). Orchestration tools to support the teacher during student collaboration: A review. *Unterrichtswiss, 47*, 143–158. https://doi.org/10.1007/s42010-019-00052-9

Van Leeuwen, A., & Rummel, N. (2020). Comparing teachers' use of mirroring and advising dashboards. *ACM International Conference Proceeding Series*, 26–34. https://doi.org/10.1145/3375462.3375471

Van Leeuwen, A., Janssen, J., Erkens, G., & Brekelmans, M. (2014). Supporting teachers in guiding collaborating students: Effects of learning analytics in CSCL. *Computers & Education, 79*, 28–39. https://doi.org/10.1016/j.compedu.2014.07.007

Van Leeuwen, A., Van Wermeskerken, M., Erkens, G., & Rummel, N. (2017). Measuring teacher sense making strategies of learning analytics: A case study. *Learning: Research and Practice, 3*(1), 42–58.

Van Leeuwen, A., Rummel, N., & Van Gog, T. (2019). What information should CSCL teacher dashboards provide to help teachers interpret CSCL situations? *International Journal of Computer-Supported Collaborative Learning, 14*, 261–289. https://doi.org/10.1007/s11412-019-09299-x

Verbert, K., Duval, E., Klerkx, J., Govaerts, S., & Santos, J. L. (2013). Learning analytics dashboard applications. *American Behavioral Scientist, 57*(10), 1500–1509. https://doi.org/10.1177/0002764213479363

Verbert, K., Govaerts, S., Duval, E., Santos, J. L., Van Assche, F., Parra, G., & Klerkx, J. (2014). Learning dashboards: An overview and future research opportunities. *Personal and Ubiquitous Computing, 18*, 1499–1514.

Voyiatzaki, E., & Avouris, N. (2014). Support for the teacher in technology-enhanced collaborative classroom. *Education and Information Technologies, 19*(1), 129–154. https://doi.org/10.1007/s10639-012-9203-2

Zhou, Q., Suraworachet, W., & Cukurova, M. (2021a). *Different modality, different design, different results: Exploring self-regulated learner clusters' engagement behaviours at individual, group and cohort activities*. https://doi.org/10.35542/osf.io/u3g4n

Zhou, Q., et al. (2021b). Investigating students' experiences with collaboration analytics for remote group meetings. In I. Roll, D. McNamara, S. Sosnovsky, R. Luckin, & V. Dimitrova (Eds.), *Artificial Intelligence in education. AIED 2021* (Lecture Notes in Computer Science) (Vol. 12748). Springer. https://doi.org/10.1007/978-3-030-78292-4_38

Part II
Applications in K-12

Chapter 9
Augmented Reality (AR) for Biology Learning: A Quasi-Experiment Study with High School Students

Christy Weng-Lam Cheong, Xingmin Guan, and Xiao Hu

Abstract In recent years, Augmented Reality (AR) has been adopted at both formal and informal learning settings. It has been found useful in training learners' core competencies, such as spatial ability, practical skills, and conceptual understanding. Various studies have highlighted its positive impact on enhancing learning outcomes (e.g., learning engagement and performance). Its impact on learners' academic emotion, however, is yet to be comprehensively investigated. The aim of this study is to explore high school students' perception of AR in biology learning, measure the impact of AR on their learning performance, and systematically examine the extent to which AR affects their academic emotion. A quasi-experiment was conducted among 103 students from a high school in mainland China. An AR learning material was developed for participants in the experiment group, while the control group used traditional learning approaches solely. Questionnaire surveys and focus-group interviews were conducted to understand participating students' perception towards AR and/or the subject of biology as well as their academic emotion during the quasi-experiment. Their learning performance before and after the intervention was also measured and compared. The findings indicate a general willingness to use AR applications in learning among the participating students. The capacity of AR was confirmed in terms of facilitating acquisition of conceptual knowledge in biology and in stimulating positive academic emotions towards the subject. These findings suggested that AR can serve as an effective means in science education to sustain students' positive learning emotion and engagement in both formal and informal learning environments.

Christy W. L. Cheong and Xingmin Guan have equally contributed in this chapter.

C. W.-L. Cheong
Academic Affairs Department, Macao Polytechnic Institute, Macao, S. A. R, China
e-mail: wlcheong@ipm.edu.mo

X. Guan · X. Hu (✉)
Faculty of Education, University of Hong Kong, Hong Kong, S. A. R, China
e-mail: xguan2@connect.hku.hk; xiaoxhu@hku.hk

Y. E. Wang et al. (eds.), *Social and Emotional Learning and Complex Skills Assessment*, Advances in Analytics for Learning and Teaching,
https://doi.org/10.1007/978-3-031-06333-6_9

Keywords Augmented reality · Academic emotion · Biology learning · High school students

9.1 Introduction

Augmented Reality (AR) refers to a combination of technologies which overlays computer-generated content onto the real world (Azuma, 1997). It has been applied in nearly all walks of life, ranging from medicine, military, manufacturing to entertainment. In education, increasing interests in AR application have equally been observed (e.g., Chen et al., 2017; Dunleavy & Dede, 2014; Tan & Waugh, 2013). In the early stage of educational AR application, the Horizon Report 2012 (Johnson et al., 2012) predicted that AR would be utilized widely in K-12 education within 4–5 years, which would produce new experiences and opportunities for information access and learning in three-dimensional (3D) space. This prediction has seemingly become reality over the past few years. Extensive discussions were reported in the literature on AR application in various educational contexts (such as experiential learning and location-based game learning) and in different academic disciplines such as mathematics (Estapa & Nadolny, 2015), chemistry (Cai et al., 2014), biology (Aivelo & Uitto, 2016), physics (Cai et al., 2017), and geography (Turan et al., 2018).

In science learning, AR has been found affordable in training students' spatial ability, practical skills, and conceptual understanding (Cheng & Tsai, 2013). These affordances of AR have been confirmed in the context of biology learning, which involves objects rarely seen in daily life and abstract concepts that students may find challenging. In Tan and Waugh's study (Tan & Waugh, 2013), a Virtual-Reality (VR) material, which enables visualization of DNA, proteins, and cellular structures in 3D space, was developed for use by secondary school students in learning molecular biology. Compared to traditional classrooms where models, diagrams and other physical tools are used to show the structure, AR can provide a realistic environment for an enhanced learning experience (Lee, 2012).

AR-facilitated learning was not only found beneficial for enhancing students' cognitive learning outcomes (e.g., learning performance) but also learning experience in the social and emotional domain (Akçayır & Akçayır, 2017). In their review of 55 published studies between 2011 and 2016, Chen et al. (2017) reported that AR in education had the trends of improving learning performance, increasing learning motivation and engagement as well as improving perceived enjoyment. Similar findings were obtained by Harley et al. (2016) in their study comparing students' emotion and learning performance in learning contexts respectively with the presence and absence of AR applications. They found that AR could enhance learning effectiveness and enjoyment (Harley et al., 2016). Such characteristics of educational AR applications provide a great opportunity to unfold the role of academic emotion in students' learning process using analytic methods.

However, compared to learning outcomes, academic emotions, i.e. emotions associated with learning (Pekrun, 2006), are a less populated area in which the impact of AR application is measured. Few studies have considered emotions comprehensively. Most of them focused on one or two emotions while there are a wide range of emotions in the learning context such as pride, anxiety, boredom, etc. (Pekrun, 2006). Emotion has been found to be an important factor shaping learning behaviors and affecting learning performance (Artino Jr et al., 2012; Mega et al., 2014). It is thus worth exploring the impact of AR in this area.

This study aims to bridge this research gap by exploring how high school students perceive AR in biology learning (Research Question 1), and to what extent AR impacts their learning performance (RQ2) and academic emotion (RQ3). Through performing analytics on high school students' self-reported academic emotion in technology-enriched learning settings, findings of this study would be helpful for (1) understanding students' academic emotion when learning with advanced technology (AR in this study); and (2) designing educational AR applications to create a better learning experience and facilitate independent learning.

9.2 Related Work

9.2.1 AR in Education

AR creates "a situation in which a real-world context is dynamically overlaid with coherent [...] virtual information" (Klopfer & Squire, 2008). It has the capacity of 3D visual presentation and real-time user-environment interaction (Azuma, 1997). It can be location-aware or vision-based (Dunleavy & Dede, 2014). Location-aware AR tracks every move of the users using GPS, which can be well utilized in learning contexts where locations matter (e.g. field trips). Vision-based AR enables users to access information by means of an object (such as QR codes and pictures) using digital devices, which is suitable for learning in classrooms, laboratories and museums.

Used in various educational contexts, AR has been suggested as one of the potential new technologies for pedagogical applications (Johnson et al., 2011). The affordances of AR as an educational application have been emphasized in existing studies in terms of (1) the opportunities available for students to observe objects and learn in three dimensions (Chen et al., 2011), (2) the facilitation of mastery of abstract concepts (Hughes et al., 2011), (3) the provision of immersive learning experience (Han et al., 2015), and (4) the bridging of formal and informal learning (Sotiriou & Bogner, 2008).

In science education, informal learning environments (e.g. museums and field trips) are perceived as partners of formal education (Ramey-Gassert et al., 1994).

AR application in informal science learning environments have been found to be effective in enhancing learning outcomes. A systematic review of published studies from 2003 to 2017 (Goff et al., 2018) reported that AR was found to be effective in enhancing students' conceptual knowledge, topic interest and engagement in science learning. Huang et al. (2016) designed an AR-based learning tool for eco-education to be used by secondary school students exploring botanical gardens. Their findings indicated that the tool was effective in stimulating positive emotions and engagement as well as enhancing learning performance. A study conducted on informal learning in science museums (Salmi et al., 2017) revealed that AR-facilitated learning was particularly helpful for enhancing preadolescent low-achievers' learning performance.

While the benefits of AR application in education are generally acknowledged, existing studies also noted its possible drawbacks. Aivelo and Uitto (2016), for example, cautioned about the digital gap among students in relation to AR operational knowledge. When insufficient knowledge or difficulties are encountered, confusion and anxiety would naturally arise to varying extents, which may discourage learning. This negative impact would become particularly strong in informal learning contexts where learning is not compulsory. To ensure effective learning after the use of AR, a comprehensive understanding of the impact of AR on academic emotion is desirable.

9.2.2 AR in Biology Learning

The discipline of biology involves a wide range of abstract concepts and objects that have little similarity to those we often see in daily life (e.g., gene and DNA). To facilitate learning, they are usually shown through verbal, visual, mathematical, dynamic or physical conceptual models (Koba & Tweed, 2009).

Given the aforementioned affordances of AR in education, AR can be an effective means to demonstrate biological conceptual models. A number of successful attempts have been reported in literature. For instance, Weng et al., (2006) developed an AR game about mitosis, meiosis and respiration for secondary school students, which were found helpful for enhancing their perception and understanding of relevant processes and phenomena. Applying AR in presenting digestive system by a flat torso of the human body in a summer school, Pribeanu et al. (2008) found that AR is successful to increase students' learning motivation and interests. On another line of research on teachers, Aivelo and Uitto (2016) examined the perception of experienced secondary and primary teachers on an AR game about parasites. They found that sustaining situational interest in the game is key to its successful utilization.

However, little research on the applications of AR to biology learning considered academic emotion in a systematic sense. Some previous studies included affective constructs such as motivation or interests, but, none of them, to the best of our knowledge, studied a range of academic emotions or differentiate emotions towards the lessons using AR applications and those towards the learning subjects (see Sect. 9.3.4).

9.2.3 Definition of Academic Emotion

Emotion is multi-component phenomena in coordinated psychological processes consisting of affective, cognitive, physiological, motivational and expressive components (Shuman & Scherer, 2015). Academic emotion refers to an organism's responses to significant stimuli during the teaching and learning process, which includes physiological activation, motivational, perceptual, evaluative and learning processes, motor expression, action tendencies, and monitoring subjective feelings (Benard, 1997).

Academic emotion is considered as a significant predictor of the learning strategies adopted by students as well as their learning outcomes, such as learning motivation, self-regulation and academic achievement (Pekrun et al., 2002). Various empirical studies have also confirmed that academic emotion affects learning behaviors and performance. An earlier study by Peper (2006) demonstrated that anxiety could reduce intrinsic motivation to learn, and at the same time increase extrinsic motivation to avoid failure. Exploring academic emotions in science education, Um et al. (2012) reported that positive emotions could facilitate cognition and enhance learning performance.

Pekrun (2000) proposed a theoretical model that categorizes discrete academic emotions into four classes defined by the activation-deactivation and positive-negative dichotomies. One of the assumptions of the framework is that discrete academic emotions make specific influences on learning and performance. Positive-activating emotions, such as enjoyment, pride and hope, help learners go on learning. Positive-deactivating emotions, such as relief and relaxation, lead learners to adjust the pace of learning but still keep them to study. Negative-activating emotions, such as anxiety, anger and shame/guilt, push learners to overcome difficulties in the learning process or work harder to avoid failure. Negative-deactivating emotions, such as boredom, hopelessness and disappointment, could make learners feel incapable of achieving the expected learning outcomes.

Despite the stated role of academic emotion in learning, efforts devoted to measure the impact of AR application on academic emotion tended to be limited and focus on few emotions. In this study, the impact of AR on academic emotion was measured using a systematic approach based on the emotion categorization defined by Pekrun's model (Pekrun, 2000).

9.3 Research Methods

To answer the three research questions, a quasi-experiment was conducted in a high school in mainland China. This section presents the research design in detail.

9.3.1 *Study Context*

The school where the study was conducted is located in a developed city of South China, offering 3-year upper secondary education (equivalent to Grade 10 to Grade 12 in the US education system). Two classes in Grade 11 participated in this study respectively as the experiment group and the control group, both of which had no prior experience with AR learning materials. The students in these two classes were academically more capable than their peers in other Grade-11 classes in the school. At the time this study was conducted, they had finished the standard high school curriculum for biology and were preparing for the National College Entrance Examination to be held at the end of Grade 12.

During the quasi-experiment, both groups were reviewing topics related to cell structure. The experiment group was provided with an AR-based material on cell structure, whereas the control group had a regular lesson reviewing the topics.

9.3.2 *AR Learning Material*

For the intervention, a vision-based AR learning material on cell structure was developed drawing upon the lesson plans and inputs from the class teacher. Developed in Unity, the material consists of two main 3D objects respectively for animal and plant cells. Each object is equipped with 3D images of the cell organelles, textual explanations of the basic concepts and features of the structure concerned, as well as interactive features.

By scanning the cell structure pictures on the textbook using smartphones or tablets provided by the school, students can get access to the AR learning material. They can interact with the 3D objects (such as turning or zooming) by touch gestures. In addition, there are 11 and 10 interactive buttons available in respective objects. These buttons, altogether, lead to textual explanations of 9 animal and 8 plant cell organelles as well as 2 conclusions for each cell structure. By clicking these buttons, students will be prompted by a note card providing the stated textual information (See Figs. 9.1 and 9.2).

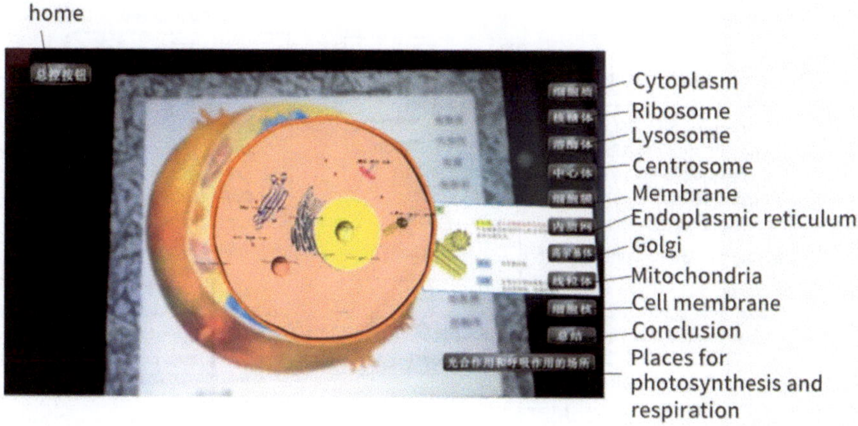

Fig. 9.1 An illustration of the AR learning material on animal cell

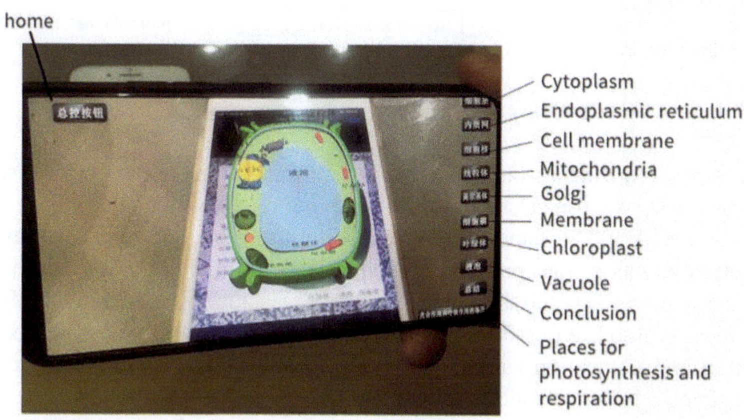

Fig. 9.2 The AR learning material on plant cell shown on a smartphone

9.3.3 Procedure

The experiment group and the control group had a biology lesson that reviewed cell structure respectively with and without access to the AR learning material. In this lesson, after the teacher introduced the topic, the experiment group was invited to use the AR learning material and given a demonstration on its operation and usage. After that, four students shared one device and explored the material for 30 min while technical assistance was provided by the teacher and the researcher if needed. During the week after the lesson, the students from the experiment group could continue using the AR learning material at home. In order to encourage them to do so, a notification was sent to their parents about this arrangement right after the lesson.

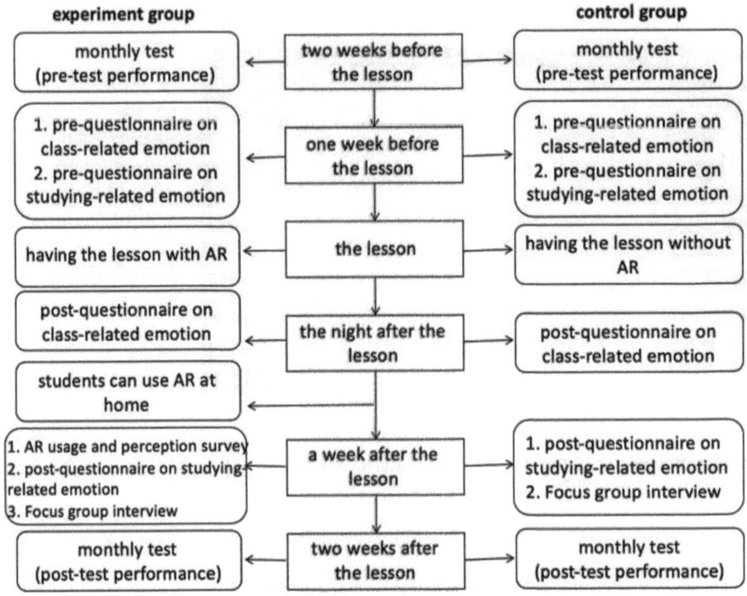

Fig. 9.3 Procedure of the quasi-experiment

One week before the lesson, both groups of students completed the pre-intervention questionnaire surveys on class-related and studying-related emotion (see Sect. 9.3.4). To measure students' emotion towards the lesson itself, the questionnaire survey on class-related emotion was conducted in both groups again after the lesson. The post-intervention survey on studying-related emotion was administered 1 week after the lesson to accommodate possible emotional changes towards the biology subject. The AR usage and perception survey was completed solely by the experiment group 1 week after the lesson to collect their views on the AR material. The academic performance of both groups before and after the lesson was tracked using the two monthly tests respectively conducted 2 weeks before and after the lesson. Figure 9.3 shows the procedure of the quasi-experiment in chronological order.

Lastly, focus-group interviews were conducted to gain a deeper understanding of relevant student perception. Five sub-groups were selected on a random basis respectively from the experiment group and the control group. In each focus-group interview, students were asked to discuss their perception of the lesson on cell structure and the subject of biology. For the experiment group, students also discussed their perceptions of AR learning.

9.3.4 Measures

To understand how students perceived AR in learning biology (RQ1), a questionnaire survey was conducted in the experiment group after the intervention. The students were asked to (1) report their frequency of after-class usage of the AR material, and (2) rate on a 5-point Likert scale their level of satisfaction towards it and relevant experience of sensory immersion. The questionnaire items were adopted and modified from those proposed and verified in a previous study on young children's perception toward AR materials (Han et al., 2015). Satisfaction was measured in terms of students' interest in and perceived user-friendliness of the AR material while sensory immersion was measured in terms of students' self-engagement and environment engagement.

In order to examine the impact of AR on learning performance (RQ2), the scores students received from their monthly tests were collected and compared to measure learning performance. The monthly tests held 2 weeks before and after the intervention were taken respectively as the pre-test and post-test. The scope of these tests covered the entire high school biology curriculum which the students had completed at the time of this study, and thus the learning performance examined is for the biology subject.

In order to measure the impact of AR on academic emotion (RQ3), students' academic emotions before and after the intervention were measured using two questionnaires adapted from the Achievement Emotions Questionnaire (AEQ) (Pekrun et al., 2002, 2005). The AEQ is a multidimensional questionnaire widely adopted for assessing students' academic emotion. For example, it was used in measuring students' negative emotions, such as anxiety, anger and boredom, in mathematics learning in lower secondary schools (Sutter-Brandenberger et al., 2018). Another study focusing on post-secondary students measured learning activity-related emotions using the AEQ, including enjoyment, anxiety and boredom (Maymon et al., 2018). The AEQ measures nine discrete emotions in all four classes defined by the activation-deactivation and positive-negative dichotomies in Pekrun's model (Pekrun, 2000). It includes enjoyment, hope, pride (positive-activating), anger, anxiety, shame (negative-activating), relief (positive-deactivating), hopelessness and boredom (negative-deactivating). The AEQ includes these emotions in three types of academic achievement situations: (1) class-related, i.e. triggered while attending lessons, (2) studying-related, i.e. towards the course or subject, and (3) test-related, i.e. towards tests and exams.

With a particular focus on the teaching and learning process, this study examined both class-related and studying-related emotions, removing the test-related emotion "relief" (positive-deactivating) from the list. Two respective questionnaires were thus designed using the AEQ manual with inter-item reliability validated in the literature. Each contained eight constructs measured on a 5-point Likert scale ranging from 1 (strongly disagree) to 5 (strongly agree). The questionnaires are presented in the Appendix.

9.3.5 Data Analysis

To answer RQ1, a descriptive statistical analysis was conducted on the data collected from the questionnaire survey about students' usage and perception of the AR material.

To answer RQ2, an independent samples t-test was conducted on the pre-test data for a learning performance comparison between the groups to examine their initial homogeneity. Another independent samples t-test was conducted on the post-test data to examine whether the two groups had significant differences in learning performance after the intervention.

To answer RQ3, after the internal reliability of each emotion construct was measured, a Mann-Whitney U test per academic emotion was conducted on responses to the pre-questionnaire to see if there were significant emotional differences between the groups before the intervention. Then the emotion constructs were separated into two groups according to the results of these tests. For the constructs where significant differences were not observed, a post-questionnaire Mann-Whitney U test was conducted per academic emotion. For the constructs where significant differences were observed, a Mann-Whitney U test was conducted within each student group to compare the ratings on each academic emotion between pre- and post-intervention questionnaires. In this way, we were able to find out which student group had significant changes in which emotion construct.

In addition, to control the false discovery rate in cases of multiple tests, the Benjamini-Hochberg procedure was conducted in all statistical tests in this study (McDonald, 2009).

9.4 Results

103 students participated in this study. Valid sets of questionnaires were collected from 88 students, forming the sample for data analysis. In this sample, 47 students were in the experiment group (26 males and 21 females) and 41 were in the control group (22 males and 19 females).

9.4.1 Students' Perception of AR

The self-reported frequencies of after-class usage of the AR material in the experiment group indicate that 70.3% of the surveyed students (33 out of 47) reported having used the material at home. Among them, 11 (23.4%) used the AR material twice or more often. Table 9.1 shows the descriptive statistics describing students' perception of AR. A median of 3.00 or above across items indicates that students were satisfied with the AR learning experience. The median for Item 1 was the highest, indicating that students were willing to use it in learning.

Findings from the focus group interviews provided some insights about the factors influencing students' use of the material. It was found that the perceived novelty of the technology may contribute to motivating the use. A student (Participant #14) noted, *"I would like to use it again at home because I think it interesting and fancy. And I even want to share this technology with my parents."* The values of AR's

Table 9.1 Statistics of students' perceptions towards AR

	Criteria	Questions	Median
Satisfaction	Interest	I would like to learn with AR frequently.	5
		I am more motivated to learning when using AR.	4
	User-friendliness	I found the AR was easy to use.	4
		AR is attractive.	4
		AR is useful for learning.	3
		AR is flexible. (I can learn at any place.)	4
Sensory immersion	Self-engagement	I felt immersed in learning while using AR	3
	Environment engagement	I felt the real cell structures were in front of me while using AR.	4

affordances in 3D visualization for learning was also acknowledged. A student (Participant #22) mentioned, *"AR shows the cell structures three-dimensionally so that we can observe them from multiple angles and understanding the structure further."* However, the technical requirement (both hardware and soft skills) for using the AR might have possibly discouraged usage, as a student (Participant #24) expressed, *"I think it inconvenient for me to download the application to learn at home."* Another student (Participants #42) also stated, *"after feeling it interesting, I found it a little hard to control so that I lost my interest soon."*

9.4.2 Learning Performance

The results of the t-test which compared the pre-test scores of the two groups indicate statistically significant differences in learning performance between the groups ($t = -2.619, p = .011$), which highlighted their initial heterogeneity. Specifically, the mean score of the experiment group was 58.81 out of 100, while that of the control group was 52.79.

Because of the groups' initial heterogeneity in learning performance, to answer RQ2, a t-test was conducted using the differences between the pre-test and post-test scores of individual students in the two groups (the means of the experimental and control groups were 19.54 and 23.76 respectively). The results indicate no statistically significant differences between the groups ($t = 1.328$, $p = .188$), rejecting the hypothesized impact of AR on learning performance.

9.4.3 Academic Emotion

Table 9.2 presents the results of the statistical analyses on data collected from the academic emotion surveys. Sufficient internal consistency for each emotion construct was confirmed using Cronbach's alpha ($\alpha \geqslant .690$) (George & Mallery, 2003).

Results of the Mann-Whitney U tests on responses to the pre-questionnaires indicated no significant differences between the experiment and control groups in most of the emotion constructs, except for studying-related hope ($\alpha = .794$). Therefore, to answer RQ3, a Mann-Whitney U test was conducted within each student group to compare the ratings on studying-related hope between pre- and post-intervention questionnaires. For other emotion constructs, a post-questionnaire Mann-Whitney U test was conducted per construct to compare the two groups' ratings.

As shown in Table 9.2, significant differences between the experiment and control groups were found in four class-related emotions. After the intervention,

Table 9.2 Statistics of student ratings on emotions

	Emotion	Group	Cronbach's alpha	Mean(pre)	Mean(post)	*p*-value(C. vs E.)
Class-related	Enjoyment	C	.921	3.14	3.27	.000**
		E		3.31	4.19	
	Pride	C	.749	3.69	3.34	.006**
		E		3.85	3.98	
	Hope	C	.903	3.61	3.56	.000**
		E		3.96	4.23	
	Anxiety	C	.690†	2.43	2.09	.491
		E		2.54	1.95	
	Boredom	C	.893	2.55	2.34	.055*
		E		2.56	1.70	
	Hopelessness	C	.935	1.58	1.63	.075*
		E		1.36	1.32	
	Shame	C	.721	2.26	2.24	.182
		E		2.62	1.83	
	Anger	C	.866	1.60	1.80	.006**
		E		1.49	1.26	
Studying-related	Enjoyment	C	.879	3.36	3.51	.050*
		E		3.72	4.14	
	Pride	C	.744	3.73	3.83	.163
		E		4.11	4.28	
	Anxiety	C	.800	2.38	2.07	.491
		E		2.23	2.02	
	Boredom	C	.905	1.71	1.83	.188
		E		1.63	1.57	
	Hopelessness	C	.792	1.67	1.56	.491
		E		1.53	1.51	
	Shame	C	.706	1.62	1.63	.491
		E		1.70	1.57	
	Anger	C	.803	1.86	1.78	.188
		E		1.57	1.45	

Note: C: control; E: experiment; †: after removing one item. **: significant at p = 0.05 level; *: significant at p = 0.10 level students in the experiment group tended to be more positive towards the lesson than their peers in the control group. In the post-questionnaire, their ratings of all positive-activating emotions, namely enjoyment, pride and hope, were significantly higher than those of students in the control group while those of anger were significantly lower. For negative-deactivating emotions (i.e., boredom and hopelessness), ratings from students in the experiment group were also lower than those from students in the control group though the differences were only significant at the $p = 0.10$ level. In terms of anxiety and shame, the two negative-activating emotions, the two student groups did not significantly differ, but we can observe that the experimental group had a larger reduction on ratings in these two emotions than the control group.

As for studying-related emotions, after the intervention, the experiment group's ratings of enjoyment were higher than those of the control group ($p = .05$). The two groups did not rate significantly differently on other studying-related emotions although a trend can be observed from the mean values of the post-questionnaire responses that the experiment group rated positive emotions higher and negative emotions lower than the control group. From the Mann-Whitney U tests comparing the pre- and post-intervention ratings on studying-related hope within the two student groups respectively, no significant differences were found ($p_e = .365$ and $p_c = .335$). In sum, the findings indicate that AR did not make a significant impact on students' emotion towards the subject of biology except for enjoyment.

Findings from the focus group interviews align with the statistical findings presented above. The perceived novelty of the AR material was found to be a stimulus of positive academic emotions. A student (Participants #12) noted that he found the AR material visually attractive and thus caught his attention. Another student (Participant #21) stated, *"I felt it was a novel technology so that I would like to try it out of curiosity."* In addition, students tended to find the opportunities of exploration in 3D space helpful for their understanding of the content concerned, making them enjoy the lesson and feeling interested and engaged. A student from the experiment group (Participant #22) said, *"my cognition of the cell structures is changing because of AR. In the past, I thought them so boring, however, I feel them interesting with AR learning so that I also have more interests in this subject."* Another one from the experiment group (Participant #42) said, *"I am looking forward to having the lesson and learning this content because I think I would be very interesting."* In contrast, a student from the control group (Participant #61) reflected, *"I am annoyed that there is too much and complex information for each [cell] structure to remember."*

9.5 Discussions

This study examined students' perceptions of using AR in biology learning as well as the impact of AR on academic emotions and learning performance. In terms of students' perception of AR, the findings indicated that students were generally willing to use AR in after-class learning and satisfied with the immersive experience provided by the technology. As Hughes et al. (2011) explained, AR is able to help students understand and master things that may not be easily seen or not at all present in the real world. It can help visualize complex concepts and models enhancing students' spatial ability (Billinghurst et al., 2005). These affordances of AR were explicitly confirmed in this study via the focus group interviews which indicated that the immersive experience enabled by AR facilitated the learning of cell structure. However, this study also found that users may be discouraged from using the AR application by its technicality. Perceived user-friendliness is thus a possible factor affecting the utilization.

No significant difference, however, was found in learning performance with the use of AR in this study. This finding is contrary to Cai et al. (2017), who found that their AR learning tool significantly improved the learning performance of junior high school students in physics. Different from this study where participating students had already completed the curriculum before the intervention and were in the process of examination preparation, it was the first time for the participants in Cai et al. (2017) to learn the content concerned. This seems to suggest that the use of AR is helpful for initial learning rather than reinforcing learning, which requires further studies to confirm. Another previous study (Cai et al., 2014) also found significant gains in learning performance through using AR materials in one chemistry lesson of Grade 11 students. It is noteworthy that there was no differentiation of experiment and control groups in Cai et al. (2014) where the performance gain was measured by comparing the test performances of the same group of students before and after a review session using the AR material. In our study, both the experiment and control groups improved their learning performances (Sect. 9.4.2), suggesting that the use of AR might not have been the reason for improved performance. More studies are called for to further examine the effect of AR in learning performances in different learning contexts.

This study found that the use of AR in learning tended to increase positive emotions and reduce negative emotions, which was partially due to the perceived novelty of the technology. This finding is in alignment with Huang et al.'s study on AR-based experiential learning (Huang et al., 2016), which found that AR increased students' sense of fun and curiosity and thus stimulated their positive emotions during learning. This study also found that the impact of AR learning on class-related emotion was more obvious than that on studying-related emotions. This is possibly because the intervention was implemented in one single lesson followed by voluntary access for only one week. As Haberman (1991) noted, people's belief and attitude are changing through emotions, which tends to be a slow process. A student from the experiment group (Participant #22) actually commented that the constant use of AR "will definitely help us to cultivate our scientific interests and experience in doing experiments."

Since the 1990s, nurturing students' literacy in Science, Technology, Engineering and Mathematics (STEM) has been on the prioritized agenda in education (Bybee, 2010). With its strengths in introducing conceptual models, facilitating understanding, and arousing learners' positive academic emotions, AR bears strong potentials in enhancing STEM education and consequent learner competencies. As suggested in this study, user-friendly AR materials can be used consistently to help develop learners' interest towards the subjects. Technical support, such as in-class or on-site instructions and demonstration, video tutorials for after-class or pre-visit reference, and on-demand assistance, can be provided to alleviate possible technology anxiety and frustration in usage. Carefully designed, the utilization of AR is likely to be an effective means to sustain learning engagement, especially in informal learning environments, where learning outcomes gained via formal education can be further fortified.

9.6 Limitations

There are several limitations in this study. First, the study was conducted on participants with prior knowledge of the topic at a time they were preparing for examinations. This research setting is helpful for examining the impact of AR on reinforcing learning but not on learning new knowledge. It would be more comprehensive to measure the impact of AR on learning performance in both of these contexts. Second, the pre-test and post-test measuring learning performance were synchronized with tests administered by the school. This approach is effective in reducing the possible interruption of school activities, but limits the contents to be tested. In this study, the two tests covered the entire high school biology curriculum rather than cell structure only. The findings of this study on learning performance thus contribute more empirical evidence to discussions about the impact of AR on learning the biology subject than specific content represented by the AR material. Third, despite the alignment with previous studies (e.g., Cai et al., 2014, 2017), students' contact time with the AR material was nonetheless limited (i.e. one lesson and one week after it) and may be insufficient for learning and emotional influences to evolve. Further studies are needed to examine possible changes in learning performance and academic emotions after the tool is used for a substantially longer period of time.

9.7 Conclusions and Future Work

This study examined the utilization of AR in biology learning from students' perspective and measured its impact on learning performance and academic emotions on the secondary education level. A general trend of willingness to use AR applications in after-class learning was found among students. The capacity of AR was confirmed in terms of facilitating acquisition of conceptual knowledge in biology and in stimulating positive academic emotions towards the learning process. These findings support the feasibility of AR application to learning in the scientific domain where individual learners' conceptualization capabilities are largely leveraged.

Diverging from its predecessor by its systematic approach to academic emotion, this study is a starting point of further research on AR-facilitated learning and its effects on social and emotional learning. Subsequent studies can be conducted by extending the research design with more AR materials and/or longer learner-material contact time. In addition, more research can be conducted to test the feasibility and impact of AR application in purely informal learning settings such as museums and libraries. Last but not least, with recent development of wearable technology, sensors (such as wristbands) can be considered for use to collect relevant psycho-physical data for the purpose of triangulating the findings in social and emotional learning.

Acknowledgments This study is supported by a General Research Fund (No. HKU 17607018) from the Research Grants Council of the Hong Kong S. A. R., China, and a Faculty Research Fund in the University of Hong Kong.

Appendix

AEQ ITEMS

Course-related emotions:

ENJ 1: I get excited about going to class.
ENJ 2:I am motivated to go to this class because it is exciting.
ENJ 3:After class, I start looking forward to the next biology class.
PRI 1: When I make good contributions in class, I get even more motivated.
PRI 2: I think that I can be proud of that I know about this subject.
PRI 3: I would like to tell my friends about how well I did in this course.
HOP 1: I hope that I will be successful in biology learning motivate me to invest a lot of effort.
HOP 2: In class, I am confident because I understand the material.
HOP 3: Before the class, I am optimistic that I will be able to keep up with the material.
ANX 1: Thinking about class makes me feel uneasy.
ANX 2: Even before class, I worry whether I will be able to understand the material.
ANX 3: When I don't understand something important in class, my heart races.
BOR 1: Because I get bored in class, my mind begins to wander.
BOR 2: Because the time drags, I frequently look at my watch.
BOR 2: I start yawning in class because I am so bored.
HOL 1: Even before class, I am resigned to the fact that I won't understand the material.
HOL 2: During class, I feel hopeless that I won't understand the material.
HOL 3: After the class, I feel hopeless continuing in this subject.

SHA 1: After the class, I'd rather not tell anyone when I don't understand something in class.

SHA 2: During the class, if the others knew that I don't understand the material I would be embarrassed.

SHA 3: During the class, I am embarrassed that I can't express myself well.

ANG 1: I wish I didn't have to attend the class because it makes me angry.

ANG 2: During the class, thinking about the poor quality of the course makes me angry.

ANG 3: When I think of the time I waste in class, I get aggravated.

Studying-related emotions:

ENJ 1: I enjoy the challenge of learning the material.

ENJ 2: Certain subjects are so enjoyable that I am motivative to do extra readings about them.

ENJ 3: I am so happy about the progress I made that I am motivated to continue studying.

PRI 1: Because I want to be proud of my accomplishments, I am very motivated in learning biology.

PRI 2: When I solve a difficult problem in my studying, my heart beats with pride.

PRI 3: I'm proud of my capacity.

HOP 1: During the class, the thought of achieving my learning objectives inspires me.

HOP 2: Before the class, I feel optimistic that I will make good progress at studying.

HOP 3: I feel confident that I will be able to master the material.

ANX 1: During the class, the subject scares me since I don't fully understand it. (dropped)

ANX 2: When I look at the books I still have to read, I get anxious.

ANX 3: After the class, I worry whether I have properly understood the material.

BOR 1: Because I am bored about this subject, I have no desire to learn.

BOR 2: While studying biology I seem to drift off because it's so boring.

BOR 3: The material bores me so much that I feel depleted.

HOL 1: My lack of confidence makes me exhausted before I even start.

HOL 2: During studying, I wish I could quit because I can't cope with it.

HOL 3: After studying I'm resigned to the fact that I don't have the ability.

SHA 1: I feel ashamed because I am not as adept as others in studying.

SHA 2: Because I have had so much troubles with the course material, I avoid discussing it.

SHA 3:I turn red when I don't know the answer to a question relating to the course material.

ANG 1: Because I get so upset over the amount of material, I don't even want to begin studying.

ANG 2: I get annoyed about having to study.

ANG3: When I sit at my desk for a long time, my irritation makes me restless.

References

Aivelo, T., & Uitto, A. (2016). Digital gaming for evolutionary biology learning: The case study of parasite race, an augmented reality location-based game. *LUMAT: International Journal on Math, Science and Technology Education, 4*(1), 1–26.

Akçayır, M., & Akçayır, G. (2017). Advantages and challenges associated with augmented reality for education: A systematic review of the literature. *Educational Research Review, 20*, 1–11.

Artino, A. R., Jr., Holmboe, E. S., & Durning, S. J. (2012). Control-value theory: Using achievement emotions to improve understanding of motivation, learning, and performance in medical education: AMEE Guide No. 64. *Medical Teacher, 34*(3), e148–e160.

Azuma, R. T. (1997). A survey of augmented reality. *Presence: Teleoperators & Virtual Environments, 6*(4), 355–385.

Benard, B. (1997). Drawing forth resilience in all our youth. *Reclaiming Children and Youth: Journal of Emotional and Behavioral Problems, 6*(1), 29–32.

Billinghurst, M., Grasset, R., & Looser, J. (2005). Designing augmented reality interfaces. *ACM Siggraph Computer Graphics, 39*(1), 17–22.

Bybee, R. W. (2010). Advancing STEM education: A 2020 vision. *Technology and engineering teacher, 70*(1), 30.

Cai, S., Wang, X., & Chiang, F.-K. (2014). A case study of augmented reality simulation system application in a chemistry course. *Computers in Human Behavior, 37*, 31–40.

Cai, S., Chiang, F.-K., Sun, Y., Lin, C., & Lee, J. J. (2017). Applications of augmented reality-based natural interactive learning in magnetic field instruction. *Interactive Learning Environments, 25*(6), 778–791.

Chen, Y.-C., Chi, H.-L., Hung, W.-H., & Kang, S.-C. (2011). Use of tangible and augmented reality models in engineering graphics courses. *Journal of Professional Issues in Engineering Education & Practice, 137*(4), 267–276.

Chen, P., Liu, X., Cheng, W., & Huang, R. (2017). *A review of using augmented reality in education from 2011 to 2016*. Springer.

Cheng, K.-H., & Tsai, C.-C. (2013). Affordances of augmented reality in science learning: Suggestions for future research. *Journal of Science Education and Technology, 22*(4), 449–462.

Dunleavy, M., & Dede, C. (2014). *Augmented reality teaching and learning*. Springer.

Estapa, A., & Nadolny, L. (2015). The effect of an augmented reality enhanced mathematics lesson on student achievement and motivation. *Journal of STEM Education, 16*, 3.

George, D., & Mallery, P. (2003). *SPSS for windows step by step: A simple guide and reference. 11.0 update* (4th ed.). Allyn & Bacon.

Goff, E. E., Mulvey, K. L., Irvin, M. J., & Hartstone-Rose, A. (2018). Applications of augmented reality in informal science learning sites: A review. *Journal of Science Education and Technology, 27*(5), 433–447.

Haberman, M. (1991). Can cultural awareness be taught in teacher education programs? *Teaching Education, 4*(1), 25–32.

Han, J., Jo, M., Hyun, E., & So, H.-J. (2015). Examining young children's perception toward augmented reality-infused dramatic play. *Educational Technology Research and Development, 63*(3), 455–474.

Harley, J. M., Poitras, E. G., Jarrell, A., Duffy, M. C., & Lajoie, S. P. (2016). Comparing virtual and location-based augmented reality mobile learning: Emotions and learning outcomes. *Educational Technology Research and Development, 64*(3), 359–388.

Huang, T.-C., Chen, C.-C., & Chou, Y.-W. (2016). Animating eco-education: To see, feel, and discover in an augmented reality-based experiential learning environment. *Computers & Education, 96*, 72–82.

Hughes, O., Fuchs, P., & Nannipieri, O. (2011). *New augmented reality taxonomy. Technologies and features of augmented environments. Handbook of augmented reality.*

Johnson, L., Smith, R., Willis, H., Levine, A., & Haywood, K. (2011). *The horizon report 2011*. The New Media Consortium.

Johnson, L., Adams, S., & Cummins, M. (2012). *The NMC Horizon Report: 2012 K-12 edition.* The New Media Consortium.

Klopfer, E., & Squire, K. (2008). Environmental detectives—The development of an augmented reality platform for environmental simulations. *Educational Technology Research and Development, 56*, 203–228.

Koba, S., & Tweed, A. (2009). *Hard-to-teach biology concepts: A framework to deepen student understanding*. NSTA Press.

Lee, K. (2012). Augmented reality in education and training. *TechTrends, 56*(2), 13–21.

Maymon, R., Hall, N. C., & Goetz, T. (2018). When academic technology fails: Effects of students' attributions for computing difficulties on emotions and achievement. *Social Sciences, 7*(11), 223.

McDonald, J. H. (2009). *Handbook of biological statistics*. Sparky House Publishing.

Mega, C., Ronconi, L., & De Beni, R. (2014). What makes a good student? How emotions, self-regulated learning, and motivation contribute to academic achievement. *Journal of Educational Psychology, 106*(1), 121.

Pekrun, R. 2000. A social-cognitive, control-value theory of achievement emotions.

Pekrun, R. (2006). The control-value theory of achievement emotions: Assumptions, corollaries, and implications for educational research and practice. *Educational Psychology Review, 18*(4), 315–341.

Pekrun, R., Goetz, T., Titz, W., & Perry, R. (2002). Academic emotions in students' self-regulated learning and achievement: A program of qualitative and quantitative research. *Educational Psychologist, 37*(2), 91–105.

Pekrun, R., Goetz, T., & Perry, R. P. (2005). *Achievement emotions questionnaire (AEQ).* User's manual. Unpublished Manuscript, University of Munich, Munich.

Peper, M. (2006). Imaging emotional brain functions: Conceptual and methodological issues. *Journal of Physiology-Paris, 99*(4), 293–307.

Pribeanu, C., Balog, A., & Iordache, D. D. (2008). *Formative user-centered usability evaluation of an augmented reality educational system.*

Ramey-Gassert, L., Walberg, H. J., III, & Walberg, H. J. (1994). Reexamining connections: Museums as science learning environments. *Science Education, 78*(4), 345–363.

Salmi, H., Thuneberg, H., & Vainikainen, M.-P. (2017). Making the invisible observable by augmented reality in informal science education context. *International journal of science education, Part B, 7*(3), 253–268.

Shuman, V., & Scherer, K. R. (2015). *Emotions, psychological structure of.* Elsevier.

Sotiriou, S., & Bogner, F. X. (2008). Visualizing the invisible: Augmented reality as an innovative science education scheme. *Advanced Science Letters, 1*(1), 114–122.

Sutter-Brandenberger, C. C., Hagenauer, G., & Hascher, T. (2018). Students' self-determined motivation and negative emotions in mathematics in lower secondary education—Investigating reciprocal relations. *Contemporary Educational Psychology, 55*, 166–175.

Tan, S., & Waugh, R. (2013). *Use of virtual-reality in teaching and learning molecular biology.* Springer.

Turan, Z., Meral, E., & Sahin, I. F. (2018). The impact of mobile augmented reality in geography education: Achievements, cognitive loads and views of university students. *Journal of Geography in Higher Education, 42*(3), 427–441.

Um, E. r., Plass, J. L., Hayward, E. O., & Homer, B. D. (2012). Emotional design in multimedia learning. *Journal of Educational Psychology, 104*(2), 485–498.

Weng, N. G., Bee, O. Y., Yew, L. H., & Hsia, T. E. (2006). An augmented reality system for biology science education in Malaysia. *International Journal of Innovative Computing, 6*, 2.

Chapter 10
Struggling Readers Smiling on the Inside and Getting Correct Answers

Garron Hillaire, Boris Goldowsky, Bart Rienties, and Samantha G. Daley

Abstract In this chapter, we explore inclusive learning analytics to predict reading comprehension outcomes in the context of middle school remedial reading classrooms for 235 students from 12 school districts in the United States. The design of the system used in this study followed the Universal Design for Learning Guidelines which have expressed the goal of supporting all learners, a common goal with inclusive learning analytics. We construct a predictive model for reading comprehension, introducing a novel method to apply the same sentiment analysis method to interpret both emotional self-reports and discussion comments, contributing new insights about the relationship between emotion and cognition during learning. The novel method of analysis was inspired by the inclusive design principles of Universal Design for Learning by treating the two data sources as multiple means of expression. The relationship the model suggests is that struggling students with positive emotional reactions coupled with neutral communication about reading materials are more likely to provide correct answers. The psychological phenomena of positive internal experience coupled with downregulated expression have previously been described as "smiling on the inside," and are considered socially desirable when outperforming peers. We explore the model in terms of bias across a range of groups legally protected from discrimination. With the limitations of the measures and the model in mind, we provide recommendations on how to support students in this psychological state of "smiling on the inside".

G. Hillaire (✉)
Massachusetts Institute of Technology, Cambridge, MA, USA
e-mail: garron@mit.edu

B. Goldowsky
CAST Inc., Wakefield, MA, USA
e-mail: bgoldowsky@cast.org

B. Rienties
The Open University, Milton Keynes, UK
e-mail: bart.rienties@open.ac.uk

S. G. Daley
The University of Rochester, Rochester, NY, USA
e-mail: sdaley@warner.rochester.edu

Y. E. Wang et al. (eds.), *Social and Emotional Learning and Complex Skills Assessment*, Advances in Analytics for Learning and Teaching,
https://doi.org/10.1007/978-3-031-06333-6_10

Keywords Inclusive learning analytics · Universal design for learning · Reading comprehension · Middle school

10.1 Introduction

The intersection of learning analytics and inclusivity considers how marginalized students are represented by data and the analysis of that data. For example, previous studies have explored the emotional expression of assistive technologies (Hillaire et al., 2019), evaluations of online learning designs (Iniesto et al., 2014), and the use of data visualization to support help-seeking behaviors of students (Daley et al., 2016). The current study uses multiple methods to triangulate the emotional states of students and looks for evidence of how those states may correlate with success in academic tasks.

Our work is based on Universal Design for Learning (UDL). Central to the UDL framework is the UDL Guidelines (CAST, 2018), which were established to guide the development of learning environments that support all learners, including struggling students (Meyer et al., 2014). The UDL Guidelines do not reference learning analytics specifically, but there are nevertheless a variety of applications of the UDL Guidelines to analytics research. Inclusive design naturally supports the aim of inclusive analytics, but it is not sufficient. To connect inclusive design to inclusive analytics, we align the analytic methods with the principles of inclusion or UDL.

In this chapter, we examined data from Udio (Boucher et al., 2015), a reading comprehension platform based on UDL. One of the key principles of supporting variability through learning design in the UDL Guidelines is to provide multiple means of action and expression (CAST, 2018). In Udio, one example of this is that students expressed their emotional reaction to reading materials in two different ways: by using a "reaction wheel" with a fixed set of words, or through an open, asynchronous discussion area (see instruments). From a design perspective, multiple means of expression are intended to support all learners with options for engaging with reading materials. The related inclusive analytics challenge that we tackle is the coordination of both of these response channels—emotional self-report and discussion comments—as two related expressions around the same reading material. We set out to see if the emotional data footprint from Udio generated by these two response options, in combination with skill-based measures, would predict success on a reading comprehension task. Based on this, we look for insights into how online environments can support diverse learners by focusing on what successful learning – in this case, reading – looks like from an emotional perspective for struggling students.

The focus on emotions in the analytic methods are not based on convenience, but on the anticipated importance of emotions in learning for struggling students. Udio was built for middle-school struggling readers, a population of learners who have often built up many years of frustrating and negative emotions associated with reading in school. Thus, when presented with a reading-based task, these students have a double hurdle to overcome: both a lack of skills and a lack of motivation, even an aversion to the task itself. As an online reading environment designed to help these students, Udio represents an opportunity to directly investigate how these interactions play out. While we have no direct access to students' emotional states, the intervention did provide us with two ways to gain insight into emotion: discussion comments that students wrote, and the words they chose from a fixed list of "reactions". We also had a proximal measure of comprehension of each reading page as well as more global measures of reading skill from standardized instruments. One novel feature of our approach is that we interpret both the self-report scale and the discussion comments with the same sentiment analysis technology (see analysis). Our motivation for using the same interpretive approach for both data sources was to align our analysis with the design principle that the two data sources provide students with multiple means of expression about the learning material. This resulted in a model in which the two emotional measures are expressed in similar scales and can both be used to predict the outcome variable. Our analysis indicates that a positive self-report in conjunction with neutral discussion comments predicts more correct reading comprehension outcomes. Generating a promising model is testament to the importance of the intertwined nature of learning design and learning analytics.

However, generating a promising model does not yet reach the goal of supporting *all* learners.

It is a goal of UDL to globally implement supports that are essential for some students with the goal of benefiting all students. This aim aligns well with the consideration of ethical concerns about student data and predictive models. We have collected demographic information about groups legally protected from discrimination, including race, gender, socio-economic status (SES), English Language Learning (ELL) status, and Individual Education Program (IEP) status. We examined the bias of the predictive model based on these groups. The bias analysis demonstrated that there are both positive and negative biases, suggesting further work is needed to improve the model before it would be able to achieve the goal of supporting all learners.

With these ethical considerations around bias explored, we finally present the implications of our results. Our model suggests that students who had (1) positive internal experiences; and (2) external presentations of neutral communication with peers are more likely to have provided correct answers on the reading comprehension check. The phenomenon of positive internal emotion with downregulated neutral expression has been previously described as "smiling on the inside" (Schall et al., 2016). Connecting these results to socio-emotional learning, we finally discuss how teachers might best support these students who are "smiling on the inside".

10.2 Related Work

10.2.1 Inclusive Learning Analytics and Universal Design for Learning

There is a known relationship between learning design and learning analytics (Macfadyen et al., 2020). The data we collect are a consequence of the design of the system used to collect the data. This relationship illustrates that the exploration of inclusive learning analytics might best be supported by exploring data from a system that was designed with inclusivity in mind.

The Universal Design for Learning (UDL) framework provides guidelines for educational goals, methods, materials, and assessments that can support all learners, including struggling students. The UDL perspective suggests that when students struggle, the criticism should first focus on the design of the system to identify if there are unnecessary barriers to student success embedded within the design, rather than assigning the deficiency to the student (Meyer et al., 2014). The UDL Guidelines are organized into three principles, encouraging the provision of (1) multiple means of engagement; (2) multiple means of action and expression; and (3) multiple means of representation (CAST, 2018).

The intersection of UDL and learning analytics was explored at the 2014 Learning Analytics Summer Institute (Hillaire et al., 2014) where the Udio platform used in this study was discussed as an example of how UDL creates an interesting opportunity to explore learning analytics. In addition, the design of "What's your reaction?" (see instruments) was used as an example of how to prototype visual learning analytics (Hillaire et al., 2016). When systems are designed up front with inclusivity in mind, the data they collect have the potential to provide insight into learning for all students including struggling students. We examine the data collected from Udio to explore what emotional footprints struggling students generate at the intersection of emotion and cognition during reading comprehension.

10.2.2 Emotion, Cognition, and Reading Comprehension

There is, of course, no simple model of how emotion and cognition interact around learning. Fiedler et al. (2014) describe "broaden and build" theory of positive emotion (Fredricks et al., 2004; Fredrickson, 1998), saying that both positive and negative emotions influence learning. Negative emotions are considered more effective for considering a complex learning task in isolation (Clore & Huntsinger, 2007), while positive emotions are considered associated with assimilative learning tasks (Fiedler & Beier, 2014). A competing perspective is that emotions in general are a deficit to learning: cognitive load theory advocates that neutrality is optimal for student performance (Fraser et al., 2012). These perspectives provide a theoretical frame to investigate student emotion during learning.

To contextualize positive emotion—or the potential lack of positive emotions—for middle school students, we consider that these students may have experienced years of affect and motivational challenges. For example, a study with 500 students in grades 2–4 that included 87 students with reading comprehension difficulties found that about half of the students appeared motivated and self-reported high levels of positive affect, while the other half appeared to lack motivation. Students with disabilities had lower motivation and more negative experiences (Sideridis et al., 2006). This result indicates that before middle school many students are already developing motivation issues for reading comprehension.

Further work illustrated that emotional responses during reading correlated with comprehension measures (with the limitation that multiple choice assessment did not have a correlation) (Daley et al., 2014). Repeated struggles lead to a lack of motivation, causing more failures, setting up a vicious cycle.

We report exploratory results on a predictive model that suggests that students who self-report positive emotions and at the same time make neutral comments in the discussion about reading are more likely to get correct answers. This internal positive emotion coupled with downregulated neutral expression has previously been described as "smiling on the inside"(Schall et al., 2016). The smiling on the inside phenomenon is considered a result of downregulated expression so that students provide socially desirable interactions when they are outperforming peers.

10.2.3 Measuring Emotion with Sentiment Analysis

There has been an increased interest in the interplay of emotion and cognition in recent years (Okon-singer et al., 2015), facilitated by the development of new emotional measures in the discipline of affective computing (Calvo & D'Mello, 2010). These new emotion measures are just starting to gain momentum in the context of education and detection of students' emotions (Herold, 2016). Sentiment Analysis is an affective computing approach to make predictions about the emotional content of text (Pang & Lee, 2006); it has been used to detect specific emotions as well as the categorical dimension of valence in text (Medhat et al., 2014). A lexical approach to sentiment analysis uses a dictionary of words that are categorized in a dimensional manner. Typically, the dimension used is valence. An example dictionary is (Warriner et al., 2013), which provides a valence score for 13,915 words. The scores are on a scale between 0 and 10, with scores 4 and below considered negative, 4–6 neutral, and above 6 positive. The scores generated for this dictionary were created in a crowd-sourced manner (Warriner et al., 2013). The way a lexical approach is applied to text is to substitute words with their associated valence scores, and then compute the average of these scores to summarize the valence of written text communication (Pang & Lee, 2006).

The affective computing method of sentiment analysis holds promise for applications in educational research (Rienties & Rivers, 2014; Sharples et al., 2015). Some recent studies, for example, have found a correlation between the calculated valence of students' course reviews with their completion rates (Wen et al., 2014).

10.2.4 Measuring Emotion by Self-Report

Perhaps the simplest way to get information on people's affective states is to ask them. This self-report technique can be implemented in a wide variety of ways, and there are several validated instruments for self-reporting emotions. For example, PANAS asks participants to rate their agreement with positive and negative words to determine the extent to which an experience is positive and the extent to which an experience is negative (Watson et al., 1988). This positive-negative axis is known as "valence" and is considered a universal dimension for categorizing emotion (Russell, 1980). The Geneva wheel places words on two dimensions, valence and arousal (the level of intensity of the response) and asks participants to rate their agreement with 12 words with intensity for each term (Sacharin et al., 2012).

All self-report methods have known limitations. From the perspective of measurement breakdown in the classroom, there are challenges for both the student accurately reporting their emotions and the teacher accurately observing emotions. For example, when a student uses a specific list of words (s)he may not interpret the word to mean the same thing as the instrument expects (Duckworth & Yeager, 2015). Furthermore, there are often challenges in the accuracy of tools used to research emotions. For example, in a recent review of publications in the journal *Emotion* (Weidman et al., 2016), 69% of studies developed an impromptu scale for emotional measures, making it difficult to compare results across studies. Given the tension between using words the target population is familiar with and using a standardized scale, there is a real challenge to get a measure that allows participants to accurately report their emotions and yield results that are comparable with other work.

In the present study, we selected a set of emotion words through formative evaluation with struggling readers (Hillaire et al., 2016). While this approach may help with the first problem of students selecting from emotion words that are interpreted in the anticipated manner, it does not solve the problem of generating results that are comparable with other studies. To address this second problem, we interpret the self-report scale we created using sentiment analysis to anchor our novel measurement scale within a more standard approach, with the intention of closing the gap between novel and standard measurements.

10.2.5 Measuring Reading Comprehension

Reading comprehension is a difficult task to measure because of the many factors involved in reading. When comparing measures that predict reading comprehension, there is evidence that the instruments used to identify struggling readers produce variable results (e.g., a student identified as a struggling reader by one diagnostic instrument may not be identified by another instrument) (Cutting & Scarborough, 2009). Difficulties in reading could be due, for example, to challenges in decoding words, in vocabulary, in background knowledge about the topic, or in understanding complex syntax (Brasseur-Hock et al., 2011). A single reading comprehension measure will be unlikely to disentangle these contributing elements and are rarely sensitive to changes over short time periods.

Previous work has illustrated a potential relationship between the emotions of the student and reading comprehension, pointing out that even the narrative of the text can evoke emotional response (Kneepkens & Zwaan, 1995). The relationship between emotion and cognition during reading comprehension may be a part of learner variability that generates inconsistent results from measures designed to predict reading comprehension. As our review on emotions and learning indicated the potential benefits of positive and neutral emotions during learning and further evidence that struggling readers often experience negative emotions, there is reason to explore how emotions relate to reading comprehension measures.

10.3 The Present Study

We explore if sentiment analysis can be used to give us comparable measures of the emotional valence of both self-report and discussion comments to explore the predictive power of these two emotional indicators: will they give us any insight into when students were motivated to read and comprehend a text? Thus, we break down our research questions as follows:

Research Question 1 (RQ1): To what extent can we use sentiment analysis to interpret self-report?

Research Question 2 (RQ2): To what extent can we use sentiment analysis to interpret students' discussion comments?

Research Question 3 (RQ3): To what extent does sentiment analysis of self-report and discussions correlate with providing correct answers in a reading comprehension activity?

10.3.1 Ethical Concerns with Predictive Modeling in Education

In learning analytics research, it is common to produce predictive models associated with learning and report accuracy of the model. When researching inclusive learning analytics, it is important to interrogate predictive models to identify model fits by considering demographic characteristics. When considering which groups to evaluate, a reasonable starting point is to consider groups protected by discrimination laws. A recent review of algorithmic bias in education points out that the most commonly studied groups are race, gender, and learner's current national location, while language learning and disability status are groups for which there is limited literature (Baker & Hawn, 2021).

We explore how emotional measures might predict reading comprehension and report both overall accuracy as well as interrogate the model in terms of fairness. To evaluate fairness we examine model predictions across groups including race, gender, Individual Education Plan (IEP) status, English Language Learning (ELL) status, and Socio-Economic Status (SES). This approach includes categories frequently studied for algorithmic bias in education (e.g., gender, race) and includes groups for which there is limited work (e.g., IEP, ELL) (Baker & Hawn, 2021). We have included these groups to anchor the analysis to groups protected by anti-discrimination laws. One of the limitations in taking this approach is how adhering to legally protected groups comes with labels and grouping approaches that are not necessarily ideal (Baker & Hawn, 2021). For example, we include ELL status, but recognize that there is a shift towards referring to students as emergent bilinguals (García et al., n.d.). We focused on contributing evidence about algorithmic bias for groups using designations frequently used by schools to connect the interpretation of results with advice for teachers.

As First Author on this paper, I am a student who was in a remedial reading class in the sixth grade, classified by others as Native American and self-identify as a member of the Lummi Nation. In the sixth grade, some could easily argue I had undiagnosed reading disabilities. I am sensitive to the problematic nature of these categories. I believe we are a long way from understanding how to adequately address bias in algorithms and engaging in this work is both difficult and necessary. Hopefully, the contribution of this chapter can improve our collective understanding of the known limitations in how we have categorized students.

10.4 Participants

The present work is part of the "Udio" project (Boucher et al., 2015), we include data from a subset of that project's participants, including 33 classrooms in 15 middle schools within 12 districts across the continental United States. The classrooms involved in this intervention were all specifically designed to serve struggling readers: designated as remedial reading, resource room, or other settings with students identified as requiring additional support in reading.

Within this overall cohort, the present study only considers data from students who consented to participate in the study and who used the three response features ("Discuss It", "What's Your Reaction?", and "Boost your Understanding") that we are analyzing, and who had at least one case of a text within which they used all three features. The demographic data for the 235 students meeting these criteria are detailed in Table 10.1.

10.5 Procedure

Udio was an online reading environment designed to support struggling readers to engage with texts in multiple genres and on a wide variety of topics. 340 different texts were available for students to choose from both fiction and non-fiction, with topics ranging from sports and entertainment, personalities to science, history, and current issues. The majority of the online texts were single-page readings, but a few spanned multiple pages. Reading supports based on the principles of Universal Design for Learning (CAST, 2018; Meyer et al., 2014) were available, such as text-to-speech for reading content aloud, English and English-to-Spanish dictionaries, the ability to highlight and collect key ideas, etc.

Students used the online Udio environment (see Fig. 10.1) on individual laptops several times a week for a period of 7 months as part of their regular school day. Although teachers provided varying degrees of direction, generally students were free to choose texts of interest to them, and to read, listen, and use any of the features to interact with those texts.

10.6 Instruments

10.6.1 What's Your Reaction?

This was an interactive self-report feature that allowed students to click one or more of the 12 emotion adjectives to express their feelings about a text (see Fig. 10.2). These were organized in a circular fashion reflecting the affective dimensions of valence and arousal, similar to the Geneva Wheel (Sacharin et al., 2012). Students were free to select any or all of the 12 words. They could also, by clicking another link "See others' reactions", see a visualization of the aggregated data showing how many of their classmates had chosen each descriptor, but others' reactions were not personally identifiable. For more information on the design of this feature, see (Hillaire et al., 2016).

10.6.2 Discuss It

This was a discussion area, presented as a sidebar next to the reading content (see Fig. 10.3). In this area, students (and teachers) could express their thoughts about the text either by writing, drawing, or leaving a voice recording. These comments, including the author's name, were visible to their classmates and teachers. It was also possible to reply to classmates' comments (in the following analysis we do not make a distinction between top-level comments and replies).

Table 10.1 Demographic information about participants

Demographics		Grade 6	Grade 7	Grade 8	Total
Gender	Male	52 63%	43 65%	55 64%	150 63.8%
	Female	31 37%	23 35%	31 36%	85 36.2%
Race	White	33 40%	20 30%	18 21%	71 30.2%
	Black or African American	24 29%	19 29%	22 26%	65 27.7%
	Hispanic or Latino	21 25%	21 32%	41 48%	83 35.3%
	Asian or Pacific islander	0 0%	2 3%	1 1%	3 1.3%
	Alaska native or American Indian	1 1%	0 0%	0 0%	1 0.4%
	Two or more races	4 5%	4 6%	3 3%	11 4.7%
	Missing	0 0%	0 0%	1 1%	1 0.4%
IEP	IEP	48 58%	37 56%	49 57%	134 57.0%
	Non-IEP	35 42%	29 44%	36 42%	100 42.6%
	Missing	0 0%	0 0%	1 1%	1 0.4%
ELL	ELL	8 10%	18 27%	25 29%	51 21.7%
	Non-ELL	75 90%	48 73%	60 70%	183 77.9%
	Missing	0 0%	0 0%	1 1%	1 0.4%
SES	Free/reduced-Price lunch	65 78%	46 70%	74 86%	185 78.7%
	Non-free/reduced-Price lunch	18 22%	20 30%	11 13%	49 20.9%
	Missing	0 0%	0 0%	1 1%	1 0.4%
Total		83 35%	66 28%	86 37%	235 100%

10.6.3 Boost Your Understanding

This feature allowed students to check their understanding of the text through a maze procedure, that is, by choosing appropriate words from a short list of options to complete a summary paragraph (see Fig. 10.4). Once a word was chosen for every blank, they could click "Check my answers". After checking, they had the opportunity to correct any mistakes, but the number of answers that they got correct

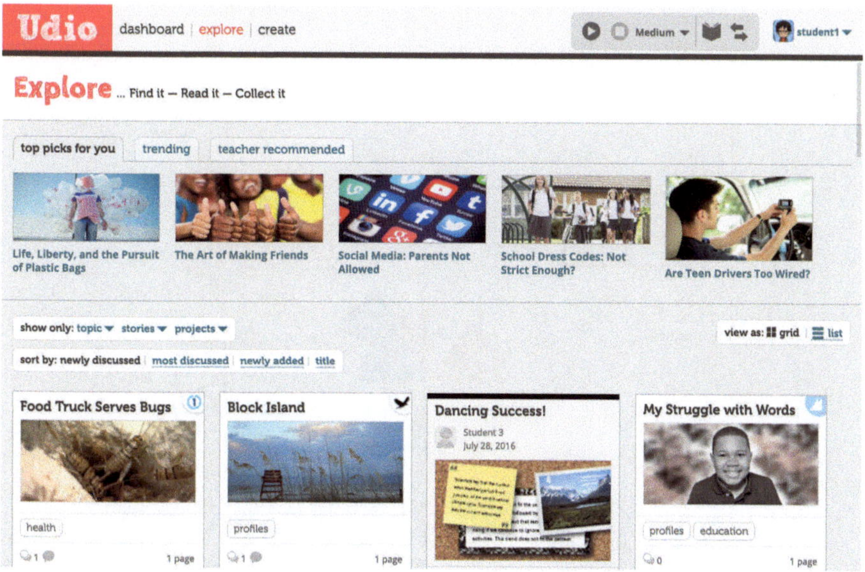

Fig. 10.1 The Udio platform

on the first try was recorded. This approach to an in-the-moment measure of reading comprehension allowed us to focus specifically on whether students could generate a reasonable summary of often very brief reading passages. The vocabulary and reading levels were similar to the passage itself to control for factors beyond making meaning from the text.

10.6.4 RAPID Probability

Prior to the study, students completed the RAPID assessment, a computer-adaptive measure developed by Lexia Learning (2016) that includes assessments of reading comprehension, vocabulary knowledge, syntactic knowledge, and word recognition. In the reading comprehension subtest, students are asked to read a passage and respond to multiple choice questions. Students are presented passages of differing levels of difficulty, depending on their performance on the other subtests. In these analyses, we use the "reading success probability" score, a percentage intended to indicate the likelihood that the student will perform at the level of reading comprehension expected for their given grade level, as defined by a commonly used standardized achievement measure. The formula used to derive the probability score incorporates not only reading comprehension, but also the other aspects of reading and language included in the RAPID, making it adaptive for the student's age, appropriate to use multiple times in a single school year, and a robust means to consider their comprehensive performance in reading.

Fig. 10.2 "Whats your reaction?" feature in Udio

10.7 The Dataset

We constructed a combined dataset by selecting student-text pairs where a student used all three features of interest in their interactions on that text, so that we had an emotion self-report, one or more discussion comments, and a comprehension check result for that student-text pair. There were 235 students who used all three features for at least one text. These students completed all three activities for between 1 and 53 texts, with the mean around 7 texts (M = 7.19; SD = 8.39) for a total of 1691 student-text pairs where the student did all three activities. Of these 1691 instances, students got a correct score on the comprehension check on 548 (32%).

10.8 Analysis

To answer Research Question 1 (RQ1):"To what extent can we use sentiment analysis to interpret self-report?", we recognize that self-report using "What's your reaction?" was designed with an ad hoc list of discrete words and then reinterpreting the

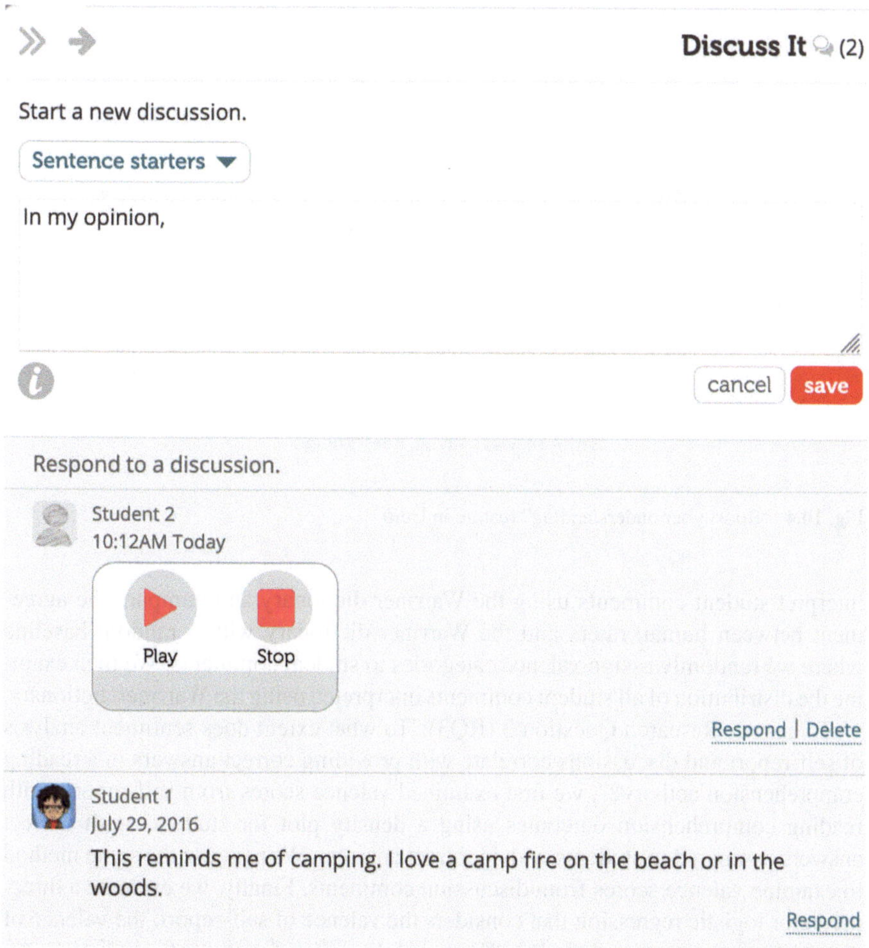

Fig. 10.3 "Discuss It" feature in Udio

data in a dimensional way through valence. Calvo et al. (2010; Weidman et al., 2016) argue that it is not ideal to reinterpret discrete emotions in dimensional terms. To address this concern, our self-report instrument was modelled including the dimensions of valence and arousal (Hillaire et al., 2016). Mapping the words to their valence values in the Warriner dictionary gives us a method to formally evaluate the degree to which this design intention was successful. We first interpret every possible response using "What's your reaction?" and then interpret the actual responses provided by students.

To answer Research Question 2 (RQ2):"To what extent can we use sentiment analysis to interpret students' discussion comments?", we first take a sample of student comments and have human raters determine the valence of the comments using the same method used when developing the Warriner dictionary. Then we

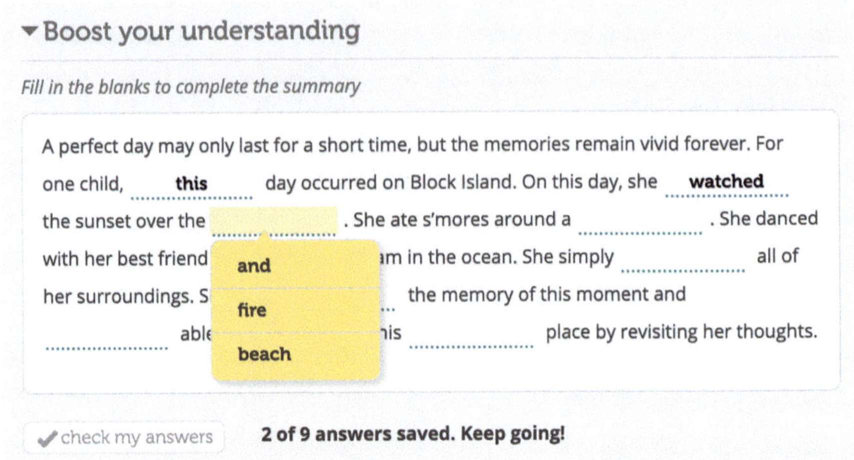

Fig. 10.4 "Boost your understanding" feature in Udio

interpret student comments using the Warriner dictionary and compare the agreement between human raters and the Warriner dictionary with a random baseline where we randomly assign valence categories to student comments. We then examine the distribution of all student comments interpreted using the Warriner dictionary.

To answer Research Question 3 (RQ3):"To what extent does sentiment analysis of self-report and discussions correlate with providing correct answers in a reading comprehension activity?", we first examined valence scores from self-reports with reading comprehension outcomes using a density plot for students with correct answers compared to students with incorrect answers. We next use the same method to examine valence scores from discussion comments. Finally, we evaluate a three-predictor logistic regression that considers the valence of self-report, the valence of discussion comments, and the RAPID probability score for students predicting student success on the reading comprehension check. After evaluating the model accuracy across all students, we examined the fairness of the model considering a range of student demographic categories.

10.9 Results

10.9.1 (RQ1) to What Extent Can We Use Sentiment Analysis to Interpret Self-Report?

To answer RQ1, we conducted a theoretical and practical analysis of the process. In the theoretical analysis, we examined all possible self-report responses and in the practical analysis we examined the actual responses from the students in this study.

When conducting a theoretical mapping of self-report words to the dimension of valence, we looked up the words from "What's Your Reaction" in the Warriner dictionary. There was a direct match for 8 of the 12 words. For the remaining 4 words, there was a word stem used as a substitution. For example, the word 'challenging' does not appear in the Warriner dictionary, so 'challenge' was used to map the option to a valence value. This mapping process was done by inspection of the dictionary. Table 10.2 shows the self-report words that were shown on the positive and negative sides of the circular word wheel and their valence scores.

The guidance for interpreting valence scores from the Warriner dictionary is that scores above 6 are positive, scores below 4 are negative, and scores between 4 and 6 are neutral. Thus, 11 of the 12 words fall into the expected category. The word "challenging" was presented as a positive word, but in the Warriner dictionary the closest match, "challenge", has a score of 5.95. While it is near the threshold value for positive, it is considered neutral based on the guidance for interpreting Warriner valence scores. However, it was close enough to the threshold value of 6 that if the mean score was calculated for a pair of reactions including 'challenging', with any other positive word the score would be positive and with any other negative word the score would be neutral.

Udio users could select between 1 and 12 words. With 12 options in a multiselect, there are 4095 combinations. For each combination of possible responses, the Warriner valence score (see Table 10.2) was substituted. (E.g., the selection of [Curious, Good] translated into [6.58, 7.89]). After the response was translated into valence scores, the arithmetic mean was calculated (e.g., 6.58 + 7.89/2 = 7.235). The result of every possible response generated the histogram in Fig. 10.5. The distribution generated had a min = 2.1; max = 7.89, std. = 0.68, mean = 4.83, median = 4.83. Data were tested (formally) for violation of normalcy by running the Shapiro-Wilk's test, which indicated that the distribution is different from a normal distribution ($p < 0.05$). The skewness of the data = 0.01 and excess kurtosis = 0.93. The kurtosis and the histogram indicate the distribution is leptokurtic, meaning it

Table 10.2 Warriner dictionary scores for react words

Polarity	React option	Warriner dictionary	Valence score
Positive	Engaging	Engaged	6.78
	Interesting	Interesting	6.78
	Challenging	Challenge	5.95
	Curious	Curious	6.58
	Calming	Calm	6.89
	Good	Good	7.89
Negative	Annoying	Annoying	3.00
	Frustrating	Frustrating	2.57
	Confusing	Confusion	3.32
	Sad	Sad	2.10
	Boring	Boring	2.71
	Dull	Dull	3.40

has more outliers than a normal distribution. This is illustrated in the heavy tailed histogram in Fig. 10.5. The skewness is low enough that the distribution is approximately symmetric. The mean of the distribution was lower than 5 (4.83), which indicates that the approximately symmetric distribution is shifted left of the neutral value by 0.17. If we were to shift the threshold values of 4 and 6 by 0.17, the median of the distribution would become neutral and the word "challenging" would move into the positive region.

When self-reporting reactions to a text, students provided between 1 and 12 words with the mean number of words selected between 4 and 5 (M = 4.45; SD = 3.09). The most common response was selecting all 12 words, occurring 132 times. The valence score for selecting all twelve words was 4.83 (the mid-point, or neutral point of the adjusted valence scale). While this response may reflect students not using the system for the intended purpose, the scoring approach results in a neutral score for responses where all 12 words are selected. The second most common reaction was to select the single positive word "good" which occurred 111 times. The third most common response was to select all six of the positive words, which occurred 106 times. The fourth most common response was to select "interesting" and "good," occurring 99 times. The fifth most common response was "interesting" alone, occurring 72 times. These common responses show up as peaks in the distribution of responses from the study, as shown in Fig. 10.6. The distribution generated had a Min = 2.1; Max = 7.89, Mean = 5.99, Median = 6.21, SD = 1.23. The distribution was left skewed with a skew of −0.85 and had an excess kurtosis of 0.56. It is not surprising that the distribution was (formally) identified as significantly different from a normal distribution as indicated by the Shapiro-Wilk's test ($p < 0.05$).

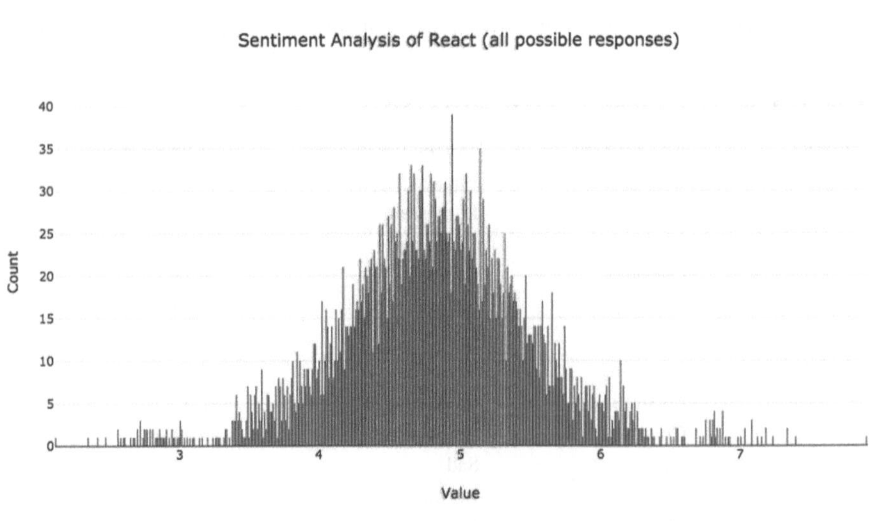

Fig. 10.5 All possible "What's your reaction?" values interpreted using the Warriner dictionary

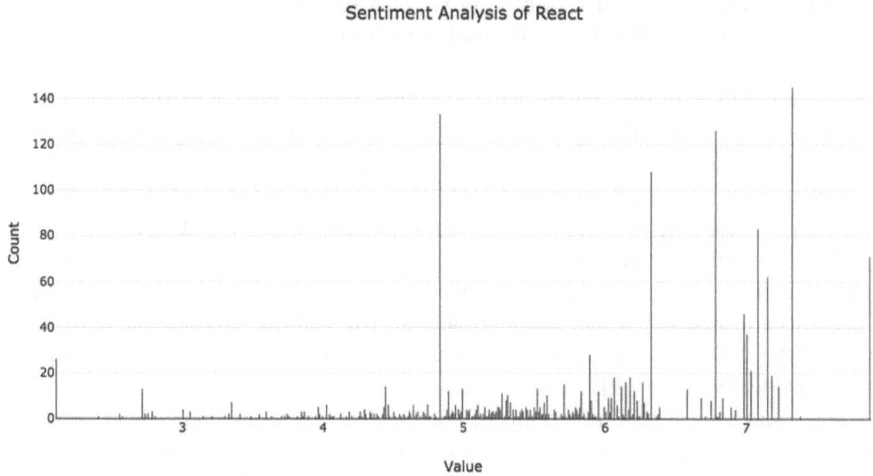

Fig. 10.6 "What's your reaction?" student responses interpreted using the Warriner dictionary.

In summary, all possible combinations of word selections did not generate a normal distribution, although it was approximately symmetrical around the midpoint 4.83. The result of students' actual use of the feature generated a distribution that was jagged and skewed to the left. Based on the most popular selections and the histogram, this result appears to be the result of students frequently selecting only positive words to describe their reaction. The minimum and maximum values were the same between the theoretically possible values and the actual outcome from use by students. While the majority of responses were the selection of some groups of positive words, there were some responses across the entire scale, as illustrated by the histogram.

From a theoretical perspective, the mapping generated a plausible distribution with a skewness that indicated there were more possible responses at the positive and negative extremities of the scale than there would be in a normal distribution. The lack of normality may actually make sense as it allows for more positive and negative responses with wide tails while maintaining symmetry so as not to nudge the user with positive or negative responses. From a practical perspective, there were some spikes in the values due to a large number of students using the same selection of words. For example, selecting just the word 'good' occurred 111 times. The common responses generated a series of spikes on the positive side of the distribution. While this generated a reasonable simulated outcome from a theoretical perspective, the spikes in the distribution indicate the measurement outcome is not exactly a continuous measure. These results indicate that it is reasonable from a theoretical perspective to translate from discrete words to the dimension of valence. In practical terms, when doing this, there are common response spikes in the resulting distribution that may influence how best to interpret the resulting data.

10.9.1.1 (RQ2) to What Extent Can We Use Sentiment Analysis to Interpret Students' Discussion Comments?

To answer RQ2, we first established a baseline data set for comparison by having human raters evaluate the emotional content of a selection of discussion comments. This allowed us to evaluate the accuracy of the sentiment analysis labels. After establishing the level of agreement with human raters, we looked at the practical outcome of mapping discussion comments to the dimension of valence.

We randomly selected 50 comments where students provided correct answers and 50 comments where students provided incorrect answers from the comprehension check to generate a 100-message sample. The 100-message sample represents 4% of the 2501 total messages in the study. For the 13 raters, in line with the approach by Warriner (2013), each message was given a score from 1 to 9 and the scores from all raters were averaged. For the 100 messages rated, there were 34 positives, 19 negatives, and 47 neutrals as indicated by the human raters. When comparing these ratings with the labels generated by using the Warriner dictionary, we also used a random baseline to compare accuracy measures against random guessing for the valence labels. We randomly assigned a label to all messages and compared those to the human raters. After repeating this 1000 times, we then averaged all accuracy measures to produce a random benchmark. The resulting overall accuracy of the random benchmark is the f-measure score of 0.16. Our sentiment analysis classifier outperformed this random benchmark, achieving an f-measure of 0.44 (see Table 10.3).

Across all comparisons, the sentiment analysis classifier outperformed the random baseline except in terms of negative recall where the classifier had a recall of 0.21 and the random classifier had a recall of 0.32. Based on these results, the predictions of negative valence are the least reliable. Given that random outperformed the sentiment analysis classifier, interpretations of negative predictions should be avoided. Negative had a precision of 1.00 and a recall of 0.21. These results indicate that the sentiment analysis classifier under-predicts negative valence by around 79% and when it does predict negative (4 times for this sample; 19*0.21 = 4), it is always correct. The results also indicate that the most suitable category for interpretation is positive with an F-Measure of 0.55. Neutral's F-Measure of 0.44 lands between Positive and Negative.

The distribution of comment valences generated had Min = 1.96; Max = 8.37, Mean = 6.07, Median = 6.15, SD = 0.91. The distribution was left skewed with a skew of −1.32 and had an excess kurtosis of 4.35 (see Fig. 10.7). It is not surprising that the distribution was significantly different from a normal distribution as indicated by the Shapiro-Wilk's test ($p < 0.05$). Again, there are spikes of values that may impact the validity of interpreting the measure as a continuous dimension.

Table 10.3 Sentiment analysis accuracy of Warriner dictionary and random baseline compared to 13 human raters

Method	Valence	Valence	Support	Precision	Recall
Warriner dictionary	Positive	34	0.43	0.76	0.55
	Negative	19	1.0	0.21	0.35
	Neutral	47	0.51	0.38	0.44
	Macro average	100	0.65	0.45	0.44
Random baseline	Positive	34	0.10	0.31	0.16
	Negative	19	0.06	0.32	0.10
	Neutral	47	0.16	0.35	0.23
	Macro average	100	0.11	0.33	0.16
Precision	$\frac{TP}{TP+FP}$	Recall		$\frac{TP}{TP+FN}$	
F-score	$\frac{TP}{TP+(1/2)(FP+FN)}$				

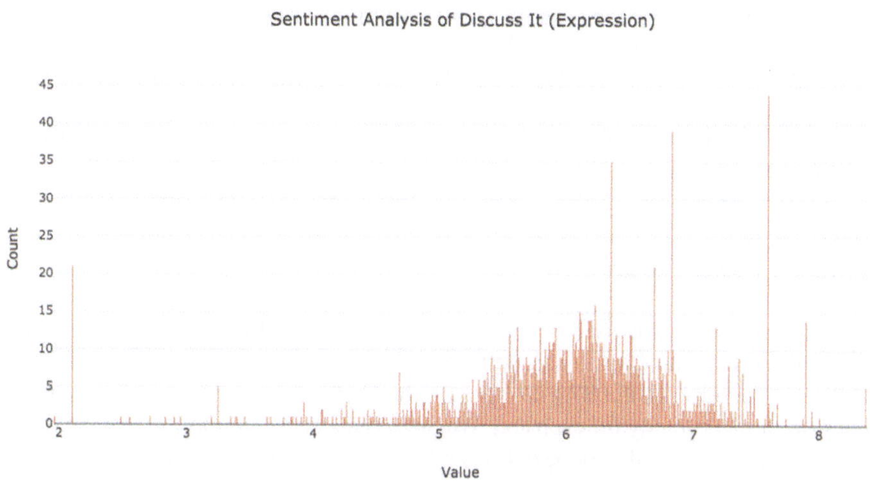

Fig. 10.7 "Discuss It" student responses interpreted using the Warriner dictionary.

10.9.2 (RQ3) to What Extent Does Using a Valence Interpretation of Self-Report and Discussion Comments Correlate with Providing Correct Answers in a Reading Comprehension Activity?

To answer RQ3, we first examined each emotion predictor in isolation to see if it predicted success on the comprehension check, and then fit a three-predictor logistic regression model to see if that would provide additional predictive power.

10.9.2.1 Sentiment of Self-Report Compared to Reading Comprehension

For self-report valence, the Mann-Whitney test indicated that the valence was different for students providing correct answers (Median = 6.32) and students providing incorrect answers (Median = 6.14), U = 276941.5, $p < 0.05$. Figure 10.8 shows the KDE density plots of the two distributions. Incorrect answers are associated with a bimodal distribution with a primary peak in the positive valence region and a secondary peak at the neutral valence. For correct answers, the neutral peak is absent and the positive peak is more pronounced. Put the other way, more positive self-report valences are associated with a larger proportion of correct answers on the comprehension check.

10.9.2.2 Sentiment of Discussion Compared with Reading Comprehension

For student comments, the Mann-Whitney test also showed a difference in the median comment valence for correct answers (Median = 6.05) and incorrect answers (Median = 6.21), U = 351837.5, $p < 0.001$. Figure 10.9 shows the KDE plots of the two distributions. In this case, both distributions show a single large peak, but the peak is in the positive valence region for incorrect answers, while for correct answers the peak is at the boundary of neutral and positive. Thus, students who participate in the discussion in a more neutral manner, as illustrated in Fig. 10.9, are more likely to get a correct answer.

10.9.2.3 Logistic Regression

Logistic regression was used to determine to what extent the three variables – self-report valence, comment valence, and the student's baseline reading comprehension ability as measured by the RAPID – predict the likelihood of answering the comprehension check correctly for each text. We first checked the assumption that the independent variables should not be collinear. The variance inflation factor (VIF) values for each independent variable were as follows: Self-Report Valence 1.03, Comment Valence 1.02, RAPID probability 1.02. These results indicated there was only a weak collinearity between the variables.

According to the model, the RAPID probability score (ranging from 0 to 99) was a strong predictor that students would provide correct answers as indicated by the odds ratio of 1.03 (95% Confidence Interval 1.03 to 1.04; $p < 0.001$). The emotional measures were both found to have a significant effect. Students' self-reported emotional reactions (ranging from 2.1 to 7.89) had a 14% increased likelihood of getting a correct answer for each valence point increase, as indicated by the odds ratio of 1.14 (95% Confidence Interval 1.05 to 1.26; $p = 0.005$). In contrast, every unit increase in the valence of their discussion comments (ranging from 1.96 to 8.37) resulted in a 13% decrease in the chance of a correct answer, as indicated by the

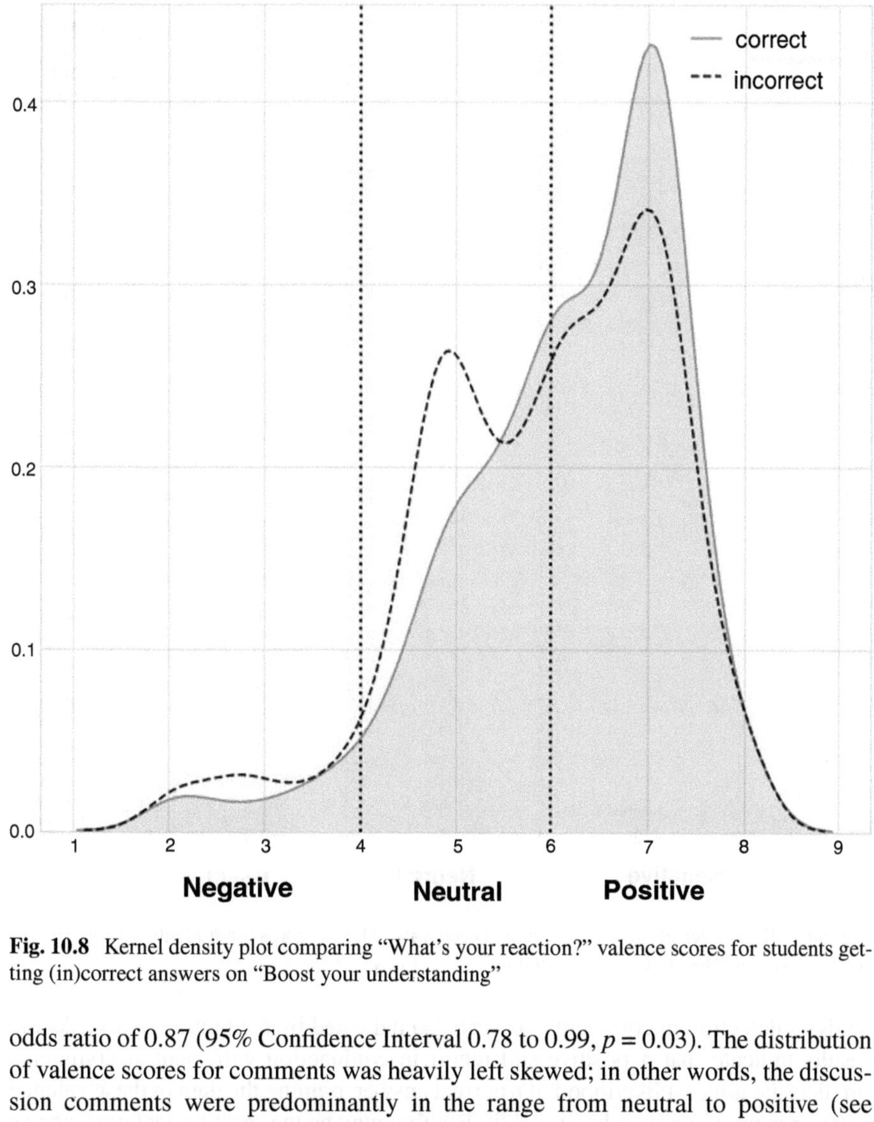

Fig. 10.8 Kernel density plot comparing "What's your reaction?" valence scores for students getting (in)correct answers on "Boost your understanding"

odds ratio of 0.87 (95% Confidence Interval 0.78 to 0.99, $p = 0.03$). The distribution of valence scores for comments was heavily left skewed; in other words, the discussion comments were predominantly in the range from neutral to positive (see Fig. 10.7). Thus, we can describe these results by saying that students who provided a positive self-report, while demonstrating a more neutral presence in the discussion, were the most likely to provide a correct answer to the comprehension check (see Table 10.4).

When comparing the logistic regression to the null model, all three tests—likelihood ratio test, score test, and Wald test—show that the three-predictor model is a better fit than the null model. The Hosmer–Lemeshow test, however, yielded $x\,2$ (8) of 57.298 and was statistically significant ($p < .05$), suggesting that the model is not a good fit to the data. While there are many options to explore this result further, one possible explanation is that there is room to improve the quality of the sentiment

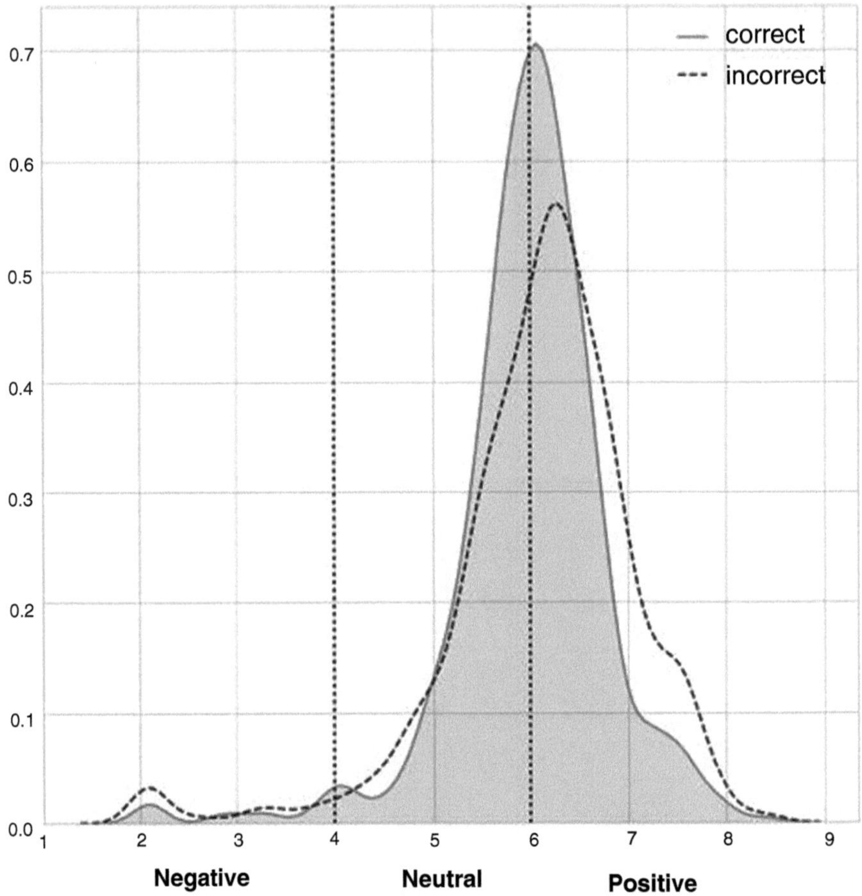

Fig. 10.9 Kernel density plot comparing "Discuss it" valence scores for students getting (in)correct answers on "Boost your understanding"

analysis detection in comments (as was also indicated by the human raters). As the results indicate that a positive self-report in conjunction with neutral comments predicts the greatest likelihood of a correct answer, binning the data in these valence categories may improve the model fit. Binning may be the most logical next step as this would directly address the spikes in the distribution found when examining both dimensions independently. However, given this investigation was exploratory in nature, the results are presented with knowledge that the model has room for improvement.

While the model in Table 10.4 indicates a 13% decrease in the probability for every point that the valence score increases, when comparing that with Fig. 10.9, the relationship appears to only exist for valence scores that range from neutral to positive. This evidence suggests that a binning strategy is likely the next best step to improve the model fit in terms of overall accuracy.

Table 10.4 Logistic regression of a three predictor model for reading comprehension

Predictor	B	SE	Wald's X 2	df	p	OR	95% CI
RAPID probability	0.0303	0.002	169.9	1	0.000***	1.03	[1.03, 1.04]
Self-report valence	0.1315	0.047	7.7	1	0.005**	1.14	[1.05, 1.26]
Comment valence	−0.1349	0.062	4.7	1	0.03*	0.87	[0.79, 0.99]
(Constant)	−1.6587	0.446	13.8	1	0.000***	–	
Test			X 2	df	p		
Overall model evaluation							
Likelihood ratio test			425.99	4	0.00***		
Score test			182.65	4	0.00***		
Wald test			341.5	4	0.00***		
Goodness-of-fit test							
Hosmer–Lemeshow			24.188	8	0.002**		

Note: * = $p < 0.05$; ** = $p < 0.01$; *** = $p < 0.001$

These results support both the relationship of positive emotions supporting assimilation and the relationship of neutral emotions supporting performance. This combination of feeling positive but behaving neutrally has been described as "smiling inside" (Schall et al., 2016). What is interesting about the potential parallel between the Schall et al. study (2016) and the results of this chapter is that the "smiling on the inside" phenomenon described high school students outperforming peers. The intent of the "smiling on the inside" study was to determine the extent and consequences of sup- pressing outward expression of positive emotions when having academic success.

In our model, the emotional state of "smiling inside" predicts success in the comprehension check: 169 true positives (TP), and 122 false positives (FP), 379 False negatives (FN), and 1021 true negatives (TN) with an overall accuracy of 70.37% (see Table 10.5).

We next look to see how the model works across the protected groups based on demographic information. We used Aequitas toolkit to compute six model accuracy metrics as well as disparity in terms of ratios compared to the majority reference group defined as the group across attributes with the most records in analysis. Our majority reference group was: Male, Hispanic, IEP, Non-ELL, Free and Reduce Price Lunch. The results of the bias analysis are reported in Table 10.6.

To interpret the bias results, we focus on FOR $\left(\frac{FN}{FN+TN}\right)$ and FDR $\left(\frac{FP}{TP+FP}\right)$ to examine bias focused on false negatives and false positives, incorporating true positives and true negatives as components of the denominators. For these statistics, when the parity scores were above 1, they were considered a negative bias meaning worse performance compared to the reference group. When the scores were below 1, they were considered a positive bias indicating better performance compared to the reference group. Significance of the difference was tested using the Aequitas package with an alpha of 0.05.

In terms of gender, there is a negative bias for females in terms of FOR indicating there are more false negatives in relation to true negatives for female students. There was no significant difference between males and females for FDR.

In terms of race, we saw a negative bias in FOR for White, Black or African American, American Indian or Alaskan Native, and students who identify with two or more races. The worst case for FOR is for the American Indian or Alaskan Native student (n = 1) with 1.0 FOR because there were no true negative records for this student. In contrast, Asian or Pacific Islander had a positive bias in FOR with a score of 0 because there were no false negative records for this population (n = 3).

For differences in FDR for race, we saw a positive bias for White, Black or African American, American Indian or Alaskan Natives, and those that identify with two or more races. The most pronounced difference was for the American Indian or Alaskan Native student (n = 1) with an FDR score of 0.0 because again there were no false positives for this student. While there was no statistical difference for Asian or Pacific Islander students for the FDR score of 1.0, it is noteworthy as the score indicates that there were no true positive records for this population (n = 3).

When considering IEP status, we see a negative bias for FOR and positive bias for FDR for students without an IEP. In contrast, students with ELL status have a positive bias for FOR and negative bias for FDR. It is worth noting that the measures involved are all language-based ("Whats your reaction?", and "Discuss it") which may be part of the underlying reason for the bias of predictions for ELL students. Students that are not on Free or Reduced Price Lunch have only a negative bias for FOR.

The biases illustrated in the model are apparent across all groups analyzed because all categories had a significant difference compared with the reference group. This would suggest an unequal impact if action were taken based on model predictions. Unequal does not necessarily mean inequitable as further analysis is required to determine if the model fits better for students that require more help with reading comprehension. It is likely that trade-off decisions would need to be considered, but analysis in Table 10.6 provides a description of disparities that merit exploration when navigating trade-offs when taking action based on the model predictions.

Table 10.5 Observed and predicted correct responses on "Boost your understanding"

		Predicted		
		Yes	No	% Correct
Observed	Yes	169 True Positive (TP)	379 False Negative (FN)	30.84
	No	122 False Positive (FP)	1021 True Negative (TN)	89.33
Total				70.37

Table 10.6 Bias analysis based on demographic information

Demographics		TPR (parity)	TNR (parity)	FPR (parity)	FNR (parity)	FOR (parity)	FDR (parity)
Gender	Male	0.62 1.00	0.72 1.00	0.28 1.00	0.38 1.00	0.09 1.00	0.72 1.00
	Female	0.54 0.87	0.74 1.04	0.29 0.9	0.46 1.21	0.14 1.63*	0.64 0.89
Race	Hispanic or Latino	0.33 1.00	0.85 1.00	0.15 1.00	0.67 1.00	0.04 1.00	0.90 1.00
	White	0.70 2.11*	0.61 0.72*	0.39 2.62*	0.30 0.44*	0.10 2.70*	0.70 0.78*
	Black or African American	0.56 1.67*	0.63 0.74*	0.37 2.48*	0.44 0.67*	0.24 6.18*	0.60 0.67*
	Asian or Pacific Islander	-	0.83 0.98	0.17 1.12	-	0.00 0.00*	1.00 1.11
	Alaska Native or American Indian	0.45 1.46*	-	-	0.55 0.82	1.00 26.10*	0.00 0.00*
	Two or more races	0.53 1.60*	0.85 1.00	0.15 0.99	0.47 0.70*	0.13 3.45*	0.50 0.56*
IEP	IEP	0.53 1.00	0.80 1.00	0.20 1.00	0.47 1.00	0.05 1.00	0.80 1.00
	Non-IEP	0.60 1.13	0.53 0.65*	0.47 2.41*	0.40 0.85	0.29 5.46*	0.60 0.75*
ELL	ELL	0.59 1.00	0.68 1.00	0.32 1.00	0.41 1.00	0.15 1.00	0.66 1.00
	Non-ELL	0.38 0.66*	0.86 1.27*	0.14 0.44*	0.62 1.49	0.02 0.16*	0.92 1.39*
SES	Free/Reduced-Price Lunch	0.56 1.00	0.75 1.00	0.25 1.00	0.44 1.00	0.10 1.00	0.70 1.00
	Non-Free/Reduced-Price Lunch	0.63 1.12	0.62 0.82	0.38 1.54*	0.38 0.85	0.16 1.60*	0.66 0.94

Negative Bias | Reference Group | Positive Bias

True Positive Rate (TPR) $\frac{TP}{TP+FN}$ True Negative Rate (TNR) $\frac{TN}{TN+FP}$

False Positive Rate (FPR) $\frac{FP}{FP+TN}$ False Negative Rate (FNR) $\frac{FN}{TP+FN}$

False Omission Rate (FOR) $\frac{FN}{FN+TN}$ False Discovery Rate (FDR) $\frac{FP}{TP+FP}$

* = Significant Difference in Parity

10.10 Discussion

Based on this analysis, there is clear evidence that mapping discrete words from self-reports onto the dimension of valence can be an effective strategy; with the caveat that common responses generate a distribution with many local maxima for common values. The distribution that comes from such an interpretation is not smooth and is likely subject to multimodal distributions, which is further suggested by the wide tails in the theoretical distribution. There are likely to be local maxima in the categories of positive, negative, and neutral valence in general; although we only had local maxima in neutral and positive for this study. This result supports the criticism that it is not easy to translate emotional measures that report discrete emotions in dimensional terms (Calvo & Kim, 2010; Weidman et al., 2016), but not easy is not the same as not possible. The results from the logistic regression suggest that binning the valence values may be appropriate. We saw a distribution with similar features in the sentiment analysis of discussion comments, as there appear to be some common phrases used in the discussion that result in identical valence scores for many comments. This indicates that emotions felt versus emotions expressed may play different roles in the learning process. While these results look promising, there are two clear limitations.

The first limitation was that comment responses in discussion and self-report resulted in many local maxima in the distribution of valence. This appears to be a surmountable challenge; the next step in improving a predictive model for this study should be considering alternatives such as multinomial logistic regression.

The second limitation was the imbalance in the accuracy of valence prediction for discussion comments. The lowest accuracy was for negative detection, which deterred interpreting the model as it relates to these responses. We therefore focused our interpretation of the model on the valence categories of positive and neutral. It has been previously noted that negative detection is perhaps the most difficult category for sentiment analysis detection in education. One explanation given for this difficulty is that the context of education has different terms than the context where sentiment analysis technologies are generally created (Wen et al., 2014).

In spite of these limitations, we did see a correlation between these measures and the learning outcome, with both emotional measures, a significant relationship even with the inclusion of the RAPID score suggesting emotions may help to explain the variance in reading comprehension performance. Moreover, the direction of these effects fits with a psychological state described as "smiling on the inside" (Schall et al., 2016). While this study was exploratory in nature, the results indicate that it would be worth explicitly examining the "smiling on the inside" phenomenon for middle school students with a wide range of motivation for reading. Given that students with disabilities are thought to have less motivation and more negative affect associated with learning (Sideridis et al., 2006) and the "smiling on the inside" phenomenon may have a long-term effect of decreasing motivation for learning (Schall et al., 2016) it is important to confirm the findings that struggling readers are smiling on the inside when getting correct answers, and gain more insights into how best to support these students to acknowledge their success when they are outperforming their peers.

10.11 Implications for Practice

We know that measures designed to predict reading comprehension do not explain all variance in performance. We identified that emotions help to explain variance in terms of reading comprehension, specifically as students generating an emotional footprint in the data that is succinctly described as smiling on the inside were more likely to get correct answers. To consider how this might influence practice, there are two things to discuss: (1) implications for teachers and (2) implications for predictive model design.

For teachers, it is important to understand that this model suggests that when struggling readers get correct answers, their emotional state of smiling on the inside combines positive internal emotion with neutral emotional expression. This psychological state is considered socially desirable when outperforming peers. If smiling on the inside is how students choose to behave, teachers that want to lean into this behavior could seek to privately acknowledge student success and avoid disrupting what students might consider a socially desirable behavior. Teachers that see smiling on the inside as a maladaptive practice could consider how they might be able to influence social norms within the classroom and allow public acknowledgement of success to be more socially desirable.

For those who work on algorithms that consider the emotional dimensions of struggling readers, the methods detailed in this study may provide guidance on designing systems that generate data capable of providing insights into emotions and learning. When considering how to take action on predictions, however, it would be advisable to work with teachers and adopt a human-in-the-loop approach towards action by observing what teachers do with these predictions and considering how systems might provide adaptive behaviors that parallel effective teaching practice. When considering how systems might become adaptive in response to emotional data, the bias analysis of the predictive model suggests that care must be taken: building a better understanding of how such an adaptive system is unequal would be required to build an equitable system.

10.12 Acknowledgements

The contents of this chapter were developed under a grant from the U.S. Department of Education (H327M11000). However, those contents do not necessarily represent the policy of the Department of Education, and you should not assume endorsement by the Federal government. We thank the many colleagues who contributed to this project: Alyssa Boucher, Linda Butler, Peggy Coyne, Kim Ducharme, Miriam

Evans, Steve Graham, Elysa Greenberger, Tracey Hall, Karen Harris, Ted Hasselbring, Rebecca Louick, Patrick Proctor, Kristin Robinson, David Rose, and Gabrielle Rappolt-Schlichtmann.

References

Baker, R. S., & Hawn, A. (2021). *Algorithmic bias in education*. https://doi.org/10.35542/OSF.IO/PBMVZ. https://edarxiv.org/pbmvz/

Boucher, A. R., Evans, M., & Graham, S. (2015). *Udio: Rich and authentic literacy experiences for struggling middle school readers*. Taylor & Francis, Routledge.

Brasseur-Hock, I. F., Hock, M. F., Kieffer, M. J., Biancarosa, G., & Deshler, D. D. (2011). Adolescent struggling readers in urban schools: Results of a latent class analysis. *Learning and Individual Differences, 21*(4), 438–452. https://doi.org/10.1016/j.lindif.2011.01.008

Calvo, R. A., & D'Mello, S. (2010). Affect detection: An interdisciplinary review of models, methods, and their applications. *IEEE Transactions on Affective Computing, 1*(1), 18–37. https://doi.org/10.1109/T-AFFC.2010.1

Calvo, R. A., & Kim, S. M. (2010). *Sentiment analysis in student experiences of learning* (pp. 111–120). Third International Conference on Educational Data Mining (EDM 2010).

CAST. (2018). Universal Design for Learning Guidelines version 2.2.

Clore, G. L., & Huntsinger, J. R. (2007). How emotions inform judgment and regulate thought. *Trends in Cognitive Sciences, 11*(9). https://doi.org/10.1016/j.tics.2007.08.005

Cutting, L. E., & Scarborough, H. S. (2009). Prediction of reading comprehension: Relative contributions of word recognition, language proficiency, and other cognitive skills can depend on how comprehension is measured. *Scientific Studies of Reading, 10*(3), 277–299. https://doi.org/10.1207/s1532799xssr1003_5. https://www.tandfonline.com/doi/abs/10.1207/s1532799xssr1003_5

Daley, S. G., Willett, J. B., & Fischer, K. W. (2014). Emotional responses during reading: Physiological responses predict real-time reading comprehension. *Journal of Educational Psychology, 106*(1), 132–143. https://doi.org/10.1037/a0033408

Daley, S. G. S., Hillaire, G., & Sutherland, L. L. M. (2016). Beyond performance data: Improving student help seeking by collecting and displaying influential data in an online middle-school science curriculum. *British Journal of Educational Technology, 47*(1), 121–134. https://doi.org/10.1111/bjet.12221

Duckworth, A. L., & Yeager, D. S. (2015). Measurement matters: Assessing personal qualities other than cognitive ability for educational purposes. *Educational Researcher, 44*(4), 237–251. https://doi.org/10.3102/0013189X15584327. http://edr.sagepub.com/cgi/content/abstract/44/4/237

Fiedler, K., & Beier, S. (2014). Affect and cognitive processes in educational contexts. In R. Pekrun & L. Linnenbrink-Garcia (Eds.), *International handbook of emotions in education, chap. 3* (pp. 36–55). Routledge.

Fraser, K., Ma, I., Teteris, E., Baxter, H., Wright, B., & Mclaughlin, K. (2012). Emotion, cognitive load and learning outcomes during simulation training. *Medical Education, 46*(11), 1055–1062. https://doi.org/10.1111/j.1365-2923.2012.04355.x

Fredricks, J. A., Blumenfeld, P. C., & Paris, A. H. (2004). School engagement: Potential of the concept, state of the evidence. *Review of Educational Research, 74*(1), 59–109. https://doi.org/10.3102/00346543074001059. http://rer.sagepub.com/cgi/doi/10.3102/00346543074001059

Fredrickson, B. L. (1998). What good are positive emotions? *Review of General Psychology, 2*(3), 300–319. https://doi.org/10.1037/1089-2680.2.3.300

García, O., Kleifgen, J. A., & Falchi, L. (n.d.). *A research initiative of the campaign for educational equity from English language learners to emergent bilinguals*. Tech. rep.

Herold, B. (2016). *Personalized learning based on students' emotions: Emerging research to know*. http://blogs.edweek.org/edweek/DigitalEducation/2016/01/personalized_learning_student_emotion_research.html

Hillaire, G., Rappolt-Schlichtmann, G., & Stahl, W. (2014). Learning analytics Summer Institute 2014 – Panel – Universal Design for Learning. https://youtu.be/Sl-6QATSw3E

Hillaire, G., Rappolt-Schlichtmann, G., & Ducharme, K. (2016). Prototyping visual learning analytics guided by an educational theory informed goal. *Journal of Learning Analytics, 3*(3), 115–142. https://doi.org/10.18608/jla.2016.33.7

Hillaire, G., Iniesto, F., & Rienties, B. (2019). Humanising text-to-speech through emotional expression in online courses. *Journal of Interactive Media in Education, 2019*(1). https://doi.org/10.5334/jime.519

Iniesto, F., Rodrigo, C., & Moreira Teixeira, A. (2014). *Accessibility analysis in MOOC platforms. A case study: UNED COMA and UAb iMOOC* (pp. 545–550). V Congreso Internacional sobre Calidad y Accesibilidad de la Formación Virtual (CAFVIR 2014) (Cafvir).

Kneepkens, E. W., & Zwaan, R. A. (1995). Emotions and literary text comprehension. *Poetics, 23*(1–2), 125–138. https://doi.org/10.1016/0304-422X(94)00021-W

Lexia Learning. (2016). Lexia RAPID Assessment: Technical Manual for RAPID Version 2.0: Grades 3–12.

Macfadyen, L. P., Lockyer, L., & Rienties, B. (2020). Learning design and learning analytics: Snapshot 2020. *Journal of Learning Analytics, 7*(3), 6–12. https://doi.org/10.18608/jla.2020.73.2

Medhat, W., Hassan, A., & Korashy, H. (2014). Sentiment analysis algorithms and applications: A survey. *Ain Shams Engineering Journal, 5*(4), 1093–1113. https://doi.org/10.1016/j.asej.2014.04.011

Meyer, A., Rose, D. H., & Gordon, D. (2014). *Universal design for learning theory and practice*. Cast Incorporated.

Okon-singer, H., Hendler, T., & Pessoa, L. (2015). The neurobiology of emotion-cognition interactions. https://doi.org/10.3389/978-2-88919-528-2. http://www.frontiersin.org/books/The_neurobiology_of_emotion-cognition_interactions/576.

Pang, B., & Lee, L. (2006). Opinion mining and sentiment analysis. *Foundations and Trends in Information Retrieval, 1*(2), 91–231. https://doi.org/10.1561/1500000001. http://www.nowpublishers.com/product.aspx?product=INR&doi=1500000001

Rienties, B., & Rivers, B. A. (2014). *Measuring and understanding learner emotions: Evidence and prospects* (pp. 1–16). Lace. http://www.laceproject.eu/publications/ learning-analytics-and-emotions.pdf.

Russell, J. A. (1980). A circumplex model of affect. *Journal of Personality and Social Psychology, 39*(6), 1161–1178.

Sacharin, V., Schlegel, K., & Scherer, K. R. (2012). Geneva emotion wheel rating study (report) p. 13. https://doi.org/10.13140/RG.2.1.2170.3527

Schall, M., Martiny, S. E., Goetz, T., & Hall, N. C. (2016). Smiling on the inside: The social benefits of suppressing positive emotions in outperformance situations. https://doi.org/10.1177/0146167216637843

Sharples, M., Adams, A., Alozie, N., Ferguson, R., Fitz Gerald, E., Gaved, M., McAndrew, P., Means, B., Remold, J., Rienties, B., Roschelle, J., Vogt, K., Whitelock, D., & Yarnall, L. (2015). *Innovating pedagogy 2015: Open University innovation report 4*. Tech. Rep., The Open University.

Sideridis, G. D., Mouzaki, A., Simos, P., & Protopapas, A. (2006). Classification of students with Reading comprehension difficulties: The roles of motivation, affect, and psychopathology. *Learning Disability Quarterly, 29*(3), 159–180. https://doi.org/10.2307/30035505

Warriner, A. B., Kuperman, V., & Brysbaert, M. (2013). Norms of valence, arousal, and dominance for 13,915 English lemmas. *Behavior Research Methods, 45*(4), 1191–1207. https://doi.org/10.3758/s13428-012-0314-x. http://www.ncbi.nlm.nih.gov/pubmed/23404613

Watson, D., Clark, L. A., & Tellegen, A. (1988). Development and validation of brief measures of positive and negative affect: the PANAS scales. *Journal of Personality and Social Psychology, 54*(6), 1063–1070. https://doi.org/10.1037/0022-3514.54.6.1063. http://www.ncbi.nlm.nih.gov/pubmed/3397865

Weidman, A. C., Steckler, C. M., & Tracy, J. L. (2016). The jingle and jangle of emotion assessment: Imprecise measurement, casual scale usage, and conceptual fuzziness in emotion research. *Emotion, 17*(2), 267–295. https://doi.org/10.1037/emo0000226. http://doi.apa.org/getdoi.cfm?doi=10.1037/emo0000226

Wen, M., Yang, D., & Rosé, C.P. (2014). *Linguistic reflections of student engagement in massive open online courses* (pp. 525–534). Proceedings of the 8th International Conference on Weblogs and Social Media ICWSM 2014 525534. https://doi.org/10.1016/j.bspc.2014.01.007. http://www.scopus.com/inward/record.url?eid=2-s2.0-84909951147&partnerID=40&md5=80f121bcfc587505feaee3a3d6675c59; http://www.researchgate.net/profile/Miaomiao_Wen2/publication/275649171_Linguistic_Reflections_of_Student_Engagement_in_Massive_Open_Online_Courses/links/554244e70cf234bdb219b2a1.pdf%5Cn; http://www.scopus.com/inward/record.url?eid=2-s2.0-84909951147&partne

Chapter 11
Exploring Selective College Attendance and Middle School Cognitive and Non-cognitive Factors Within Computer-Based Math Learning

Maria Ofelia Z. San Pedro, Ryan S. Baker, Alex J. Bowers, and Neil T. Heffernan

Abstract Middle school is a key juncture in the processes that influence whether a student will have a successful post-secondary outcome such as going to a selective college, but research on factors leading to this choice does not yet utilize the extensive fine-grained data now becoming available on middle school learning and engagement. Leveraging recent methodological advances in measurement and educational data mining, we apply automated detectors which can infer student learning, academic emotions, and engagement, to data from middle school mathematics software usage. We then use the measures derived to predict which students will go to selective colleges several years later. The result is a model that can distinguish whether a student will eventually go to either a selective or a non-selective college 77.4% of the time. The resulting model can also run in real-time, creating the potential for providing actionable data quickly to teachers and guidance counselors.

Keywords Selective college · Post-secondary institution · Engagement · Affect · Academic emotions · Intelligent tutoring system

M. O. Z. San Pedro (✉)
Applied Research, ACT, Inc., Iowa City, IA, USA
e-mail: sweet.san.pedro@act.org

R. S. Baker
Teaching, Learning, and Leadership Division, University of Pennsylvania, Philadelphia, PA, USA

A. J. Bowers
Teachers College, Columbia University, New York, NY, USA
e-mail: Bowers@tc.edu

N. T. Heffernan
Worcester Polytechnic Institute, Worcester, MA, USA
e-mail: nth@wpi.edu

Y. E. Wang et al. (eds.), *Social and Emotional Learning and Complex Skills Assessment*, Advances in Analytics for Learning and Teaching,
https://doi.org/10.1007/978-3-031-06333-6_11

Attending a more selective post-secondary institution has been shown to be associated with higher-quality learning, higher likelihood to graduate, and improved career prospects and economic gains (Bowen et al., 2009; Carnevale & Rose, 2003; Ovink et al., 2018; Shamsuddin, 2016; Thomas, 2000). Schools that are more selective tend to have higher access to financial resources, more faculty attention that can increase a student's success in college, more career counseling, better access to internships, and better preparation for application to graduate schools (Carnevale & Rose, 2003; Hoxby, 2009). However, access to selective colleges is skewed by race, ethnicity and socioeconomic status. Overall, white and Asian students are found to be more likely to enroll in four-year colleges, especially in highly selective colleges (Goldrick-Rab, 2007; Reardon et al., 2012), while African- American and Hispanic students are less represented in highly selective colleges (Carnevale & Rose, 2003; Reardon et al., 2012). Students from low SES families usually lack the economic resources necessary to pursue postsecondary education (Ellwood & Kane, 2000; Karabel & Astin, 1975; Zhou & Bowers, 2020). While demographics appear to be a significant part of the gaps in access to a selective college, they do not illuminate all the possible reasons why students fail to attend college, let alone a selective college. In particular, some students may not attend a selective college due to experiences that occur much early on in their lives. Many students effectively drop out of the pipeline towards academic success well before reaching college (Balfanz, 2009; Balfanz et al., 2007; Bowers, 2010; Bowers & Sprott, 2012a, b; Bowers & Zhou, 2019; Neild, 2009). Such change occurs both in terms of decreasing motivation (Anderman & Maehr, 1994) or greater degree of academic failure that can begin to manifest in middle school (NMSA, 2002; Neild, 2009). This often results in extreme forms of disengaged behavior such as non-attendance and classroom misconduct (Tobin & Sugai, 1999; Tobin et al., 1996).

11.1 Social Cognitive Career Theory and Pathway to College

Due to the possibility of this kind of early school disengagement, school counselors are encouraged to support students in developing the cognitive and non-cognitive skills necessary to being college-ready (Conley, 2008; Conley et al., 2009), and help them transition to postsecondary education (Gibbons et al., 2006). If students who are at risk could be spotted early, better-targeted interventions could be developed for these students (Bowers, 2010, 2021). Several of these potential actionable factors are seen in Social Cognitive Career Theory (SCCT, Lent et al., 1994, 2000). According to SCCT, higher levels of interest in an activity emerge within contexts where the individual has higher self-efficacy and outcome expectations, leading to the development of intentions or goals for further exposure and engagement with that activity (Lent et al., 1994) (Fig. 11.1).

Recent SCCT research has focused on high school or college students, and relatively few studies have analyzed hypotheses related to SCCT in middle school students (but see Fouad & Smith, 1996; Gibbons & Borders, 2010; Turner & Lapan,

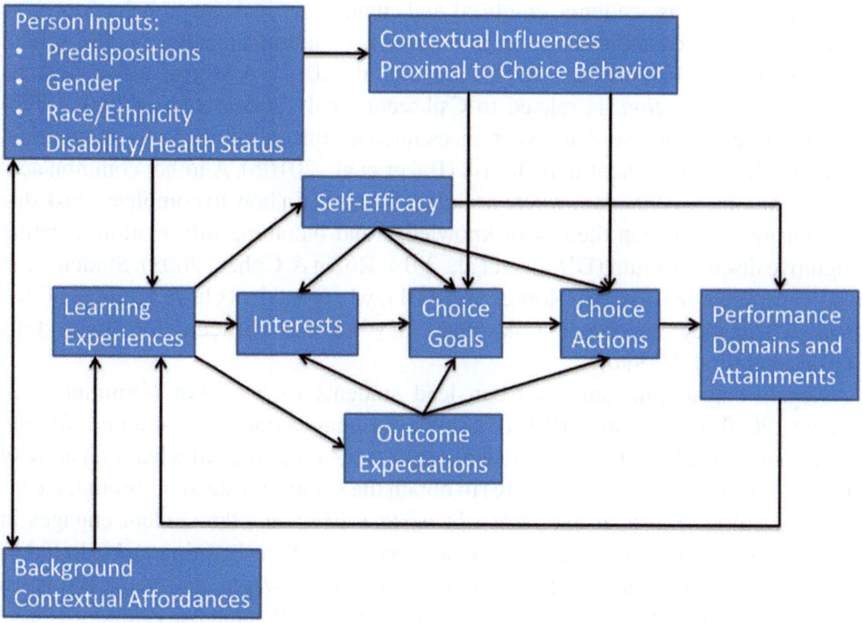

Fig. 11.1 Social cognitive career theory

2002). However, it is in middle school where students start to develop their abilities and interest in pursuing their studies and advanced careers (Cabrera et al., 2001; Camblin, 2003). During middle school, students begin to develop academic abilities, interests, and choices that will have a strong influence on later academic outcomes (Cupani & Pautassi, 2013), and become engaged or disengaged from school and learning, driven in part by changes in self-perception such as whether they see themselves as intelligent and capable of succeeding academically (Camblin, 2003; NMSA, 2002).

Hence, there have been increasing recommendations that college planning begin as early as sixth grade (Allensworth et al., 2014). Students who start thinking about college as early as middle school tend to become interested in achieving a good academic record. They may plan to take appropriate courses once they are in high school or choose to be involved in extracurricular activities that will contribute to their college applications (Roderick et al., 2008, 2011).

11.2 Cognitive and Non-cognitive Factors in Academic Settings

Understanding students' long-term outcomes such as selective college attendance necessitates looking beyond their academic performance and individual abilities, towards "non-cognitive factors" (Farrington et al., 2012) in their learning experiences such as academic emotions and engaged or disengaged behaviors. One example of an academic emotion is *boredom*, common in many middle school classrooms (e.g. Rowe et al., 2009; Pardos et al., 2013). A second affective state, *engaged concentration*, is related to Csikszentmihalyi's construct of flow (1990) and describes when a student experiences intense concentration, focused attention, and complete involvement in their task (Baker et al., 2010b). Another common academic emotion is *confusion*, where a student is uncertain how to complete a task due to a mismatch between their prior knowledge and incoming information, creating cognitive disequilibrium (D'Mello et al., 2014; Rozin & Cohen, 2003). Students can also experience *frustration* (Kort et al., 2001), where students have feelings of distress when they encounter tasks that may be too difficult for their skills (Csikszentmihalyi, 1990).

Negative academic emotions can lead students to zone out (Drummond & Litman, 2010; Feng et al., 2013) or exhibit disengagement in classrooms. *Gaming the system* is a behavior when a student exploits the properties of a learning activity (i.e., within an educational software) to obtain the solution instead of through meaningful learning (Baker et al., 2004). In *off-task behavior*, the student engages in extraneous activities and completely disengages from their learning tasks. In learning activities, students also exhibit *careless behavior* when they make errors on questions despite knowing how to successfully answer (Clements, 1982).

These disengaged behaviors, together with boredom, have been found to be associated with poorer learning, lower self-efficacy (Narciss, 2004; Schunk, 1989), diminished interest in educational activities, negative attitudes toward math content (Baker, 2007; Baker et al., 2008), poorer performance on standardized examinations (Pardos et al., 2013), and, most importantly, increased attrition and dropout rates (Craig et al., 2004; Daniels et al., 2009; Goodman, 1990; Mann & Robinson, 2009; Pekrun et al., 2010). By contrast, students who are more engaged in school tend to have higher academic motivation and achievement (Fredricks et al., 2004; Pardos et al., 2013). Academic emotions and disengaged behaviors are also associated with college enrollment (San Pedro et al., 2013); students who frequently experience engaged concentration in middle school mathematics are more likely to go to college, while students who frequently experience confusion and boredom or who game the system are less likely to go to college (San Pedro et al., 2013). Hence, engagement and academic emotions in middle school learning appear to play an essential early role in students' educational experiences.

11.3 Educational Technology in Assessing Cognitive and Non-cognitive Factors

Researchers in recent years have used educational technologies to study academic emotions and engagement, both in laboratory settings and in actual classrooms, in fine-grained detail. Educational data mining (EDM; Baker & Yacef, 2009) researchers have developed automated models (using a combination of interaction data and classroom observations of students) that can infer students' academic emotions, engagement, and knowledge in real time, and have found evidence that the constructs these models infer are associated with differences in student outcomes. These recent advances have progressed in large measure due to the expansion of computer-based learning environments usage in schools, providing a rich source of data that helps us understand students' learning processes (Canfield, 2001; Heffernan & Heffernan, 2014; Koedinger & Corbett, 2006). .

Assessments or measures derived from these models are different from the questionnaire responses and coarse-grained variables (such as demographic information or test scores) typically used in research on educational outcomes. Assessments developed using EDM predict educational outcomes such as learning gains (Baker et al., 2004; Cocea et al., 2009; Sabourin et al., 2011) and standardized exams (Pardos et al., 2013), and have been widely used within online learning environments that produce rich student interaction data, such as intelligent tutoring systems (Baker et al., 2010b; Pardos et al., 2013; Walonoski & Heffernan, 2006) and educational games (Shute et al., 2015; Bosch et al., 2015)..

Despite these advances, there has been limited research on whether these fine--grained measures can predict long-term student outcomes – in particular, attending a selective college. In this paper, we evaluate and predict whether a student will attend a selective college or not, five to six years later, based on their interaction with an educational software system, the ASSISTments system, during middle school. We assess key aspects of student emotion, engagement, and knowledge by leveraging existing machine-learned detectors of student affect, knowledge, and engaged/disengaged behaviors previously developed for the ASSISTments system. We investigate in particular, the following research questions:

1. How are middle school student knowledge, academic emotions, and disengaged behaviors associated with going to a selective college?
2. Are middle school student knowledge, academic emotions, and disengaged behaviors predictive of going to a selective college?

We conclude with a discussion of potential implications for the design and interventions of interactive educational systems for sustained attendance and engagement in school.

11.4 Methods

We investigate student knowledge, performance, affect and engagement through students' interaction with the ASSISTments system (Heffernan & Heffernan, 2014) when they were in their middle school years (7th or 8th grade). We conduct this research in a data set of 5472 students who used the ASSISTments system, between 2004 and 2008. Enrollment records in a post-secondary institution for the 5742 students were obtained in 2013 from the National Student Clearinghouse (http://

www.studentclearinghouse.org). For purposes of focusing on college selectivity, students not found to be enrolled in a post-secondary institution were excluded from our sample. Out of the 5742 students, 2810 students enrolled in a post-secondary institution and were considered in the current study. Also, for the purposes of the analyses in the present study, we only considered the last post-secondary institution the student enrolled in, using this as basis for assessing whether the student attended a selective college.

11.4.1 The ASSISTments System

The ASSISTments system (Fig. 11.2) (Heffernan & Heffernan, 2014) is a tutoring system for middle school mathematics provided by Worcester Polytechnic Institute (WPI) which serves as the data source for our independent variables. This free web--based educational system delivers mathematics problems and questions, assesses student performance, provides hints and suggestions, provides targeted feedback on common errors, and scaffolds the development of improved answers by breaking

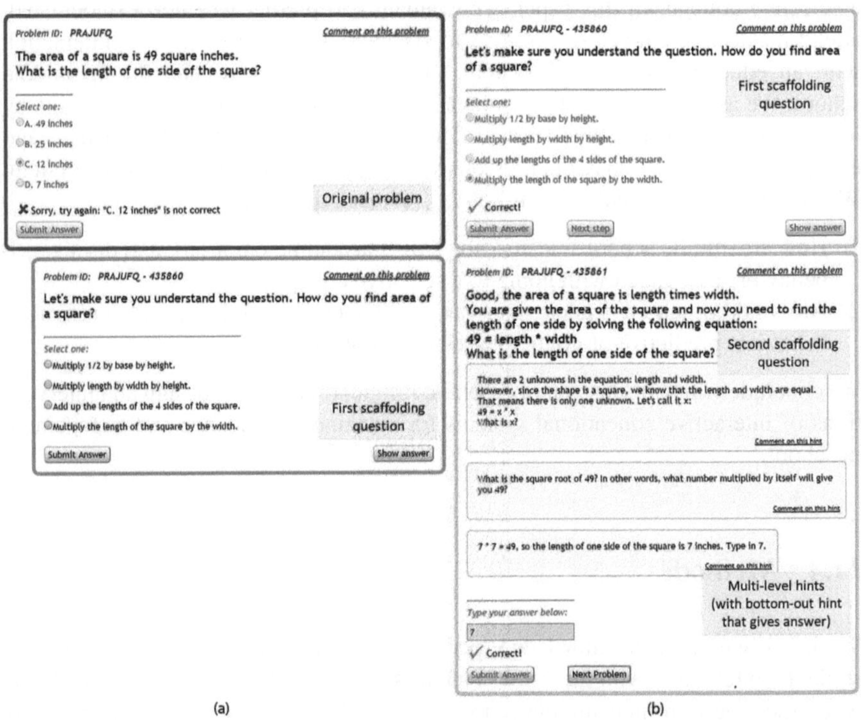

Fig. 11.2 Example of a problem in ASSISTments. (**a**) If a student gets a problem incorrect, hints and scaffolding problems are there to aid the student in eventually getting the correct answer. (**b**) Example of Scaffolding and Hints in ASSISTments

complex problems into simpler steps. When students working on an ASSISTments problem answer correctly, they proceed to the next problem. If they answer incorrectly, they are provided with scaffolding questions where the problem is broken down into its component steps in order to concretize the systematic thinking needed to solve the problem. The intention is to identify which part of the student's thinking is incorrect. This information about the student's problem solving is then provided to teachers as detailed reports and summaries for assessment and diagnostic purposes.

Interaction log data from the ASSISTments system were obtained for the sample population of 2810 students from middle schools in the Northeastern United States. The students used the system at various times starting from school years 2004–2005 to 2007–2008 (with a few students continuing tutor usage until 2008–2009). These students were drawn from four districts that used the ASSISTments system at various times throughout the course of the year. Two districts were urban with large proportions of students requiring free or reduced-price lunches due to poverty, relatively low scores on state standardized examinations, and large proportions of students learning English as a second language. The other two districts were suburban, serving generally middle-class populations, with relatively higher scores on state standardized examinations. In general, students in our sample used ASSISTments three to four times a month in classes held in their school's computer lab. Students were guided and instructed by teachers trained in formative assessment. These teachers used ASSISTments in their math curricula for review of concepts and test preparation. Overall, the students in the sample made 2,024,893 actions within the software (where an action consisted of making an answer or requesting help), within 1,021,272 mathematics problems (counting both original and scaffolding problems).

11.4.2 Dependent Variable: College Selectivity

College selectivity measures are generally determined by an aggregate index computed across several factors, including: the median SAT or median composite ACT entrance exam score; the average high school class rank of the student; the average student GPA in high school; and the percentage of students accepted (Carnevale & Rose, 2003). Each of the 270 post-secondary institutions attended by our sample of students was classified in terms of selectivity. The most commonly-used measure of college selectivity (c.f., Carnevale & Rose, 2003; Griffith & Rothstein, 2009; Schmidt et al., 2011) is the annual Barron's index (College Division of Barron's Education Series, 2012), which classifies colleges into ten categories from most selective or 'Most Competitive' to 'Noncompetitive' and 'Special', which consists of specialty institutions such as schools of music, culinary schools, automotive training schools, and art schools.

Of the 2810 students, 32 students attended an institution with a 'Special' classification and 46 students attended an institution unclassified in Barron's. We excluded these students from our sample, leaving us with data from 2732 students with 9 selectivity classifications to use for our analyses.

Barron's index makes fine distinctions between degrees of selectivity, as shown in Table 11.1. In this paper, we analyze enrollment in either a selective college or a non-selective college, in a binary fashion, rather than attempting to treat this scale as numerical.

As seen in Table 11.1, our sample (like the national population of students) is skewed towards the 'Non-Competitive' end of the scale; our sample also has relatively few students attending universities in 'Very Competitive+' and 'Competitive+' classifications. Simplifying our DV can make it more evenly distributed and reflect more meaningful and practical distinctions between a selective school and a not selective school. We examined four different ways to split into selective/non--selective (see Table 11.2): 4+. vs. 3−, 6+. vs. 5−, 8+. vs. 7−, 10+. vs. 9−. We used these binary splits to label post-secondary institutions as selective or non-selective and used the resultant variable as the predicted variable in the analysis below. Figure 11.3 shows the number of students in each binary split for each cut-off. For the 4/3 cut-off, there are more students who went to a selective college (n = 1540 students) than not (n = 1192 students). For the 6/5 cut-off, 690 students went to a

Table 11.1 Barron's college selectivity rating

Selectivity rating	Selectivity description	Required GPA	Required SAT	Example institution(s)	Number of students	Number of institutions in sample
10	Most competitive	B or higher	1240 or higher	Columbia, Harvard, Stanford	122	31
9	Highly competitive+	B or higher	1240 or higher	Cornell University	109	15
8	Highly competitive	B or higher	1240 or higher	Fordham University	108	23
7	Very competitive+	B- or higher	1146 to 1238	Yeshiva University	25	11
6	Very competitive	B- or higher	1146 to 1238	Hunter College	326	29
5	Competitive+	C or higher	1000 or higher	Buffalo State College	30	7
4	Competitive	C or higher	1000 or higher	St. Joseph's College	820	73
3	Less competitive	C or below C	Below 1000	Berkeley College	72	15
2	Non--competitive	C or below C	Below 1000	College of Staten Island	1120	66
1	Special			Julliard School	32 (excluded)	8
	(Unclassified)			Glendale Community College	46 (excluded)	42

Table 11.2 Cut-offs for classes of 'selective' and 'not selective' from Barron's selectivity rating

Selectivity rating	Selectivity description	Cut-off 1 (I)	Cut-off 2 (II)	Cut-off 3 (III)	Cut-off 4 (IV)
10	Most competitive	Selective	Selective	Selective	Selective
9	Highly competitive+	Selective	Selective	Selective	Not selective
8	Highly competitive	Selective	Selective	Selective	Not selective
7	Very competitive+	Selective	Selective	Not selective	Not selective
6	Very competitive	Selective	Selective	Not selective	Not selective
5	Competitive+	Selective	Not selective	Not selective	Not selective
4	Competitive	Selective	Not selective	Not selective	Not selective
3	Less competitive	Not selective	Not selective	Not selective	Not selective
2	Non-competitive	Not selective	Not selective	Not selective	Not selective

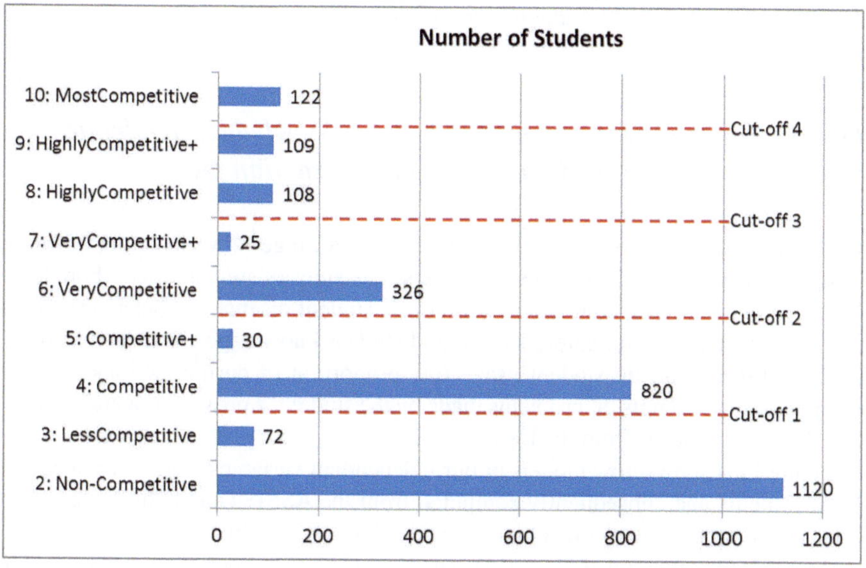

Fig. 11.3 Number of students in 'selective' and 'not selective' class for each cut-off

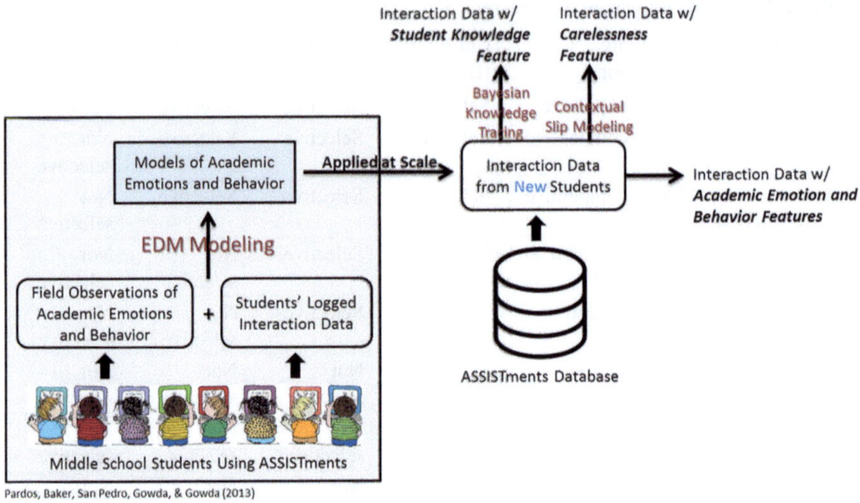

Fig. 11.4 Feature generation in ASSISTments interaction data

selective college compare to the 2042 students went to a not selective college. For the 8/7 cut-off, there were only 339 students who went to a selective college, and in the 10/9 cut-off, only 122 students went to a selective college.

11.4.3 Independent Variables: Student Knowledge, Academic Emotions and Behavior from Interaction Data

We predict and analyze college selectivity using a range of variables or features computed from the log files of ASSISTments. Measures of student affect (boredom, engaged concentration, confusion, frustration), student disengaged behaviors (off--task, gaming the system, carelessness), and student knowledge were derived from models. Information on student usage (the proportion of correct actions and the number of first attempts on problems made by the student, a proxy for overall usage) was directly extracted from the logs.

Figure 11.4 shows how models of our independent variables were developed for ASSISTments and subsequently computed from the ASSISTments interaction log data. The models of academic emotions and behavior were first reported in (Pardos et al., 2013; Ocumpaugh et al., 2014). These models were applied to every student action within the system, producing a sequence of predictions of the students' knowledge, academic emotions and behavior across the history of each student's use of ASSISTments. These could then be aggregated into a set of single overall assessments for each student.

Once the models of academic emotions, behavior and knowledge are applied to the dataset of our sample students, producing values for these independent variables, they were then used for our final model of college selectivity. This process is sometimes referred to as "discovery with models" (e.g. Baker & Yacef, 2009) where existing models are used as a component in a new and different analysis or model.

11.4.4 Modeling Student Knowledge

Student knowledge was derived from tutor usage in ASSISTments by applying Corbett and Anderson's (Corbett & Anderson, 1995) Bayesian Knowledge Tracing (BKT) model to the data (Fig. 11.5). BKT is a knowledge-estimation model which is used in many online learning systems. BKT and infers students' latent knowledge from their performance on problems. In the case of student interaction with ASSISTments, student knowledge is assessed from each student's first attempt to answer each problem. Each time a student attempts a problem or problem step for the first time, BKT calculates (and recalculates on next problem) the estimates of that student's knowledge for the skill involved in that problem or problem step. Knowledge estimations for each skill are made using four parameters: (1) L_0, the initial probability that the student knows the skill, (2) T, the probability of learning the skill at each opportunity to use that skill, (3) G, the probability that the student will give the correct answer despite not knowing the skill, and (S) the probability that the student will give an incorrect answer despite knowing the skill. Brute-force grid search was used to fit the model to the data (see Baker et al., 2010a).

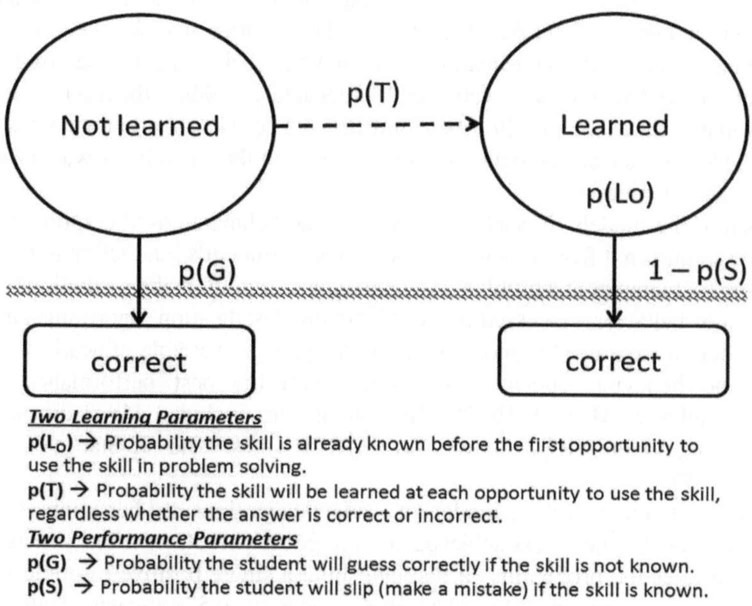

Fig. 11.5 Bayesian knowledge tracing

11.4.5 Modeling Academic Emotions and Disengaged Behavior

The academic emotions modeled within ASSISTments consist of boredom, confusion, frustration, and engaged concentration. Disengaged behaviors modeled include gaming the system, off-task behavior, and carelessness. With our student sample belonging to urban and suburban districts, two sets of detectors were used: models optimized for students in urban schools were used to label data from students who attended urban schools (Pardos et al., 2013), and models optimized for students in suburban schools were used to label data from students who attended suburban schools (Ocumpaugh et al., 2014). This choice is based on evidence that urban and suburban students manifest their emotions differently in online learning (Ocumpaugh et al., 2014). The same detectors were used for gaming the system and for off-task behavior, across contexts, as these constructs manifest more consistently across populations.

These detectors were initially developed (in Pardos et al., 2013; Ocumpaugh et al., 2014) using a three-stage process: first, field observers noted down student engagement and academic emotions while students used ASSISTments using the BROMP protocol for quantitative field observation of emotion and engagement (Baker et al., 2020) and the HART field observation app for Android (Ocumpaugh et al., 2015); second, those field notes were synchronized with the log files generated by student interaction with ASSISTments at a precision of around 1–2 second error, using an internet time server; and third, data mining was used to create models that could predict the field observations (i.e. student academic emotions and engagement) from the log files. This process resulted in automated detectors of academic emotions and engagement that can be applied to log files at scale, specifically different log data from the same learning environment, such as the data set used in this project. These detectors were validated by repeatedly building them on a subset of the available data (4/5 of 229 urban students; 4/5 of 243 suburban students), and testing them on unseen students (the other 1/5), and their goodness was measured using standard metrics.

Each of the models of academic emotions and behaviors used combinations of features engineered from raw information about a student's interaction (e.g. action is a hint, first attempt at a problem is a help request, etc.) to make predictions of that emotion or behavior, discussed below. Common classification algorithms and feature selection were used in modeling each independent variable of academic emotions and behavior, choosing the model with the best performance (AUC ROC– Hanley & McNeil, 1982). These algorithms included J48 decision trees, logistic regression JRip, Naïve Bayes, REP-Trees, and K-Star (Witten & Frank, 2005)..

The effectiveness of these models of academic emotions and behaviors is shown in Table 11.3. The detectors achieved an average AUC ROC of 0.702, where AUC ROC indicates the probability of distinguishing a single positive example from a single negative example. An AUC ROC value of 0.5 indicates chance-level

Table 11.3 Model performances (AUC ROC) of urban and suburban detectors of academic emotions and behaviors

	Boredom	Confusion	Engaged concentration	Frustration	Off-- task	Gaming
Urban detector AUC ROC	0.632	0.736	0.678	0.743	0.819	0.802
Suburban detector AUC ROC	0.666	0.744	0.631	0.589	0.819	0.802

performance, and 1.0 indicates the model performs perfectly. For example, the gaming detector had an AUC ROC of 0.802; as such, it could distinguish a gaming student from a non-gaming student 80.2% of the time.

Compared to gaming the system and off-task behavior and academic emotions, assessment of the disengaged behavior carelessness was generated differently. Instead of using models trained from field observations, the instance of carelessness was assessed with a model that infers whether a student error for each student action are due to not knowing the skill or due to being careless (i.e., careless errors or "slips", answering incorrectly despite actually knowing how to answer it correctly) (Baker et al., 2008; San Pedro et al., 2011).

Modeling carelessness or slip in the context of educational software is derived from BKT where we use the "contextual slip" model from (Baker et al., 2008; San Pedro et al., 2011) in operationalizing carelessness. To model carelessness, we apply BKT to our data to generate initial estimations of whether the student knew the skill at each problem step. Bayesian equations are then used with these estimations to compute the probability that incorrect actions were slips, based on the correctness or student performance on succeeding attempts to use the skill (Baker et al., 2008; San Pedro et al., 2011). These probability values are then used to create a model that can predict slip or carelessness contextually at each practice opportunity, from data such as response time, past history, and the pattern and type of errors, without any future information.

11.4.6 Modeling College Selectivity

We applied the detectors to measure student knowledge, academic emotions (boredom, confusion, engaged concentration, frustration), behavior (off-task behavior, gaming the system, carelessness), as well as obtaining measures of overall student correctness (a proxy for short-term academic success), and the number of actions made by the student, a proxy for overall usage (see Table 11.4). We then fit a logistic regression model predicting whether a student in the data set attended a selective or non-selective college, using the student average for each of the predictor values across the year (i.e., average boredom per student).

We used logistic regression analysis since we have (a set of) dichotomous outcomes resulting in a non-linear relationship between our predictors and outcome

Table 11.4 Predictors used in logistic regression model

	Mean	Standard deviation	Minimum	Maximum
Boredom	0.224	0.071	0.023	0.466
Engaged concentration	0.642	0.064	0.341	0.937
Confusion	0.082	0.050	0.000	0.371
Frustration	0.144	0.086	0.000	0.514
Off-task	0.216	0.080	0.065	0.837
Gaming	0.132	0.137	0.004	0.777
Knowledge	0.347	0.213	0.035	0.940
Carelessness	0.206	0.135	0.010	0.799
Correctness	0.459	0.150	0.000	0.946
Number of actions	722.52	822.38	2	14,378

variable. Choosing logistic regression allows for relatively good interpretability, while matching the statistical approach used in much of the other work predicting enrollment and success in higher education and educational pathways (Cabrera, 1994; Eccles et al., 2004; Núñez & Bowers, 2011; Stephan & Rosenbaum, 2013; San Pedro et al., 2013).

The final model created for each cut-off was cross-validated at the student level (six-fold), e.g. the models were repeatedly trained on 5/6 of the students and then tested on the remaining 1/6 of the students. This procedure estimates how well the models can be expected to perform when applied to entirely new students. The model's quality was assessed using two metrics, AUC ROC (described above) and Cohen's Kappa (see Table 11.8). Cohen's Kappa assesses the degree to which a model is better than chance at predicting a particular category (Cohen, 1960), and is a common metric for assessing categorical predictions. Cox and Snell (1989) and the Nagelkerke's (1991) pseudo-R^2 are also used to evaluate how useful the explanatory variables are in predicting the response, quantifying the amount of variance explained by the models, following recommendations in (Bowers & Lee, 2013).

All predictor variables were standardized (using z-scores), in order to increase interpretability of the resulting odds ratios and to show a clear indication of each predictor's contribution to the class variable (college is selective).

11.5 Results

11.5.1 Correlational Analyses

Before developing our college selectivity model, we looked at our original, non--standardized predictors or independent variables and examined their relationships with each other. From Table 11.5, the strongest positive associations were found between student knowledge and carelessness ($r = 0.956$, $p < 0.001$), student knowledge and correctness ($r = 0.807$, $p < 0.001$), and confusion and boredom ($r = 0.710$,

Table 11.5 Correlations of independent variables

	Boredom	Engaged Concentration	Confusion	Frustration	Off-task	Gaming	Knowledge	Carelessness	Correctness	Number of Actions
Boredom	1									
Engaged concentration	−0.347**	1								
Confusion	0.710**	−0.070**	1							
Frustration	0.405**	0.126**	0.317**	1						
Off-task	0.453**	−0.397**	0.170**	−0.049*	1					
Gaming	−0.351**	0.390**	−0.208**	0.173**	−0.503**	1				
Knowledge	−0.390**	0.085**	−0.491**	−0.069**	0.048*	−0.269**	1			
Carelessness	−0.475**	0.145**	−0.500**	−0.003	−0.030	−0.155**		1		
Correctness	−0.134**	−0.055*	−0.348***	−0.230**	0.211**	−0.586**	0.807**	0.673**	1	
Number of actions	−0.493**	0.477**	−0.332**	0.090**	−0.389**	0.546	0.067	0.178	−0.192	1

*p < 0.05; **p < 0.001

Table 11.6 Correlations of going to a selective college to independent variables in different cut-offs

	Selective college (I)	Selective college (II)	Selective college (III)	Selective college (IV)
Boredom	−.093**	−.119**	−.057	−.062*
Engaged concentration	.118**	.159**	.106**	.093**
Confusion	−.239**	−.236**	−.161**	−.108**
Frustration	−.174**	−.176**	−.126**	−.079**
Off-task	.086**	.068**	.073**	.048*
Gaming	−.247**	−.234**	−.197**	−.129**
Knowledge	.408**	.408**	.302**	.204**
Carelessness	.365**	.361**	.263**	.177**
Correctness	.448**	.439**	.331**	.227**
Number of actions	−.0002	.009	−.026	−.018

*p < 0.05; **p < 0.001

$p < 0.001$).The strongest negative associations were between correctness and gaming ($r = -0.586$, $p < 0.001$), off-task and gaming ($r = -0.503$, $p < 0.001$), and confusion and carelessness ($r = -0.500$, $p < 0.001$). Significant correlations among the predictors were evident, an indication of the existence of collinearity in a full--featured model (all predictors included in the model) for predicting whether a student will go to a selective college. Hence, we present reduced models below, rather than combining all features in a single model.

We also computed the correlations between each of our predictors and the dependent variable, whether the student attended a selective college. From Table 11.6, selectivity or students going to a selective college, across all cut-offs, is significantly correlated to each of our predictors except for number of actions, engaged concentration in the first cut-off, and off-task behavior in the second and fourth cut-offs. We conduct this analysis across all cut-offs in order to establish that the findings are stable for different cut-offs.

For the most part, going to a selective college is positively associated with engaged concentration, student knowledge, carelessness, and correctness, while negatively associated with boredom, confusion, frustration and gaming. Surprisingly, the first and third cut-offs resulted in a weak but significant positive correlation between going to a selective college and off-task behavior.

11.5.2 Differences of Predictors Between Going to a Selective College and Not Going to a Selective College

After analyzing the correlations from Table 11.6, we can look at the difference in mean values for each independent variable for students who attended selective colleges and students who attended a non-selective college in each cut-off. With the exception of number of actions and off-task for two cut-offs behavior, a statistically significant difference in means for each independent variable was found between the two groups for all cut-offs (Table 11.7).

For all cut-offs, engaged concentration, student knowledge, percentage of correct answers, and carelessness had higher mean values for *students who attended selective colleges*. The difference in engaged concentration accords with studies relating this affective state to effective learning (Craig et al., 2004; D'Mello et al., 2008; Rodrigo et al., 2009), as well as to evidence that engaged concentration with academic subjects is related to interest (Csikszentmihalyi & Schneider, 2000). In terms of student performance and learning, the differences in student knowledge and correctness indicate that successful demonstration of skill in ASSISTments during middle school is more common in students who attended a selective college. Looking at carelessness by itself, there was more carelessness for students who went to a selective college. It may seem counter-intuitive that a disengaged, careless student is more likely to go to a good college, but this finding aligns with past research that not only found carelessness to be positively associated with college enrollment (San Pedro et al., 2013), but was also more common in successful, confident students (Clements, 1982). Carelessness may be a result of overconfidence, and thus as a disengaged behavior of generally successful students.

On the other hand, boredom, confusion, frustration, and gaming the system had higher mean values for those who did *not attend a selective college*, for all cut-offs. These differences can be attributed to the fact that when boredom, confusion and frustration are not addressed properly, they may have negative influences in student learning. This is in line with previous findings that associate boredom with poorer learning outcomes (Craig et al., 2004; Pekrun et al., 2010; Pardos et al., 2013) and high school dropout (Farrell, 1988; National Research Council & Institute of Medicine, 2004; Rumberger, 1987). While confusion can sometimes lead to learning, when confusion is not addressed it is known to be associated with poorer learning (D'Mello & Graesser, 2012). Students who experience frustration and remain in that affective state are less likely to learn (D'Mello et al., 2008), and can even become bored (D'Mello & Graesser, 2012). It is also not surprising that gaming the system was more frequent among students who attended a non-selective college, since gaming the system is known to be associated with poorer learning (Cocea et al., 2009; Fancsali, 2015), poorer performance on standardized state exams (Pardos et al., 2013), and a lower chance of attending college (San Pedro et al., 2013).

Table 11.7 Features for students who attended selective college (1) and who did not attend selective college (0)

Boredom	Selective college		Mean	SD	t-value	Cohen's d	Engaged concentration	Selective college		Mean	SD	t-value	Cohen's d
	Cut-Off 1	0	0.237	0.055	9.026**	0.333		Cut-Off 1	0	0.640	0.053	−1.906^{MS}	0.071
		1	0.214	0.079					1	0.644	0.072		
	Cut-Off 2	0	0.231	0.064	8.043**	0.406		Cut-Off 2	0	0.640	0.060	−3.204*	0.157
		1	0.203	0.084					1	0.650	0.074		
	Cut-Off 3	0	0.226	0.068	4.238**	0.290		Cut-Off 3	0	0.641	0.062	−1.862^{MS}	0.125
		1	0.206	0.085					1	0.649	0.075		
	Cut-Off 4	0	0.225	0.070	3.418*	0.379		Cut-Off 4	0	0.641	0.063	−2.596*	0.241
		1	0.198	0.085					1	0.657	0.074		
Confusion	**Selective college**		**Mean**	**SD**	**t-value**	**Cohen's d**	**Frustration**	**Selective college**		**Mean**	**SD**	**t-value**	**Cohen's d**
	Cut-off 1	0	0.095	0.047	13.254**	0.508		Cut-off 1	0	0.159	0.074	*8.317**	0.312
		1	0.071	0.050					1	0.132	0.093		
	Cut-off 2	0	0.088	0.049	12.706**	0.560		Cut-off 2	0	0.152	0.082	7.912**	0.373
		1	0.061	0.047					1	0.120	0.094		
	Cut-off 3	0	0.084	0.050	8.241**	0.478		Cut-off 3	0	0.148	0.085	5.788**	0.358
		1	0.061	0.046					1	0.117	0.092		
	Cut-off 4	0	0.083	0.050	5.282**	0.489		Cut-off 4	0	0.145	0.086	4.118**	0.382
		1	0.058	0.046					1	0.0122	0.091		

Off-task	Selective college		Mean	SD	t-value	Cohen's d	Gaming	Selective college		Mean	SD	t-value	Cohen's d
	Cut-off 1	0	0.211	0.084	−2.469*	0.097		Cut-off 1	0	0.167	0.148	11.647**	0.461
		1	0.219	0.076					1	0.105	0.121		
	Cut-off 2	0	0.214	0.083	−1.713	0.070		Cut-off 2	0	0.149	0.144	13.122**	0.488
		1	0.220	0.071					1	0.083	0.101		
	Cut-off 3	0	0.214	0.081	−2.575*	0.149		Cut-off 3	0	0.141	0.140	11.781**	0.503
		1	0.226	0.073					1	0.073	0.092		
	Cut-off 4	0	0.215	0.081	−1.554	0.120		Cut-off 4	0	0.136	0.138	9.505**	0.527
		1	0.225	0.066					1	0.064	0.078		
Knowledge	**Selective college**		**Mean**	**SD**	**t-value**	**Cohen's d**	**Carelessness**	**Selective college**		**Mean**	**SD**	**t-value**	**Cohen's d**
	Cut-off 1	0	0.249	0.159	−24.052**	0.893		Cut-off 1	0	0.151	0.095	−21.137**	0.775
		1	0.423	0.218					1	0.248	0.146		
	Cut-off 2	0	0.294	0.186	−23.063**	1.082		Cut-off 2	0	0.176	0.114	−18.634**	0.944
		1	0.503	0.211					1	0.294	0.153		
	Cut-off 3	0	0.322	0.201	−17.631**	1.024		Cut-off 3	0	0.191	0.126	−12.932**	0.878
		1	0.528	0.206					1	0.306	0.155		
	Cut-off 4	0	0.337	0.208	−12.092**	1.121		Cut-off 4	0	0.200	0.131	−8.763**	0.976
		1	0.569	0.202					1	0.329	0.161		

(continued)

Table 11.7 (continued)

Correctness	Selective college		Mean	SD	t-value	Cohen's d	Number of actions	Selective college		Mean	SD	t-value	Cohen's d
	Cut-off 1	0	0.385	0.112	−26.424**	0.984		Cut-off 1	0	698.97	738.47	−1.347	0.051
		1	0.517	0.150					1	740.75	881.66		
	Cut-off 2	0	0.420	0.128	−25.210**	1.184		Cut-off 2	0	715.35	806.29	−0.784	0.035
		1	0.577	0.146					1	743.74	868.52		
	Cut-off 3	0	0.439	0.003	−19.982**	1.160		Cut-off 3	0	727.42	838.63	0.827	0.048
		1	0.601	0.008					1	687.94	697.04		
	Cut-off 4	0	0.451	0.144	−14.149**	1.311		Cut-off 4	0	727.48	834.55	1.459	0.135
		1	0.640	0.152					1	616.39	484.81		

$*p < 0.05$; $**p < 0.001$; *MS* marginally significant

Table 11.8 Goodness-of-Fit and performance values of selective college enrollment model

	R^2 (Cox & Snell)	R^2 (Nagelkerke)	Kappa	AUC ROC
Cut-off 1	0.221	0.296	0.419	0.774
Cut-off 2	0.212	0.313	0.386	0.801
Cut-off 3	0.122	0.230	0.142	0.793
Cut-off 4	0.063	0.204	0.029	0.821

11.5.3 Logistic Regression Model of Going to a Selective College

After looking at our individual variables and their relation to selective college attendance, we then built a logistic regression model that integrates multiple features and is predictive of selective college attendance. Goodness of fit metrics are given in Table 11.8.

Our final models achieved a cross-validated AUC ROC across cut-offs ranging from 0.774 to 0.821 and cross-validated Kappa values from 0.029 to 0.419 (we discuss the low Kappa below). All the models across cut-offs were statistically significantly better than a null model, and achieved a fit of R^2 (Cox & Snell) ranging from 0.063 to 0.221 and R^2 (Nagelkerke) values from 0.204 to 0.313. These values indicate that for example in cut-off 1, the final model's predictors explain 22.1% to 29.6% of the variance of those who attended a selective college.

As can be seen in Table 11.9, engaged concentration, confusion, frustration, gaming, student knowledge and correctness maintained the same directionality as in Tables 11.6 and 11.7 as predictors in a final model, while off-task and boredom switched direction in the overall model. Despite not having a significant correlation to attending a selective college by itself, number of actions became a significant predictor of going to a selective college when controlling for other predictors.

For the first cut-off, the final model of going to a selective college ($\chi^2(df = 7, N = 2732) = 680.752, p < 0.001$) included engaged concentration, confusion, frustration, gaming, carelessness, correctness and number of actions as predictors. Controlling for other predictors, each unit increase in correctness increased the odds of a student going to a selective college by 2.2. Similarly, the more engaged concentration, carelessness, or usage of ASSISTments a student showed, the greater the likelihood of that student going to a selective college. On the other hand, when controlling for other predictors, the more a student exhibits confusion, frustration and gaming, the odds of the student going to a selective college reduces.

The final model for the second cut-off had engaged concentration, frustration, off-task behavior, gaming, student knowledge, correctness and number of actions for its predictors ($\chi^2(df = 7, N = 2732) = 650.892, p < 0.001$). It is interesting to note that the resulting set of significant predictors and their relations to going to a selective college was similar to the final model in the first cut-off, with the exception of confusion being replaced by off-task behavior. When controlling for other predictors, off-task behavior is negatively associated with going to a selective college

Table 11.9 Selective college enrollment model

Cut-off1	Features	Coefficient	Standard error	Chi--Square	p-value	Odds Ratio
	Engaged Concentration	.119	.060	3.956	.047	1.127
	Confusion	−.153	.064	5.710	.017	.858
	Frustration	−.206	.053	14.907	<.001	.814
	Gaming	−.186	.077	5.862	.015	.830
	Carelessness	.275	.081	11.628	.001	1.316
	Correctness	.835	.098	72.805	<.001	2.305
	Number of Actions	.200	.064	9.870	.002	1.222
	Constant	.404	.046	76.681	<.001	1.497
Cut-off2	**Features**	**Coefficient**	**Standard error**	**Chi--Square**	**p-value**	**Odds ratio**
	Engaged concentration	.171	.056	9.327	.002	1.186
	Frustration	−.182	.051	12.672	<.001	.834
	Off-task	−.122	.067	3.321	.068	.885
	Gaming	−.230	.104	4.840	.028	.795
	Student knowledge	.312	.096	10.485	.001	1.366
	Correctness	.831	.123	45.593	<.001	2.296
	Number of actions	.195	.062	9.841	.002	1.215
	Constant	−.1387	.056	611.729	<.001	.250
Cut-off3	**Features**	**Coefficient**	**Standard error**	**Chi--Square**	**p-value**	**Odds ratio**
	Boredom	.267	.102	6.809	.009	1.306
	Engaged concentration	.160	.066	5.885	.015	1.174
	Frustration	−.277	.091	9.333	.002	.758
	Student knowledge	.422	.145	8.537	.003	1.526
	Correctness	.728	.136	28.742	<.001	2.071
	Number of actions	.150	.080	3.497	.061	1.162
	Constant	−2.394	.079	913.297	<.001	.091
Cut-off4	**Features**	**Coefficient**	**Standard error**	**Chi--Square**	**p-value**	**Odds ratio**
	Engaged concentration	.229	.094	5.978	.014	1.257
	Correctness	1.159	.097	144.072	<.001	3.186
	Constant	−3.730	.143	678.318	<.001	.024

(different than its non-significant relation when considered alone), aligning with prior studies that find off-task behavior to be associated with poorer learning outcomes (Goodman, 1990; Cocea et al., 2009).

The third cut-off resulted in a final model (χ^2(df = 6, N = 2732) = 353.994, $p < 0.001$) that had boredom, engaged concentration, frustration, student knowledge, correctness, and number of actions as predictors. Changes in a student's engaged concentration, frustration, student knowledge, correctness, or number of actions had a similar effect on the likelihood of the student going to a selective college when controlling for other predictors as for the other cut-offs.. However, in this model, once we control for other variables in the model, boredom is significant positively associated with college attendance. It is possible that once we control for students who are both bored and unsuccessful, all that remains are students who are bored with the material because it is too easy (cf. Pekrun et al., 2010).

The fourth cut-off resulted in a final model with two predictors – engaged concentration and correctness ($\chi^2(df = 2, N = 2732) = 176.375, p < 0.001$), each of them positively associated with going to a selective college when controlling for the other predictor.

Comparing the final models of going to a selective college in the different cut--offs the model for the first cut-off performed well overall (across the R^2 values, Kappa and AUC ROC), while the model for the fourth cut-off performed the worst in terms of R^2 values and Kappa (but performed best in terms of AUC ROC). These values may be attributed to the extreme data imbalance in the fourth cut-off, where only 122 students were labeled as attending a selective college out of 2732 students. Based on its performance, we choose the final model from the first cut-off for discussion below.

11.6 Discussion and Conclusion

In this paper, we investigated a set of malleable and actionable factors that occur during a student's learning experience, outside grades, tests and demographic information: student knowledge, performance, academic emotions and behavior within a middle school learning environment. Taking data from 2732 students who used ASSISTments over the course of a year or more in middle school, we used a combination of features of student success and engagement while using the system to develop a logistic regression model that can distinguish whether a student will eventually enroll in a selective college in four different instances (i.e. cut-offs in labeling selective and not selective colleges).

Our best-performing model (using cut-off 1) can distinguish 77.4% of the time whether a student will eventually enroll in a selective college, with engaged concentration, confusion, frustration, gaming the system, carelessness, correctness and number of actions to be significant predictors of going to a selective college. The positive connection between academic performance and attending a selective college is consistent with past research using other indicators of academic performance (cf. Baron & Norman, 1992; Carnevale & Rose, 2003; Griffith & Rothstein, 2009), studies that identify college readiness to be linked to high performance during schooling (Roderick et al., 2009), as well as studies that predict that college enrollment is correlated with indicators of aptitude (Christensen et al., 1975; Eccles et al., 2004).

This final model also sheds light on the impact of emotional and behavioral factors experienced by students in classrooms. As our results here show, academic emotions and disengagement are associated with a student's choice of whether to attend a selective college or not, even after controlling for student performance and learning. Hence, affect and engagement or disengagement with school appear to be another key factor influencing these processes. Affect and engagement develop early in schooling and become particularly prominent during the middle school years. When compared to student behaviors such as school violence, fighting in class, or disrupting class (Kellam et al., 1998; Reinke & Herman, 2002), the academic emotions and disengaged behaviors explored in this study are very mild in nature. Nonetheless, they are associated with long-term student outcomes. While researchers have studied disengaged behavior of an intensity that leads to disciplinary referrals, the behaviors studied in this paper are more frequent, and likely more actionable than the highly problematic behaviors which result in disciplinary referrals.

Academic emotions and student behavior are likely to play an important role inthe development of academic and career self-efficacy and interests, and can thus serve as additional information and predictors in current models for college and career pathways. This richer information can also be included in reports (in software dashboards or evaluation assessments) that may assist educators in identifying at--risk students and encourage those students to participate in educational activities and programs tailored to their specific learning needs, and help them remain in the academic pipeline. In career guidance counseling studies, questionnaire-based measures are currently used to evaluate a student's career choice (cf. Betz et al., 1996; Campbell et al., 1992) and attitudes toward career domains (Tapia & Marsh, 2004). As established in this study, online learning environments create a valuable opportunity to keep students from dropping out of the academic pipeline. In assessing students' learning experiences as early as middle school—through academic emotions, engaged and disengaged behavior—there is a potential for more effective interventions based on rich and meaningful information.

There have been growing efforts to develop software that automatically provides support when students are disengaged or experiencing negative affect while interacting with the software (D'Mello et al., 2007; Forbes-Riley & Litman, 2011; Rowe et al., 2009; Woolf et al., 2010). Results presented in this paper provide supporting evidence for which academic emotions and disengaged behaviors need to be addressed or promoted in middle school, to support long-term student achievement. For example, confused students can be given learning support to help resolve their confusion – resolved confusion is associated with better learning outcomes than never being confused at all (D'Mello & Graesser, 2012; Lee et al., 2011). Students with prolonged confusion can also transition to become bored or frustrated, another reason to address this academic emotion. Frustrated students can be provided with hints that aid in student learning or with motivational comments (D'Mello & Graesser, 2012; DeFalco et al., 2018). Students who game the system can be given supplementary materials that help them learn skills bypassed through gaming (Baker et al., 2006).

While boredom and off-task behavior did not enter into this final model, it does not mean that they cannot and should not be addressed, since they are still predictive of going to a selective college on their own. Bored students can be provided with problems that are more interesting, with greater novelty and challenge to reduce boredom or to support their emotional self-regulation (Acee et al., 2010; Pekrun et al., 2010). Similarly, it may be worth exploring the addition of data on engagement and affect to formative assessment systems used by teachers, for example when students encounter frustration when completing their homework. These indicators, can inform educators as early as middle school about whether a student is at-risk of being disengaged with learning and potentially be unable to attend a selective college down the line. Such early indicators may be used to track students' progress, creating the potential for more effective and earlier guidance for students, targeted towards the factors that often prevent students from attending selective colleges despite having excellent qualifications (cf. Hoxby & Avery, 2012).

To the degree that these models can give information not just on whether a student will attend a selective college but also on which factors reduce the probability of that occurring, these models may help both teachers and guidance counselors create more targeted and individual interventions, potentially helping open the doors of selective colleges to a wider diversity of students. Research has indicated that school guidance counselors are receptive and understand the importance of using data analytics (Young & Kaffenberger, 2011). An early warning system for counselors that provides data on learning, emotions and engagement during classroom activities could supplement student information from teachers and parents to aid them in their academic program planning for students. In coordination with teachers, guidance counselors can use this information on middle school learning, academic emotions and engagement to identify students who may be in need of counseling services – for example, persistent negative emotions during online learning may be a symptom of a broader problem. We believe that further research is needed to determine exactly how to best use data from online learning to drive support for learners. As this research goes forward, counseling efforts that consider both cognitive and non-cognitive skills during learning will have the opportunity to aid in providing adequate opportunities in college preparation. Ultimately, our goal as a society should be in preparing every student in their middle school and high school years to take full advantage of the opportunities that our society can afford them; helping students get past challenges of all kinds.

Acknowledgements This research was supported by grants NSF #OAC-1636782, NSF #DRL-1252297, NSF #DRL-1031398, NSF #SBE-0836012, and grant #OPP1048577 from the Bill & Melinda Gates Foundation.

References

Acee, T. W., Kim, H., Kim, H. J., Kim, J. I., Chu, H. N. R., Kim, M., … Wicker, F. W. (2010). Academic boredom in under-and over-challenging situations. *Contemporary Educational Psychology, 35*(1), 17–27.

Allensworth, E. M., Gwynne, J. A., Moore, P., & De la Torre, M. (2014). *Looking forward to high school and college: Middle grade indicators of readiness in Chicago public schools*. University of Chicago Consortium on Chicago School Research.

Anderman, E. M., & Maehr, M. L. (1994). Motivation and schooling in the middle grades. *Review of Educational Research, 64*(2), 287–309.

Baker, R. S. (2007, April). Modeling and understanding students' off-task behavior in intelligent tutoring systems. In *Proceedings of the SIGCHI conference on human factors in computing systems* (pp. 1059–1068). ACM.

Baker, R. S., & Yacef, K. (2009). The state of educational data mining in 2009: A review and future visions. *JEDM-Journal of Educational Data Mining, 1*(1), 3–17.

Baker, R. S., Corbett, A. T., Koedinger, K. R., & Wagner, A. Z. (2004, April). Off-task behavior in the cognitive tutor classroom: When students game the system. In *Proceedings of the SIGCHI conference on human factors in computing systems* (pp. 383–390). ACM.

Baker, R. S., Corbett, A. T., & Aleven, V. (2008, January). More accurate student modeling through contextual estimation of slip and guess probabilities in Bayesian knowledge tracing. In Intelligent tutoring systems (pp. 406–415). Springer.

Baker, R. S., Corbett, A. T., Gowda, S. M., Wagner, A. Z., MacLaren, B. A., Kauffman, L. R., … Giguere, S. (2010a). Contextual slip and prediction of student performance after use of an intelligent tutor. In *User modeling, adaptation, and personalization* (pp. 52–63). Springer.

Baker, R. S., D'Mello, S. K., Rodrigo, M. M. T., & Graesser, A. C. (2010b). Better to be frustrated than bored: The incidence, persistence, and impact of learners' cognitive–affective states during interactions with three different computer-based learning environments. *International Journal of Human-Computer Studies, 68*(4), 223–241.

Baker, R. S., Ocumpaugh, J. L., & Andres, J. M. A. L. (2020). BROMP quantitative field observations: A review. In R. Feldman (Ed.), *Learning science: Theory, research, and practice*. McGraw-Hill.

Balfanz, R. (2009). *Putting middle grades students on the graduation path: A policy and practice brief*. Everyone Graduates Center & Talent Development Middle Grades Program. Johns Hopkins University.

Balfanz, R., Herzog, L., & Mac Iver, D. J. (2007). Preventing student disengagement and keeping students on the graduation path in urban middle-grades schools: Early identification and effective interventions. *Educational Psychologist, 42*(4), 223–235.

Baron, J., & Norman, M. F. (1992). SATs achievement tests, and high-school class rank as predictors of college performance. *Educational and Psychological Measurement, 52*, 1047–1047.

Betz, N. E., Borgen, F. H., & Harmon, L. W. (1996). *Skills confidence inventory*. Consulting Psychologists Press.

Bosch, N., D'Mello, S., Baker, R., Ocumpaugh, J., Shute, V., Ventura, M., Wang, L., & Zhao, W. (2015). Automatic detection of learning-centered affective states in the wild. In *Proceedings of the 2015 international conference on intelligent user interfaces*. ACM.

Bowen, W. G., Chingos, M. M., & McPherson, M. S. (2009). *Crossing the finish line: Completing college at America's public universities*. Princeton University Press.

Bowers, A. J. (2010). Grades and graduation: A longitudinal risk perspective to identify student dropouts. *The Journal of Educational Research, 103*(3), 191–207.

Bowers, A. J. (2021). Early warning systems and indicators of dropping out of upper secondary school: The emerging role of digital technologies. In S. Vincent-Lancrin & R. Baker (Eds.), *Smart data and digital technology in education: Learning analytics, AI and beyond*. Organisation for Economic Co-Operation and Development (OECD) Publishing.

Bowers, A. J., & Lee, J. (2013). Carried or defeated? Examining the factors associated with passing School District bond elections in Texas, 1997–2009. *Educational Administration Quarterly, 49*(5), 732–767.

Bowers, A. J., & Sprott, R. (2012a). Examining the multiple trajectories associated with dropping out of high school: A growth mixture model analysis. *The Journal of Educational Research, 105*(3), 176–195.

Bowers, A. J., & Sprott, R. (2012b). Why tenth graders fail to finish high school: A dropout typology latent class analysis. *Journal of Education for Students Placed at Risk (JESPAR), 17*(3), 129–148.

Bowers, A. J., & Zhou, X. (2019). Receiver operating characteristic (ROC) area under the curve (AUC): A diagnostic measure for evaluating the accuracy of predictors of education outcomes. *Journal of Education for Students Placed at Risk (JESPAR), 24*(1), 20–46.

Cabrera, A. F. (1994). Logistic regression analysis in higher education: An applied perspective. *Higher Education: Handbook of Theory and Research, 10*, 225–256.

Cabrera, A. F., La Nasa, S. M., Burkum, K., & R. (2001). *Pathways to a four-year degree: The higher education story of one generation*. Penn State University.

Camblin, S. (2003). *The middle grades: Putting all students on track for college*. Pacific Resources for Education and Learning.

Campbell, D. P., Hyne, S. A., & Nilsen, D. (1992). *Manual for the Campbell interest and skill survey*. National Computer Systems.

Canfield, W. (2001). ALEKS: A web-based intelligent tutoring system. *Mathematics and Computer Education, 35*(2), 152–158.

Carnevale, A. P., & Rose, S. J. (2003). *Socioeconomic status, race/ethnicity, and selective college admissions*. Century Foundation.

Christensen, S., Melder, J., & Weisbrod, B. A. (1975). Factors affecting college attendance. *Journal of Human Resources*, 174–188.

Clements, M. A. (1982). Careless errors made by sixth-grade children on written mathematical tasks. *Journal for Research in Mathematics Education, 13*, 136–144.

Cocea, M., Hershkovitz, A., & Baker, R. S. J. D. (2009). The impact of off-task and gaming behaviors on learning: Immediate or aggregate? In *Proceedings of the 14th international conference on artificial intelligence in education* (pp. 507–514).

Cohen, J. (1960). A coefficient of agreement for nominal scales. *Educational and Psychological Measurement, 20*(1), 37–46.

College Division of Barron's Education Series (Ed.). (2012). *Barron's profiles of American colleges* (30th ed.). Barron's Educational Series.

Conley, D. T. (2008). *College knowledge: What it really takes for students to succeed and what we can do to get them ready*. Wiley.

Conley, D., Lombardi, A., Seburn, M., & McGaughy, C. (2009). *Formative assessment for college readiness: Measuring skill and growth in five key cognitive strategies associated with postsecondary success*. Online Submission.

Corbett, A. T., & Anderson, J. R. (1995). Knowledge tracing: Modeling the acquisition of procedural knowledge. *User Modeling and User-Adapted Interaction, 4*(4), 253–278.

Cox, D. R., & Snell, E. J. (1989). *Analysis of binary data* (Vol. 32). CRC Press.

Craig, S., Graesser, A., Sullins, J., & Gholson, B. (2004). Affect and learning: An exploratory look into the role of affect in learning with AutoTutor. *Journal of Educational Media, 29*(3), 241–250.

Csikszentmihalyi, M. (1990). *Flow: The psychology of optimal experience*. Harper-Row.

Cupani, M., & Pautassi, R. M. (2013). Predictive contribution of personality traits in a sociocognitive model of academic performance in mathematics. *Journal of Career Assessment, 21*(3), 395–413.

D'Mello, S., & Graesser, A. (2012). Dynamics of affective states during complex learning. *Learning and Instruction, 22*(2), 145–157.

D'Mello, S., Picard, R., & Graesser, A. (2007). Towards an affect-sensitive autotutor. *IEEE Intelligent Systems, 22*(4), 53–61.

D'Mello, S. K., Craig, S. D., Witherspoon, A., Mcdaniel, B., & Graesser, A. (2008). Automatic detection of learner's affect from conversational cues. *User Modeling and User-Adapted Interaction, 18*(1–2), 45–80.

Daniels, L. M., Stupnisky, R. H., Pekrun, R., Haynes, T. L., Perry, R. P., & Newall, N. E. (2009). A longitudinal analysis of achievement goals: From affective antecedents to emotional effects and achievement outcomes. *Journal of Educational Psychology, 101*(4), 948.

DeFalco, J. A., Rowe, J. P., Paquette, L., Georgoulas-Sherry, V., Brawner, K., Mott, B. W., Baker, R. S., & Lester, J. C. (2018). Detecting and addressing frustration in a serious game for military training. *International Journal of Artificial Intelligence and Education, 28*(2), 152–193.

D'Mello, S. K., Lehman, B., Pekrun, R., & Graesser, A. C. (2014). Confusion can be beneficial for learning. *Learning & Instruction, 29*(1), 153–170.

Drummond, J., & Litman, D. (2010, January). In the zone: Towards detecting student zoning out using supervised machine learning. In *Intelligent tutoring systems* (pp. 306–308). Springer.

Eccles, J. S., Vida, M. N., & Barber, B. (2004). The relation of early adolescents' college plans and both academic ability and task-value beliefs to subsequent college enrollment. *The Journal of Early Adolescence, 24*(1), 63–77.

Ellwood, D., & Kane, T. J. (2000). Who is getting a college education? Family background and the growing gaps in enrollment. In *Securing the future: Investing in children from birth to college* (pp. 283–324).

Fancsali, S. (2015). Confounding carelessness? Exploring causal relationships between carelessness, affect, behavior, and learning in cognitive tutor algebra using graphical causal models. In *Proceedings of the international conference on educational data mining* (pp. 508–511).

Farrell, E. (1988). Giving voice to high school students: Pressure and boredom, Ya know what I'm Sayin'? *American Educational Research Journal, 4*, 489–502.

Farrington, C. A., Roderick, M., Allensworth, E., Nagaoka, J., Keyes, T. S., Johnson, D. W., & Beechum, N. O. (2012). *Teaching adolescents to become learners: The role of noncognitive factors in shaping school performance--A critical literature review*. Consortium on Chicago School Research.

Feng, S., D'Mello, S., & Graesser, A. C. (2013). Mind wandering while reading easy and difficult texts. *Psychonomic Bulletin & Review, 20*(3), 586–592.

Forbes-Riley, K., & Litman, D. (2011). Benefits and challenges of real-time uncertainty detection and adaptation in a spoken dialogue computer tutor. *Speech Communication, 53*(9), 1115–1136.

Fouad, N. A., & Smith, P. L. (1996). A test of a social cognitive model for middle school students: Math and science. *Journal of Counseling Psychology, 43*(3), 338.

Fredricks, J. A., Blumenfeld, P. C., & Paris, A. H. (2004). School engagement: Potential of the concept, state of the evidence. *Review of Educational Research, 74*(1), 59–109.

Gibbons, M. M., & Borders, L. D. (2010). Prospective first-generation college students: A social--cognitive perspective. *The Career Development Quarterly, 58*(3), 194–208.

Gibbons, M. M., Borders, L. D., Wiles, M. E., Stephan, J., & Davis, P. E. (2006). Career and college planning needs of ninth graders--As reported by ninth graders. *Professional School Counseling, 10*, 168–178.

Goldrick-Rab, S. (2007). What higher education has to say about the transition to college. *The Teachers College Record, 109*(10), 2444–2481.

Goodman, L. (1990). *Time and learning in the special education classroom*. SUNY Press.

Griffith, A. L., & Rothstein, D. S. (2009). Can't get there from here: The decision to apply to a selective college. *Economics of Education Review, 28*(5), 620–628.

Hanley, J., & McNeil, B. (1982). The meaning and use of the area under a receiver operating characteristic (ROC) curve. *Radiology, 143*, 29–36.

Heffernan, N. T., & Heffernan, C. L. (2014). The ASSISTments ecosystem: Building a platform that brings scientists and teachers together for minimally invasive research on human learning and teaching. *International Journal of Artificial Intelligence in Education, 24*(4), 470–497.

Hoxby, C. (2009). The changing selectivity of American colleges. *Journal of Economic Perspectives 2009, 23*(4), 95–118.

Hoxby, C. M., & Avery, C. (2012). *The missing "one-offs": The hidden supply of high-achieving, low income students* (No. w18586). National Bureau of Economic Research.

Karabel, J., & Astin, A. W. (1975). Social class, academic ability, and college "quality". *Social Forces, 53*(3), 381–398.

Kellam, S. G., Ling, X., Merisca, R., Brown, C. H., & Ialongo, N. (1998). The effect of the level of aggression in the first grade classroom on the course and malleability of aggressive behavior into middle school. *Development and Psychopathology, 10*(02), 165–185.

Koedinger, K. R., & Corbett, A. T. (2006). Cognitive tutors: Technology bringing learning science to the classroom. In K. Sawyer (Ed.), *The Cambridge handbook of the learning sciences*. Cambridge University Press.

Kort, B., Reilly, R., & Picard, R. (2001). An affective model of interplay between emotions and learning: Reengineering educational pedagogy—Building A learning companion. In *Proceedings IEEE international conference on advanced learning technology: Issues, achievements and challenges* (pp. 43–48). IEEE Computer Society.

Lee, D. M. C., Rodrigo, M. M. T., & d Baker, R. S., Sugay, J. O., & Coronel, A. (2011). Exploring the relationship between novice programmer confusion and achievement. In *Affective computing and intelligent interaction* (pp. 175–184). Springer.

Lent, R. W., Brown, S. D., & Hackett, G. (1994). Toward a unifying social cognitive theory of career and academic interest, choice and performance. *Journal of Vocational Behavior, 45*(1), 79–122.

Lent, R. W., Brown, S. D., & Hackett, G. (2000). Contextual supports and barriers to career choice: A social cognitive analysis. *Journal of Counseling Psychology, 47*(1), 36–49.

Mann, S., & Robinson, A. (2009). Boredom in the lecture theatre: An investigation into the contributors, moderators and outcomes of boredom amongst university students. *British Educational Research Journal, 35*(2), 243–258.

Nagelkerke, N. J. (1991). A note on a general definition of the coefficient of determination. *Biometrika, 78*(3), 691–692.

Narciss, S. (2004). The impact of informative tutoring feedback and self-efficacy on motivation and achievement in concept learning. *Experimental Psychology, 51*, 214–228.

National Middle School Association. (2002). *Research summary #12: Academic achievement. 2002*. Retrieved January 14, 2003, from www.nmsa.org/research/ressum12.htm

National Research Council & Institute of Medicine. (2004). *Engaging schools: Fostering high school students' motivation to learn*. National Academy Press.

Neild, R. C. (2009). Falling off track during the transition to high school: What we know and what can be done. *The Future of Children, 19*(1), 53–76.

Núñez, A. M., & Bowers, A. J. (2011). Exploring what leads high school students to enroll in Hispanic-serving institutions A multilevel analysis. *American Educational Research Journal, 48*(6), 1286–1313.

Ocumpaugh, J., Baker, R., Gowda, S., Heffernan, N., & Heffernan, C. (2014). Population validity for educational data mining models: A case study in affect detection. *British Journal of Educational Technology, 45*(3), 487–501.

Ocumpaugh, J., Baker, R. S., Rodrigo, M. M. T., Salvi, A., van Velsen, M., Aghababyan, A., & Martin, T. (2015). HART: The human affect recording tool. In *Proceedings of the ACM Special Interest Group on the Design of Communication (SIGDOC)*.

Ovink, S., Kalogrides, D., Nanney, M., & Delaney, P. (2018). College match and undermatch: Assessing student preferences, college proximity, and inequality in post-college outcomes. *Research in Higher Education, 59*(5), 553–590.

Pardos, Z.A., Baker, R.S.J.D., San Pedro, M. O. C. Z., Gowda, S. M., & Gowda, S.M. (2013) Affective states and state tests: Investigating how affect throughout the school year predicts end of year learning outcomes. In *Proceedings of the 3rd international conference on learning analytics and knowledge* (pp. 117–124).

Pekrun, R., Goetz, T., Daniels, L. M., Stupnisky, R. H., & Perry, R. P. (2010). Boredom in achievement settings: Exploring control–value antecedents and performance outcomes of a neglected emotion. *Journal of Educational Psychology, 102*(3), 531.

Reardon, S., Baker, R., & Klasik, D. (2012). *Race, income, and enrollment patterns in highly selective colleges, 1982–2004*. Center for Education Policy Analysis, Stanford University. Retrieved from http://cepa.stanford.edu/content/race-income-and-enrollmentpatterns-highly-selective-colleges-1982-2004

Reinke, W. M., & Herman, K. C. (2002). Creating school environments that deter antisocial behaviors in youth. *Psychology in the Schools, 39*(5), 549–559.

Roderick, M., Nagaoka, J., Coca, V., & Moeller, E. (2008). *From high school to the future: Potholes on the road to college* (Research Report). Consortium on Chicago School Research.

Roderick, M., Nagaoka, J., & Coca, V. (2009). College readiness for all: The challenge for urban high schools. *The Future of Children, 19*(1), 185–210.

Roderick, M., Coca, V., & Nagaoka, J. (2011). Potholes on the road to college high school effects in shaping urban students' participation in college application, four-year college enrollment, and college match. *Sociology of Education, 84*(3), 178–211.

Rodrigo, M. M. T., Baker, R. S., Jadud, M. C., Amarra, A. C. M., Dy, T., Espejo-Lahoz, M. B. V., … Tabanao, E. S. (2009). Affective and behavioral predictors of novice programmer achievement. *ACM SIGCSE Bulletin, 41*(3), 156–160.

Rowe, J. P., McQuiggan, S. W., Robison, J. L., & Lester, J. C. (2009, July). Off-task behavior in narrative-centered learning environments. In *AIED* (pp. 99–106).

Rozin, P., & Cohen, A. B. (2003). High frequency of facial expressions corresponding to confusion, concentration, and worry in an analysis of naturally occurring facial expressions of Americans. *Emotion, 3*(1), 68–75.

Rumberger, R. W. (1987). High school dropouts: A review of issues and evidence. *Review of Educational Research, 57*(2), 101–121.

Sabourin, J., Mott, B., & Lester, J. (2011). Modeling learner affect with theoretically grounded dynamic Bayesian networks. In *Proceeding of the ACII 2011* (pp. 286–295).

San Pedro, M. O. C., Baker, R., & Rodrigo, M. M. (2011). Detecting carelessness through contextual estimation of slip probabilities among students using an intelligent tutor for mathematics. In *Proceedings of 15th international conference on artificial intelligence in education* (pp. 304–311).

San Pedro, M. O. Z., Baker, R. S. J. D., Bowers, A. J., & Heffernan, N. T. (2013). Predicting college enrollment from student interaction with an intelligent tutoring system in middle school. In *Proceedings of the 6th international conference on educational data mining* (pp. 177–184).

Schmidt, W., Burroughs, N., Cogan, L., & Houang, R. (2011). Are college rankings an indicator of quality education? In *Forum on public policy online* (Vol. 3). Oxford Round Table.

Schunk, D. H. (1989). Self-efficacy and achievement behaviors. *Educational Psychology Review, 1*, 173–208.

Shamsuddin, S. (2016). Berkeley or bust? Estimating the causal effect of college selectivity on bachelor's degree completion. *Research in Higher Education, 57*(7), 795–822.

Shute, V. J., Moore, G. R., & Wang, L. (2015). Measuring problem solving skills in plants vs. Zombies 2. In *Proceedings of the 8th international conference on educational data mining (EDM 2015)*. Madrid, Spain.

Stephan, J. L., & Rosenbaum, J. E. (2013). Can high schools reduce college enrollment gaps with a new counseling model? *Educational Evaluation and Policy Analysis, 35*(2), 200–219.

Tapia, M., & Marsh, G. (2004). An instrument to measure mathematics attitudes. *Academic Exchange Quarterly, 8*(2), 1–8.

Thomas, S. L. (2000). Deferred costs and economic returns to college major, quality, and performance. *Research in Higher Education, 41*(3), 281–313.

Tobin, T. J., & Sugai, G. M. (1999). Using sixth-grade school records to predict school violence, chronic discipline problems, and high school outcomes. *Journal of Emotional and Behavioral Disorders, 7*(1), 40–53.

Tobin, T., Sugai, G., & Colvin, G. (1996). Patterns in middle school discipline records. *Journal of Emotional and Behavioral Disorders, 4*(2), 82–94.

Turner, S., & Lapan, R. T. (2002). Career self-efficacy and perceptions of parent support in adolescent career development. *The Career Development Quarterly, 51*(1), 44–55.

Walonoski, J. A., & Heffernan, N. T. (2006, January). Detection and analysis of off-task gaming behavior in intelligent tutoring systems. In *Intelligent tutoring systems* (pp. 382–391). Springer.

Witten, I. H., & Frank, E. (2005). *Data mining: Practical machine learning tools and techniques.* Morgan Kaufmann.

Woolf, B. P., Arroyo, I., Muldner, K., Burleson, W., Cooper, D. G., Dolan, R., & Christopherson, R. M. (2010, January). The effect of motivational learning companions on low achieving students and students with disabilities. In *Intelligent tutoring systems* (pp. 327–337). Springer.

Young, A., & Kaffenberger, C. (2011). The beliefs and practices of school counselors who use data to implement comprehensive school counseling programs. *Professional School Counseling, 15*(2), 67–76.

Zhou, X., & Bowers, A. (2020). A typology of parental involvement in student experience: A latent class analysis. *The High School Journal, 103*(2), 99–131.

Part III
Applications in Adult and Professional Education

Chapter 12
Single-Case Learning Analytics to Support Social-Emotional Learning: The Case of Doctoral Education

Luis P. Prieto, María Jesús Rodríguez-Triana, Paula Odriozola-González, and Yannis Dimitriadis

Abstract High rates of dropout and mental health problems in doctoral education hint that social and emotional learning (SEL) support could help doctoral students face the challenges of such an arduous, lengthy, and unstructured learning experience. The uniqueness of each individual student, doctoral process and contextual influences, makes it hard for researchers to study this kind of setting. These same peculiarities make it difficult for traditional (cohort-based) learning analytics (LA) approaches to measure, understand and support these students' learning experience, leaving doctoral education an under-researched area in LA. In this chapter, we present a novel approach to LA that is specifically designed to deal with these peculiarities, and that could support single doctoral students in developing SEL skills, without the need to compare within a cohort. We illustrate the potential of this "single-case learning analytics" (SCLA) approach, using data from a pilot (diary-based) intervention with N = 9 doctoral students using a simple LA tool over a period of several weeks. Our results highlight the added value of the insights obtained from the analysis of (quantitative and qualitative) data from a single learner, collected over time. Our results also showcase the improved performance of such single-case models over cohort-based ones. An overview of challenges defining an agenda for future research on SCLA to support for SEL in doctoral education closes the chapter.

L. P. Prieto (✉) · M. J. Rodríguez-Triana
Tallinn University, Tallinn, Estonia
e-mail: lprisan@tlu.ee; mjrt@tlu.ee

P. Odriozola-González
Universidad de Cantabria, Santander, Spain
e-mail: paula.odriozola@unican.es

Y. Dimitriadis
Universidad de Valladolid, Valladolid, Spain
e-mail: yannis@tel.uva.es

Y. E. Wang et al. (eds.), *Social and Emotional Learning and Complex Skills Assessment*, Advances in Analytics for Learning and Teaching,
https://doi.org/10.1007/978-3-031-06333-6_12

Keywords Doctoral education · Social and emotional learning · Learning analytics · Single-case learning analytics · Diary studies

12.1 Introduction

Doctoral education represents one of the most challenging and lengthy learning processes humans engage in (barring the whole process of lifelong learning itself). Yet, the study of this kind of learning process remains a comparatively under-researched area within education (Boud & Lee, 2006), along with special education or lifelong learning itself (Field, 2012).

What do all these areas of education have in common? Much of the difficulty of studying these educational processes comes from their inherently individual nature (i.e., cohorts do not play such a prominent role in the learning process, or do not exist altogether). This, in turn, makes *individual differences* heavily influence the learning process and its outcomes (Jonassen & Grabowski, 2012). Similarly, lacking the joint context of a cohort doing the same set of learning activities, the *contextual* (Boyle & Ravenscroft, 2012) and situated (Winn, 1996) nature of learning processes influences them even more strongly than in other areas of education.

In the case of doctoral education, the *length* of the doctoral process itself (taking three, four or more years) and the fact that it seamlessly crosses boundaries of space and context (Bronkhorst & Akkerman, 2016) poses an additional challenge for researchers attempting to study it (leading to many cross-sectional but few longitudinal studies focusing on its process). The lack of a clearly-defined structure of activities and the relative vagueness of doctoral learning outcomes (which are operationalized quite differently even within the same research discipline) poses an additional difficulty for its study and for any potential interventions to support doctoral learning. These difficulties are also a probable reason behind the scarcity of technological (and, specifically, learning analytics) proposals aimed at supporting doctoral education (see, e.g., Di Mitri et al., 2017 for an exception).

In recent years, two challenges have emerged within the doctoral education research literature: (a) the comparatively high rates of dropout in doctoral degrees, which many studies situate around 50% (Bair & Haworth, 2004; Wollast et al., 2018) – and even higher in online degrees (Terrell et al., 2012); and (b) the high incidence of emotional disorders like stress, depression or anxiety, with an increasing prevalence now estimated at 30–50% (Levecque et al., 2017), leading to warnings of a "mental health crisis" in doctoral education (Evans et al., 2018). Further, there is now evidence that these two phenomena (attrition and emotional distress) often overlap (González-Betancor & Dorta-González, 2020). These inter-related challenges highlight that socio-emotional and motivational aspects of the doctoral process need further attention and support and could be crucial to degree completion and learner wellbeing, as they are for school-level education (Zins & Elias, 2007).

The scale of these problems (there are close to three million doctoral students worldwide, according to latest estimates, Taylor, 2021) and the level of personalization seemingly required to support social and emotional learning (SEL) in doctoral education, makes us wonder whether technological support could be beneficial. More concretely, learning analytics (LA), which have been applied to promote self-reflection and critical thinking (e.g., Kovanović et al., 2018), could be supportive in the development of SEL competences like self-awareness and self-management (CASEL, 2003). Yet, the specific characteristics of doctoral education still remain a challenge for traditional, cohort-based methods in LA. Fortunately, new approaches to LA that specifically try to address time series data paying attention to single learners, are starting to emerge (Prieto et al., 2020b; Saqr & López-Pernas, 2021).

This paper explores the conjecture that LA could support doctoral students' SEL competencies. Given the aforementioned challenges and specificities of doctoral education, we argue that a new approach to LA is needed for such personalized support, which we have labeled "single-case learning analytics" (SCLA). But, *what is the added value of an SCLA approach over more typical cohort-based LA?* This is the research question that this chapter aims to start answering.

After a revision of related literature in the areas of SEL, LA and doctoral education, this chapter describes the main characteristics of the SCLA approach, and a simple example of SCLA technology. Later, to illustrate the potential usefulness of SCLA for doctoral SEL, we draw from data gathered in an exploratory case study, an intervention pilot with N = 9 doctoral students using that simple SCLA platform during 6–8 weeks. The mixed-methods analysis of such pilots' data showcases the differential insights and objective performance of cohort-based versus SCLA models of a SEL-related phenomenon, in the particular context of two countries (Estonia and Spain) during the first months of the current COVID-19 pandemic. The chapter is closed with an outline of outstanding challenges and a research agenda for this new area of SCLA.

12.2 Related Work

12.2.1 Social and Emotional Learning (SEL) and Doctoral Education

The learning (and teaching) of social and emotional skills that go beyond the traditional academic curricula, has garnered a lot of attention in recent educational research and practice (Collie et al., 2012). SEL has shown effects not only on academic performance, but also on extracurricular variables such as workplace success, more positive relationships and improved mental health and personal wellbeing (Bar-On et al., 2006), as implemented in multiple primary and secondary education programs (CASEL, 2013). According to the Collaborative for Academic, Social, and Emotional Learning (CASEL)'s main conceptual framework, SEL encompasses five core competencies: self-awareness, self-management, responsible decision making, social awareness and relationship skills.

Yet, there exist relatively few programs promoting SEL specifically in higher education (Conley, 2015). Conley's review of SEL in higher education notes that other programs that focus on such non-cognitive skills do exist, even if they are not labelled 'SEL': programs based on cognitive-behavioral techniques, mindfulness, relaxation programs, or social skills training programs. Among these SEL-oriented programs, it seems that the evidence for effectiveness is strongest among mindfulness-based programs, followed by cognitive-behavioral, relaxation and social skills training. Conley also notes that programs that have a supervised practice component seem to be more effective than those that do not incorporate this aspect. Additionally, the review highlights that much research is still needed to elucidate what programs (or components within programs) are most effective for different kinds of higher education students, and that stronger evaluations of such programs using validated instruments are needed. Very recent research has attempted to advance in this direction (e.g., the Social Emotional Learning Scale –SELS, see Thomas et al., 2021), albeit the scale has only been validated with primary/secondary school populations.

In terms of doctoral-level education which, as we have seen, has strong social and emotional demands for students, SEL programs are still rare (Edwards et al., 2019). Similar to other higher education areas, certain SEL-inflected programs/interventions do exist. For instance, Barry et al.'s (2019) mindfulness-based program showed improvements in doctoral students' psychological capital. Howell et al.'s (2017) qualitative case study around a gratitude-focused intervention, is another such example: its qualitative findings suggest an improvement in communication, social and emotional wellbeing. Yet, these examples of interventions that resemble SEL training are still difficult to find.

12.2.2 Learning Analytics for Social and Emotional Learning

The use of technology to support SEL is still in its infancy (as expected, given that SEL as a field is still relatively new), and it largely encompasses the same technologies used to teach and learn cognitive/academic skills, from SEL-oriented games, mobile apps or teacher professional development platforms (Stern et al., 2015). In schools, a widely-used direct-to-learner SEL technology is the Ripple Effects Whole Spectrum Learning Intervention®, an adaptive (expert system) that includes skills-building and motivational counseling, which has shown significant effects against a number of outcome variables (Perry & Bass, 2008).

A number of learning analytics (LA) systems and proposals targeting SEL-related or non-cognitive aspects of learning exist for a variety of educational contexts. As usual in the field of LA, maybe the most common learning setting is higher (typically, undergraduate) education (Beardsley et al., 2019; Liu & Huang, 2017; Walton et al., 2020), although proposals aimed at schools also exist (see, e.g., Williamson, 2017a critique of ClassDojo). Less frequent are LA proposals that target other settings where cohorts are not available, such as lifelong learning (see Prieto, Rodríguez-Triana, et al., 2020b for an exception) or doctoral education (see the next subsection).

The purpose and nature of these SEL-LA proposals is quite varied, with many of them aiming to develop some sort of SEL competence or motivation strategy. For instance, Beardsley et al.'s (2019) ClassMood app tries to make visible the emotional atmosphere of the classroom, to provide awareness and help instructors shift it towards those favorable to learning. The rest of the chapters in this book showcase further examples in this area.

Many of the proposals so far are exploratory in nature, and often comprise observational or illustrative case studies using authentic data, such as the assessment of the impact of technology on learner wellbeing by Hakami et al. (2021), or the use of computational methods to trace motivational strategies over time by mining unstructured text (Liu & Huang, 2017). In most of these proposals, the main target stakeholder (i.e., the one consuming the analytics) is the teacher/instructor (Beardsley et al., 2019; Liu & Huang, 2017; Walton et al., 2020), while learner-facing LA proposals are rather uncommon (e.g., Prieto et al., 2020b).

In terms of the data sources used as input for the analytics, proposals often draw from logs of digital platforms and LMSs (Walton et al., 2020; Williamson, 2017a). It is worth noting that quantitative and qualitative questionnaires are also often used (Hakami & Hernandez-Leo, 2021; Liu & Huang, 2017), sometimes in the form of diaries (Prieto et al., 2020b). Albeit scarce, the use of more fine-grained behavioral and physiological signals (e.g., gaze, facial expressions) is not unheard of in this area (D'Mello et al., 2017; Williamson, 2017b). Data analytic techniques in these proposals are in line with other areas of LA: descriptive and exploratory statistics (Beardsley et al., 2019; Liu & Huang, 2017; Prieto et al., 2020b; Walton et al., 2020), automated content analysis and statistical machine learning models (Liu & Huang, 2017), social network analysis (Williamson, 2017a) or epistemic network analysis (Prieto et al., 2020b).

These initial LA proposals around SEL and related areas also note several gaps that need to be addressed by future SEL-LA research: the need to develop and empirically evaluate interventions that use LA to promote SEL (Liu & Huang, 2017; Walton et al., 2020); the challenges of scaling up these SEL-oriented analytics (which still feature manual tasks like coding or labeling for later use in supervised machine learning) (Prieto et al., 2020b; Walton et al., 2020); or the development of suitable, reliable, accurate (and unobtrusive) tools and methods to measure these SEL competencies (Duckworth & Yeager, 2015) and their evolution over time and at different stages of learner development (from early ages to adulthood).

12.2.3 Learning Analytics for Doctoral Education

The use of LA to support doctoral education (be it SEL-oriented or otherwise) is even more scarce in the literature, and to the best of our knowledge only a handful of proposals in this area exist. Di Mitri et al. (2017) studied the effectiveness of a machine learning prediction model regarding different performance indicators (abilities, challenge, productivity, stress and flow) of the daily learning activities of nine doctoral students at the Open University of the Netherlands. The proposed prediction model used multimodal learning analytics based on data collected through a laptop tool and a fitness wristband, and aimed at comparing the predicted performance and the self-rated performance on typical research tasks, categorized as reading, writing, meeting, communicating and others. The authors stressed the importance of the diversity of research topics and the learning tasks of doctoral students (despite the fact that all students were sampled from the same area of technology-enhanced learning), and suggested that the prediction model might help students to reflect on the predicted performance. However, the authors did not discuss the contextual influences.

Cahyani et al. (2019) involved a small set of doctoral students (N = 19) of different programs (24% of a larger sample comprising undergraduate, masters and doctoral students), in their study on the assessment of the usefulness of multiple learning analytics. They found out that PhD students do not perceive sufficient usefulness of social network and classroom-related data, while they assessed positively other research-oriented indicators, such as time spent accessing online campus libraries, news/opinion websites, and reference/learning websites. Thus, the perceived usefulness of LA indicators seems to depend on the special features of doctoral learning activities, heavily focused on non-classroom-based research activities (which also vary widely across doctoral disciplines and stages during the PhD).

From this overview of areas related to SEL, doctoral education and LA, we can hence conclude that, despite the great need for social and emotional skills during the long and strenuous process of finishing a PhD, there exist few evidence-based interventions that target this educational level. Furthermore, despite the promise of SEL interventions to impact both academic and wellbeing variables (as demonstrated in school programs in the US and around the world), there is a lack of technological proposals (and LA proposals) that both aim at developing SEL skills and are designed with the specific challenges of doctoral education in mind (e.g., uniqueness of learners, their contexts and each dissertation's specific research tasks). Even fewer proposals have been evaluated across multiple countries, higher education institutions and disciplines (to the point of non-existence). These gaps represent an opportunity for a new approach to LA that is specifically designed for these peculiarities of doctoral education. The following sections describe such an approach and illustrate its usefulness through the example of an authentic case study in doctoral education, using a simple technological platform.

12.3 Single-Case Learning Analytics

As learning analytics researchers, we have two main goals: to increase our understanding of learning processes, and to support learners in enacting them (SoLAR, 2011). Given the challenges that doctoral education poses (the unusual length of the learning process, the importance of individual differences, and the uniqueness of each learning context), we suggest that a new approach to LA is needed. These challenges are also a hallmark of SEL, which is a lifelong learning process, in which our individual peculiarities and context also play a critical role.

In our attempt to tackle the three aforementioned issues that make the study of doctoral education (and doctoral SEL) challenging, we can define three key characteristics of a new approach to LA, which we have labeled "single-case learning analytics" (SCLA):

1. The focus on analyzing and *understanding single cases/processes (N = 1) in depth*, without necessarily comparing them with a cohort or other external reference. Rather, an SCLA approach would compare learners against themselves in the past (i.e., against a certain baseline/initial state).
2. The *gathering and analysis of learning data over extended periods of time*, in the form of (often, multivariate) time series. This enables the aforementioned within-learner comparisons (i.e., evolution over time).
3. The *tracking of multiple contextual variables over time*, to understand each learner's unique context and how it (along with learners' individual differences) influences the learning process. Given the breadth of contextual variables that could be tracked (often impossible to predict during the design of an LA system), SCLA can resort to the analysis of unstructured data sources (e.g., coding/labeling of open-ended answers provided by the learner) as a way to glean which contextual aspects may be relevant and track them (see the illustrative case study below).

Yet, within this general framework of SCLA, different kinds of SCLA proposals are possible, with different purposes: a) to observe, *explore* and inductively understand a learning context or process (and eventually generate potential theories or hypotheses); b) to deliver particular *interventions* and understand their effects on individual learners; c) to deductively *validate* pre-existing theories or models of learning, understanding how they apply under a particular context/learner. These different purposes will in turn drive the choice of data science methods to be incorporated into the SCLA system (exploratory data analysis vs. inferential vs. predictive).

Taking SEL as a specific field of application of SCLA, we hypothesize that this kind of LA approach can be especially useful to develop non-cognitive competences related to the individual, such as *self-awareness* of emotions and patterns of behavior, as well as their *self-management* (see CASEL, 2003; Eklund et al., 2018).

There are many other aspects that should be taken into account within a single-case approach to LA, which we cannot cover here due to space limitations: the limited scope of the insights and inferences that can be extracted from SCLA (we will not be able to claim generalizability of such models of a single learner to another learner); the need for rich data about the single learners (to enable the use of computational and statistical methods); the richness or multiplicity of theories that can be used to analyze and triangulate the findings of such single-case analyses; or the need for interpretable models over black-box ones (as we will use them to understand a particular learning situation as researchers, or to help learners themselves understand it). In this sense, analysis methods that are theoretically flexible and mix quantitative and qualitative approaches (such as quantitative ethnography methods, see Shaffer, 2017) can be especially useful within an SCLA approach (see Prieto et al., 2020b for an example of the added value of such methods in the context of a lifelong learning process).

Finally, it is worth noting that SCLA can eventually be useful for another common goal of any educational researcher: to produce generalizable theories or models of learning and contextual influences that are applicable across multiple learners and contexts. Although this is not the main goal of SCLA, once enough data is collected and analyzed for single cases/learners in multiple contexts, we can also start making more usual cohort-based analyses, to find "average effects" of contextual variables or interventions. While these average trends may not necessarily apply to every single learner and their context, they are still useful as a starting point for factors or interventions to be tried out by single learners; a later application of SCLA would find out exactly how these initial factors or interventions play out in the learner context (or which others may be worth adding or removing).

12.4 A Simple SCLA Platform: LAPills

As we set out to explore the SCLA approach and how it could be applied to SEL in doctoral education, we wondered what would be the simplest possible implementation of an SCLA system that still enables us to explore these ideas. Drawing from prior work in another field where personalized, longitudinal data collection was deemed useful for reflection (teacher professional development, see Prieto et al., 2020a), we developed the LAPills platform[1] as a "minimum viable SCLA system".

LAPills allows teachers (or doctoral students) to gather custom data on their own, over time (storing it into what we call *data gathering sessions*, see Fig. 12.1, top), and apply different analyses to understand the data gathered. LAPills is based on the principle that LA is most useful when aligned with the learning design (LD) of the experience (see, e.g., Lockyer et al., 2013; Rodríguez-Triana et al., 2015), as taking such learning design (which is an important part of the context in which the learning process takes place) into account can help guide the analysis and interpret its results. In LAPills, this alignment between LD and LA is achieved through the definition and use of data gathering session *templates*, which can be reused again and again, to enact an LA-enriched set of learning activities. The templates define not only the learning design (i.e., the sequence of learning activities and resources/materials to be used in them), but also the points and instruments where relevant data gathering takes place (e.g., a questionnaire to be filled in after a certain activity). The templates also define at which moments during the activities the data

[1] LAPills platform: https://web.htk.tlu.ee/lapills/

introduced should be fed back to the learner via an LA dashboard (e.g., for reflection purposes, see Fig. 12.1, bottom).

LAPills is currently implemented as a simple web platform (based on the Drupal content management system). As data gathering technologies, LAPills currently supports only questionnaires (including both open and closed questions) and activity timers (for, e.g., simple classroom observations, see Saar et al.). Yet, other data gathering means available in a web browser (e.g., audio, video, logs of the LAPills

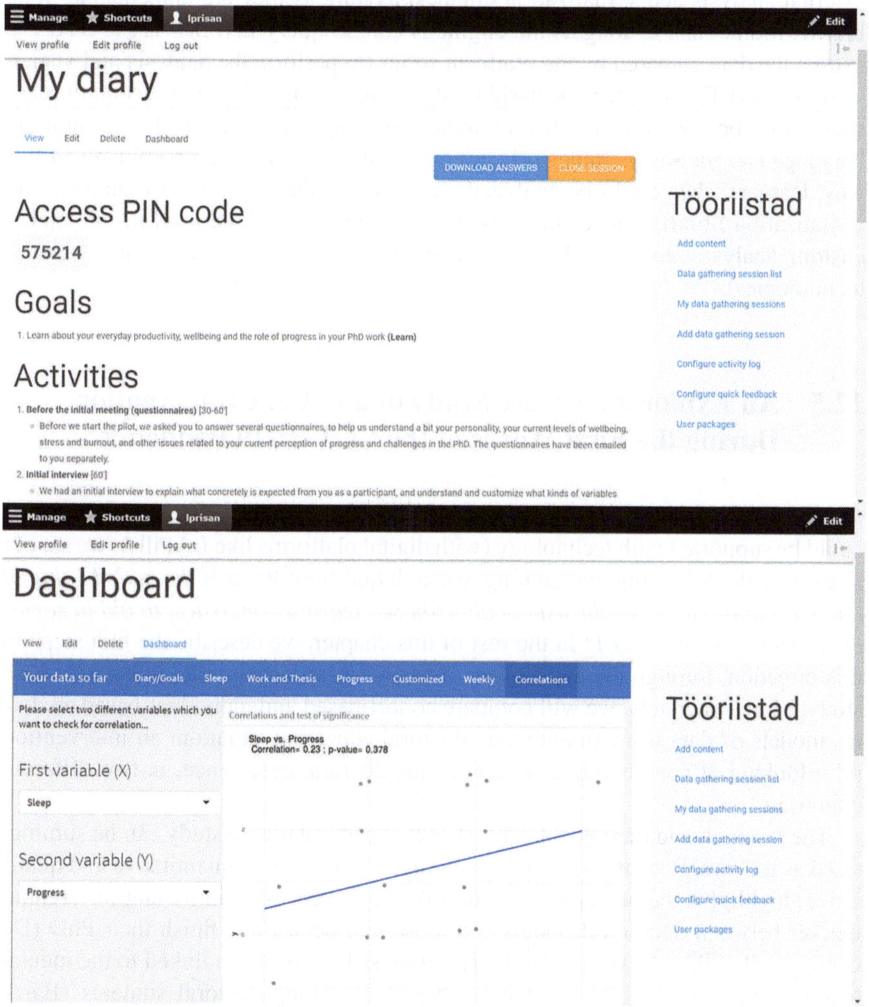

Fig. 12.1 Screenshots of the LAPills platform. Top: data gathering session page, detailing the learning design and including links to the different data gathering instruments (e.g., questionnaires) and to download the data gathered so far. Bottom: custom dashboard with analyses customized on the basis of the particular learning design (in this case, interactive visualization to explore the correlations between quantitative variables in the diary data being gathered during the session)

platform itself or other platforms) could be used in extensions of the LAPills platform.

In terms of data visualization, LAPills incorporates by default a dashboard that simply summarizes the responses gathered so far for each of the questions in the questionnaires defined in a data gathering session template (similar to the summaries available in questionnaire services like Google Forms). Yet, to fulfill the potential of an SCLA approach, context-specific analyses of the learners' data are needed (e.g., in the form of custom LA dashboards). LAPills also allows the integration of external analysis and visualization engines, as long as they are embeddable in an HTML iframe, and as long as the engine is able to query LAPills' HTTPS API to extract the data gathered by the platform so far (to perform the analysis and visualization tasks). For instance, the dashboard featured in Fig. 12.1 (and used in the case study described below) has been implemented using the R statistical programming language and its Shiny framework and hosted data visualization platform.[2] In this way, learners' data could be analyzed using any of the vast array of analysis and visualization libraries in R (and similar mechanisms could be used to implement custom analyses and visualizations using Python or other major data science technologies).

12.5 An Exploratory Case Study of an SCLA Intervention During the First Waves of the COVID-19 Pandemic

Against the backdrop of what single-case learning analytics (SCLA) is and how it could be supported with technology (with digital platforms like LAPills), we set out to explore the following overarching research question: *What is the added value of an SCLA approach over the usual cohort-based learning analytics, to aid in socio-emotional learning (SEL)?* In the rest of this chapter, we describe the first steps in this direction, through the application of an SCLA approach in an illustrative case study. More concretely, we will compare cohort-based and individual-based analytics models of data from an authentic doctoral education situation: an intervention pilot looking at non-cognitive aspects of the doctoral experience, in two different countries.

The research and theoretical context of this particular case study can be summarized as follows. Recent research in doctoral education (both quantitative and qualitative) highlights the perception of one's own progress as an important differential marker between doctoral dropouts and those that persist and finish their PhD (De Clercq et al., 2021; Devos et al., 2017). Progress has also been linked to the mental health symptoms that are worryingly prevalent among doctoral students (Barry et al., 2018; Milicev et al., 2020). Such perception of progress also appears prominently in studies of productivity and emotional wellbeing in the innovation and

[2] Shiny platform: https://www.shinyapps.io/

knowledge work industry (e.g., Amabile & Kramer, 2011). Given that previous progress-related research in doctoral education used (retrospective) interviews or questionnaires removed from students' everyday experience, we aimed at better understanding the everyday experiences of doctoral progress (and what contextual factors may be related to them) using diaries as the main data gathering device. We further hypothesized that an SCLA approach analyzing such data and reflecting it back at doctoral students themselves, could help doctoral students develop several of the core SEL skills (CASEL, 2003; Eklund et al., 2018), especially self-awareness and self-management.

The educational context, methods and results in the following sections describe our attempts at answering the following case-specific research questions: *In what ways are the insights from SCLA modeling of progress experiences and their contextual relations, different from a cohort-based analysis of the same diary data?* (RQa); and *how do SCLA models/analyses of progress (based on the diary data) compare to cohort-based ones, in terms of objective performance?* (RQb).

12.5.1 Context

The intervention pilot took place in the Spring of 2020, framed within a collaboration between Tallinn University (TLU, Estonia) and the University of Valladolid (UVA, Spain). Albeit both countries are at opposite sides of the European spectrum (both geographically and culturally), and the institutions themselves were also quite different (TLU being relatively new and comparatively small, while UVA is medium-sized and its history dates back several centuries), both shared common concerns: the high rates of dropout (or long completion times, over the four years allotted to the degree in both institutions) and the increasing prevalence of mental health issues among their doctoral students.

In the frame of this collaboration, doctoral workshops on the topic of progress and wellbeing were organized in both institutions in the winter of 2019–2020 (in which practices like journaling or self-tracking were recommended, based on the work of Amabile and colleagues). After the workshops, nine students agreed to try out a follow-up pilot intervention featuring those practices, as described below.

The temporal context of the pilot intervention is also noteworthy. The pilot intervention took place in the months of April–June 2020, coinciding (by chance, not by design) with the first waves of the recent COVID-19 pandemic, as different levels of restriction were put in place in both Estonia and Spain. Such restrictions forced all doctoral students to work from home for most of the length of the intervention, but did not have a major impact in the delivery of the intervention itself (through a digital platform). Yet, these events make the results, insights and the data themselves quite context-specific and not generalizable to other contexts or time periods.

12.5.2 Methods

The study around this intervention took the form of an exploratory mixed-methods (Creswell & Clark, 2007) case study (Yin, 1994), with the goal of exploring the potential effects and insights that could be garnered from (quantitative and qualitative) diary data taken longitudinally on a close-to-daily frequency by students themselves.

Participants Participants were a self-selected group of nine (N = 9) doctoral students that had attended the aforementioned workshops on doctoral progress and wellbeing. All nine students were female and, as we can see in Table 12.1, five of them were from a Spanish university, and four of them from an Estonian one (all of them were of local nationality). Students were from different disciplines (from Educational Sciences to Biomedicine or Humanities) and were at different stages of their PhD (from first-year students to those finishing after four years of studies).

Data Gathering The main data source of this study was a structured diary questionnaire which included both quantitative and open data inputs. Quantitative questions included an assessment of the students' own progress on that day (from −3, very unsatisfactory, to +3, very satisfactory) the number of hours slept, and the number of hours worked on the dissertation materials and in other work tasks not related to their thesis topic. Open-ended inputs were a narrative description of the events of the day, as well as the feelings and reflections that they prompted; and a list of goals to be achieved on the next day. This diary was implemented through the LAPills platform, and participants were encouraged to fill it daily (at least on the workdays) for at least six weeks (could be filled in for as long as they desired). After the second week of the intervention, a custom dashboard was activated within LAPills showing the time series for the different quantitative variables in the diary, a reproduction of the open-ended questions, as well as graphs showing correlations between different (quantitative) variables and simple (SCLA) linear/tree models of the "perceived progress" variable as a function of other quantitative variables, based on the data of that participant (see the results section for examples). Aside from the diaries, the study also included pre-, mid- and post-intervention interviews and

Table 12.1 Basic demographic characteristics of the cohort of participants in the pilot study

Participant #	Gender	Nationality	Discipline	Nr. diary entries
P1	F	Estonian	Technology	17
P2	F	Spanish	Technology	20
P3	F	Spanish	Literature	46
P4	F	Spanish	Nursing	49
P5	F	Estonian	Technology	23
P6	F	Estonian	Humanities	14
P7	F	Estonian	Education	17
P8	F	Spanish	Chemical Eng.	45
P9	F	Spanish	Physics	36

questionnaires to further understand doctoral students' mental health and progress experiences. The rest of the chapter, however, will focus on the diary data only, which is most amenable to the kinds of longitudinal analysis that SCLA is concerned with.

Data Analysis The responses to the narrative open-ended question in the diary were analyzed through what Hsieh and Shannon (2005) would call "conventional content analysis" (i.e., using an inductive coding process), performed by the first author after segmenting at the sentence-level. This manual content analysis was performed to glean different contextual factors from the data themselves, rather than from existing theories. Later on, different epistemic network analyses (ENA) (Shaffer & Ruis, 2017) were performed on the coded diary sentences, thus mapping each of the days of each of the participants to a two-dimensional "epistemic space".

To answer the research question about the (subjective) differential insights of single-case vs. cohort-based analyses of data (RQa), simple stepwise linear regression models (using AIC to select the "best model") of the "perceived progress" variable were developed. This kind of model was chosen due to their easy interpretability, both for researchers and for doctoral students themselves. We built these models using different sets of variables as predictors: (a) the quantitative variables in the diary itself (hours of sleep, hours of work, hours of thesis-related work); (b) the code counts of the different codes from the content analysis of the narrative open-ended responses in the diaries (Prieto, Rodríguez-Triana, et al., 2020b); and (c) the scores of each day in the two first dimensions of the ENA epistemic space defined by the diary entries (similar to what Prieto, Rodríguez-Triana, et al., 2020b; Swiecki & Shaffer, 2020 do). Predictors (b) and (c) were added to the modeling in order to understand the value of unstructured data (key in the SCLA approach) over using solely quantitative predictors. To compare the insights of the cohort-based and SCLA approaches, these regression models (and their interpretation) were built first for the whole cohort of nine doctoral students, and then separately for each participant, using only her data.

To answer the research question about the (objective) performance of the different models of perceived progress developed for the previous questions (RQb), models were evaluated in terms of their ability to predict a day's perceived progress on the basis of that day's predictor variables. We triangulated three evaluation metrics commonly used in regression models: mean absolute error (MAE), root mean squared error (RMSE) and R-squared. Given the inherent time-series nature of the diary data, we used backtest cross-validation (also known as walk-forward cross-validation, Hyndman & Athanasopoulos, 2018), in which the models are trained on the first n elements of the time series (in this case, the data of the first n diary entries), and the error is measured on trying to predict the next m elements (unseen by the model so far). This process of training and testing is repeated while increasing the training window ($n + 1$, $n + 2$…) and predicting the following m elements. In this case, given the relatively small number of data points (see Table 12.1), the models were evaluated from a minimum of $n = 7$ training data points, and using $m = 3$ testing points.

12.5.3 *Results*

The nine PhD students that participated in the SCLA pilot intervention entered a total of 267 journal entries in the system.

12.6 Cohort-Based Models of Progress

To start understanding the everyday experiences of progress across participant doctoral students (i.e., for the whole cohort), and how different kinds of factors seemed to relate to it, we used linear regression models trying to relate students' self-assessment of progress (in a −3 to +3 Likert scale), with the different quantitative variables recorded daily (hours of sleep, hours worked in general, hours worked on their thesis research). The results can be seen in Table 12.3: the best stepwise linear regression model (i.e., balancing predictive power with simplicity) showed that both hours of sleep and hours dedicated to thesis research were significant (and positive) predictors of progress. This model, however, explained only about 17% of the variance in the whole cohort's dataset.

These initial quantitative models seem logical (e.g., dedicating time to develop the dissertation contributions should help feel progress in the PhD) but not very enlightening. To start delving into other contextual factors that could also relate to progress, we could look at what kinds of events and circumstances were mentioned in journal entries, as per our (inductive) qualitative content analysis. Table 12.2 shows the most common codes encountered across all participants.

Yet, do these codes appear differently on days with good progress (vs. other days)? Figure 12.2 below shows an epistemic network analysis (ENA) comparing the epistemic space of journal entries in which the reported progress was positive (i.e., from +1 to +3), versus those days in which it was not (i.e., from −3 to 0). We can see that the networks of positive-progress days are significantly different from that of non-positive ones. Positive progress days feature heavily both thesis work and other types of work (many of the doctoral students had other work obligations unrelated to the thesis, like teaching or project work), as well as learning activities. On the other hand, non-positive progress days tended to feature mentions to not only these "other work" tasks, but also meetings, emails or calls, taking time off, feeling tired or sick, or feeling like they didn't have enough time.

We can also use these code counts in different days in exploratory stepwise regression models as predictor variables, to try to disentangle and assess the strength of their relationships to the level of progress perceived by students in a given day. The results of this modeling across students (see Table 12.3) suggest that, aside from doing thesis-related activities, spending time learning and applying productivity techniques (like the Pomodoro method, Cirillo, 2006) seemed to be positively associated with good progress. Conversely, interruptions, feelings of time pressure or having to dedicate time to household or family obligations, were negatively

Table 12.2 Most commonly encountered codes in the inductive qualitative content analysis of (unstructured) diary entries across all participants

Code	Definition	Examples	Count
ThesisWork	Working on materials that will contribute to the dissertation (e.g., reading, writing, data analysis, etc.)	*"I also worked on a PhD report for my next paper and started data collection."*	207
OtherWork	Other work obligations that do not contribute to thesis materials (e.g., teaching, admin or unrelated project work)	*"Search information in the internet about a new piece of equipment that my boss wants me to purchase"* *"I only prepared classes"*	109
LackingTime	Feelings of time pressure, not having enough time or other time-related stress	*"I had a sensation of overwhelm, since we decided to write the paper a few days ago and the deadline is very close by"*	38
FreeTimeRest	Spending time or whole days resting or enjoying leisure activities.	*"I took the day for resting, since I had barely slept [...]"*	31
ExhaustionSickness	Feelings of exhaustion (physical or emotional), as well as sickness.	*"Migraine day."* *"In the night I started with the thesis and it was hard to focus, I was a bit tired."*	29

associated with positive progress days. This model, enhanced with qualitative code counts, accounts for about 31% of the variance in the cohort dataset (a substantial improvement over the model including just the quantitative variables of the diary), hinting at the value of such contextual variables extracted from unstructured data.

We could also build similar models of progress using the ENA coordinates of a day's epistemic network (as done, e.g., in the iSENS data analysis method, Swiecki & Shaffer, 2020). Yet, while the stepwise linear models had some of these dimensions as significant predictors, their predictive power (in terms of variance explained) and ease of interpretability did not match the code-count-based ones (hence, not reported in Table 12.3 for brevity's sake).

12.7 Individual (SCLA-Based) Models of Progress

To contrast the models and insights across all participants of our pilot cohort, and understand the added value of a single-case learning analytics (SCLA) approach, we can now examine similar analyses. Due to space limitations we will focus our analysis on three of the participants, at different ends of the spectrum of diary data provided (see also Table 12.1): P3 was the doctoral student that introduced the most sentences in the diary, while P6 was the least prolific participant (since she had the least number of entries, often also refusing to provide an unstructured text entry),

Table 12.3 Parameters and variance explained (in terms of R-squared) of different cohort-based and individual (SCLA) stepwise linear regression models. First two columns of parameters: cohort-based models; third and fourth columns: individual models for P3; fifth and sixth columns: Individual models for P1; seventh and eigth columns: Individual models for P6

	Quanti vars only	Quanti+codes	Quanti vars only	Quanti+codes	Quanti vars only	Quanti+codes	Quanti vars only	Quanti+codes
Sleep	0.23 **	0.19 *			1.07 *	1.18 *	0.44 *	0.50 **
	(0.08)	(0.08)			(0.46)	(0.37)	(0.19)	(0.11)
Thesis	0.31 ***	0.21 ***	0.35 ***	0.31 ***		0.33	0.20 *	0.17 **
	(0.04)	(0.05)	(0.05)	(0.04)		(0.29)	(0.09)	(0.05)
Effort		−1.03 **						
		(0.32)						
GenericTasks		0.72		−0.84 *				
		(0.37)		(0.37)				
HouseworkFamilyPersonal		−0.97 ** (0.30)		−2.67 *** (0.52)		−1.82 (2.27)		
InterruptionsInterferences		−1.37 * (0.59)				−2.83 (2.10)		
LackingTime		−0.48 *				−0.67		
		(0.20)				(0.55)		
Learning		0.90 **						
		(0.30)						
ProductivityTechniques		1.51 ** (0.57)		1.15 * (0.51)				
ThesisWork		0.51 ***				−1.01		
		(0.13)				(0.67)		

Work			0.17 **	0.10	0.52 **	0.46 *		
			(0.06)	(0.05)	(0.16)	(0.15)		
ExhaustionSickness				−0.29 (0.18)				−1.29 *** (0.26)
EmailCalls						1.42		
						(1.56)		
FreeTimeRest						0.92 (0.87)		
PostponeProcrastinate						−1.42 * (0.54)		
OtherWork						−0.26		0.63 *
						(0.24)		(0.21)
N. Obs.	267	267	46	46	17	17	14	14
Adjusted R^2	0.17	0.31	0.73	0.84	0.38	0.83	0.36	0.79

* = 0.05; ** = 0.01; *** = 0.001

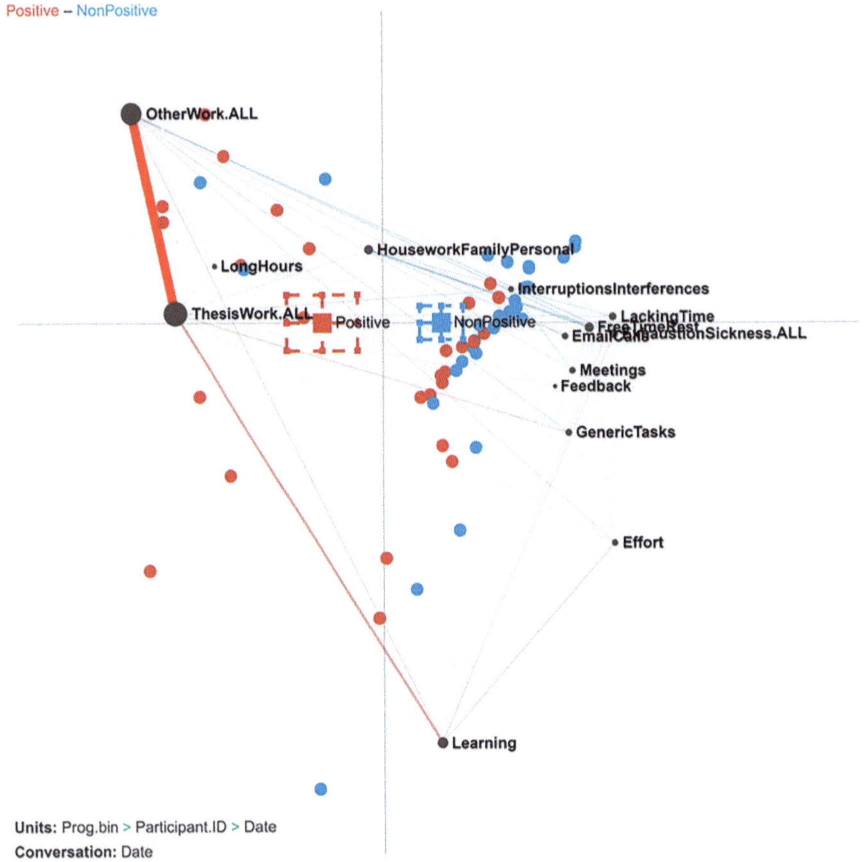

Fig. 12.2 Epistemic network analysis (ENA) of codes present in the whole-cohort dataset, rotated by the means of positive (vs. non-positive) progress days

and P1 had also rather few entries, but her unstructured entries were longer. This will help us explore the limits of the SCLA approach in terms of the volume of quantitative and qualitative data needed to obtain insights about the participant student's situation and progress trends.

By looking at stepwise linear models for the three aforementioned doctoral students (Table 12.3) that take into account only the quantitative variables, we can observe quite different models in terms of what predictors are considered significant and their relative strength, both amongst particular students (e.g., P3's model considers both hours dedicated to the thesis and hours dedicated to work in general, while P6's considers thesis hours and sleep, but not work hours) and with the cohort-based model (P6's model shows sleep as a stronger predictor, while the cohort-based model considered thesis-related time a stronger predictor of positive progress). Even more notably, we can see how the variance explained by these individual, quantitative variable models (in terms of their adjusted R-squared) is also higher than the cohort-based one.

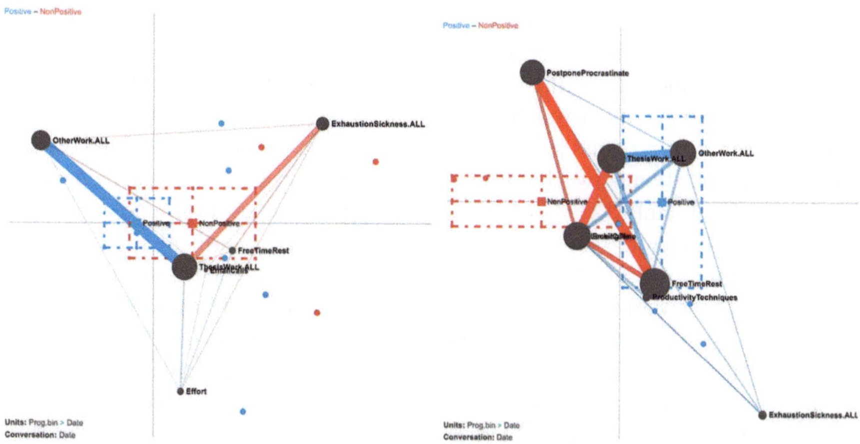

Fig. 12.3 Epistemic network analysis (ENA) of codes present in P3's (left) and P1's (right) unstructured diary data, rotated by the means of positive (vs. non-positive) progress days

If we then take into account the inductive qualitative content analysis of the unstructured narrative text accompanying each diary entry, we can see that the factors mentioned in these entries vary widely between different doctoral students. For instance, for P3, mentions to thesis work (especially, reading and writing) were very common, but also were other work mentions (specifically, to teaching duties) and exhaustion or sickness mentions. However, for P1, aside from thesis and other work tasks, mentions to postponing tasks (including procrastination) or feeling like there is not enough time were more common, as were mentions to dedicating time to emails and calls. Figure 12.3 below shows the two different ENA spaces generated by the codes of P3 and P1's narrative journal entries, comparing positive and non-positive progress days. We can see that, for P3, managing to work on both her thesis and her other work duties was the clearest marker associated with good progress days, even if that required some effort: *"Today I did not do much about the thesis, I was grading and sending emails. Yet, I have determined what direction to follow now that I have changed [the] section [I'm writing]."* Not so productive days were marked by the appearance of sensations of sickness or exhaustion, even if some thesis-related work was done: *"Despite being very tired, I have managed to read something, I have given my last literature class of the year, and I have read some pages of the [thesis-related] novel."* In contrast, P1's epistemic space suggests that postponing thesis-related tasks to take free time was strongly associated with non-positive progress days, while taking time off in the days where thesis-related activities were also present was rather associated with positively perceived progress (*"Finished first draft of the article as planned. Now I'm going to rest for 2 days from writing."*). Interestingly, days in which thesis work was combined with time emailing or in calls, tended to be perceived as less productive: *"Wrote one paragraph. Didn't progress very well. I had many interruptions (calls)"*. As we can see, many of these particular patterns were not noticeable in the cohort-based analyses, but could be beneficial for the doctoral students to be pointed out and reflected upon, to maybe change their ways of working.

As we did in the case of the cohort-based models, we could also add the code counts of these frequently mentioned factors to stepwise regression models trying to predict the reported progress perceived by students in a given day, to see the relative strength of association between different factors in a particular doctoral student. As we can see from Table 12.3, such models are quite different from one another, in terms of the predictors that seem meaningful. It is also worth noting that these machine learning models enhanced with qualitative code counts explain a much higher share of the variance in the perception of progress. For instance, the individual model of P3's progress that considers not only time dedicated to the thesis and other work as predictors, but also the presence of family/personal obligations or the use of productivity techniques, explains up to 84% of the variance in that student's dataset (as per the model's adjusted R-squared). As with the case of cohort-based models, ENA scores could also be added to predictive models of progress, but these indicators did not provide an appreciable advantage over these "naive" code-count models.

12.8 Objective Performance of Cohort- and SCLA-Based Models of Progress

So far, we have provided illustrations of different learning analytics models that can be built with the longitudinal data of a simple intervention in a doctoral education setting. Although we have already seen how SCLA (i.e., individual-based) models provide insights about associations between factors that are arguably more directly applicable to a particular student's situation, we could also ask ourselves whether these individual models are really more predictive than the cohort-based ones, within their scope of application (a whole cohort for cohort-based models, an individual for the SCLA ones). Taking into account that the diary data we are focusing on is inherently a time series, the most appropriate way to evaluate it is through backtesting (also known as walk-forward or time series cross-validation, Hyndman), rather than the usual (time-independent) cross-validation. In this kind of backtest, we use models like those presented in the previous section, to progressively try to predict the next data point (or few data points), as we accumulate diary entries over time.

How well do SCLA-based models (i.e., trained on a single individual, trying to predict a single individual's progress score) perform, compared with an equivalent "classic" LA model of the same kind (i.e., trained on a cohort of individuals)? We could consider two slightly different types of cohort model evaluation, depending on how the cohort-based LA system is expected to work: a) one in which we train a cohort model as we accumulate data about the cohort over time, and try to predict the next data points of that same cohort (what we called "cohort-on-cohort"); and b) one in which we train a cohort model with the data of a cohort, and then try to increasingly train and predict the data of a new participant never seen in the training phase (which we labeled "cohort-on-participant").

As we can see in Fig. 12.4, in our doctoral student diary dataset, the participant (SCLA-based) models performed consistently better than the cohort-based ones as data was accumulated, across the three predictive performance metrics used (mean absolute error, R-squared and root mean squared error). It is also interesting to note that the variance in performance for these SCLA models was higher than that of the cohort-based ones (which is to be expected - cohort models are expected to perform more reliably, given the larger number of data points used for training). Yet, it is noteworthy that the median performance of these individual participant-based models was substantially better than their cohort-based counterparts (see Table 12.4).

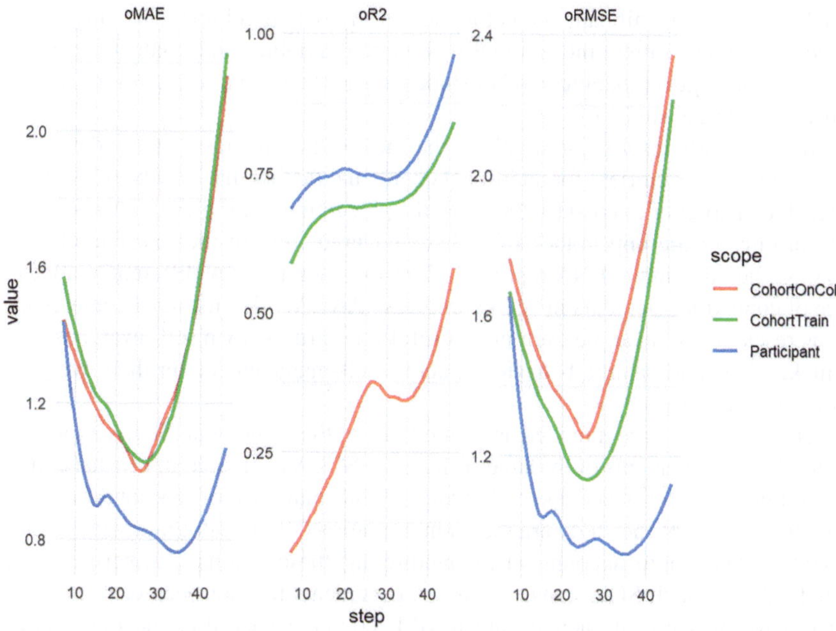

Fig. 12.4 Predictive performance metrics from backtest (i.e., walk-forward) validation of different kinds of models based on the diary data, as data accumulates over time (step), smoothed using a LOcally reWEighted ScatterPlot Smoothing (LOESS) method. MAE = Mean absolute error (smaller is better); R2 = R-squared (higher is better); RMSE = Root mean squared error (smaller is better)

Table 12.4 Median, mean and standard deviation of different evaluation metrics from the backtest (i.e., walk-forward) validation of different kinds of progress models based on the diary data

	MAE		R^2		RMSE	
Type of model scope/ evaluation	Median	Mean(stdev)	Median	Mean(stdev)	Median	Mean(stdev)
Cohort-on-cohort	1.27	1.31(0.37)	0.26	0.31(0.21)	1.55	1.59(0.36)
Cohort-on-participant	1.06	1.29(0.80)	0.83	0.68(0.34)	1.19	1.39(0.81)
Participant (SCLA)	**0.80**	**0.94(0.87)**	**0.92**	**0.75(0.33)**	**0.94**	**1.09(0.97)**

MAE Mean absolute error (smaller is better), R^2 R-squared (higher is better), *RMSE* Root mean squared error (smaller is better)

12.9 Discussion and Conclusions

The results from the case study analysis above illustrate the objective advantage of single-case statistical models over cohort-based ones – if sufficient data from the single learning process are gathered. Part of this performance advantage stems from these single-case models' narrower scope (i.e., it is easier to predict the data for a single learner than for a whole cohort), but also from the fact that the machine learning models chosen (stepwise linear regression) already perform a selection of relevant predictors within the whole array of variables in the dataset, of which there are bound to be fewer applicable in the scope of prediction, in the cohort models (few variables will be significant predictors across the whole cohort). The single-case models are able to pinpoint relationships among variables and contextual factors that may not be generalizable to other students – nor they need to be, as long as they apply to the target student.

The case study results also highlight the value of mining unstructured data (e.g., the narrative diary entries) for additional contextual factors that may be relevant for the SEL construct of interest (in this case, the perception of progress). This is apparent not only in the appearance of code-count predictors in the linear models of progress, but also in the ENA representations that clearly show differential appearance of contextual elements in "good progress" days. While this may be interesting for us researchers, it can be even more useful for learners' own self-awareness and to make plans and change behavior patterns (self-management) in their everyday learning experience.

Yet, these results are intended as illustrations of the value of an SCLA approach, not as blanket statements of the superiority of an SCLA approach (or of what counts as good progress for every doctoral student). The small size of our cohort and the limited timespan of the study are the main limitations of our pilot intervention, and should be taken into account when interpreting these results. Furthermore, the actual value of such SCLA analyses for PhD students has not been evaluated yet, nor did we measure the actual gain in SEL competencies after the intervention (something that remains difficult, given the lack of validated research instruments targeting doctoral students, or adults in general). The current research design also does not allow to understand which components of the SCLA experience could be responsible for any potential gains (i.e., is it the looking at the dashboard, or the writing of diaries, or both? Etc.).

Despite these limitations, our initial work already uncovered several challenges that LA researchers working on SEL should be mindful of, which make up an initial research agenda for SCLA to support doctoral SEL:

- *Dealing with the unstructured tasks/contexts of the doctorate.* As shown in our small dataset of everyday doctoral experiences, the learning tasks of a doctoral student are very heterogeneous, unstructured, and cross contextual boundaries (at home, in the lab, the office, etc.) seamlessly. This makes the interpretation of LA outputs challenging, as absolute values of quantitative metrics (e.g., I worked four hours today on my thesis) have very different meanings according to students' individual differences and situations (is it a weekend? Do I have teaching duties along with my research work? Is it now evaluation season? Etc.). Novel techniques of mining such contextual variables from different data sources across different spaces and contexts, will be a critical part of any SCLA effort attempting to support doctoral SEL. In this sense, recent work on rules-based (Cai et al., 2019) or natural language processing (NLP)-based automated coding of unstructured text data (Prieto et al., 2021) represent a promising direction to provide the kinds of insights showcased in our small case study, at scale.

- *Temporal analytics for doctoral SEL.* The simple machine learning models used in the case study above still do not exploit the inherently time-series nature of the data to be gathered in a SCLA approach (about a single person, over long periods of time). Although temporal techniques like sequence analysis or Markov processes are still not widely applied in LA (Knight et al., 2017), the study of doctoral SEL using SCLA seems like a perfect match for techniques that enable the understanding of the changing dynamics and evolution of learning processes (see also the notion of ENA trajectories, Brohinsky et al., 2021). One critical aspect that such time-aware models will need to take into account, is their interpretability by end users (see the following point).
- *Privacy, data ownership, interpretability, personalization and other ethical dilemmas.* As we saw in our overview of related work, LA platforms often take a hierarchical top-down approach in which learner data is analyzed for the benefit of instructors or administrators. In an SCLA approach to support doctoral SEL, a much larger emphasis is put into learners' use and understanding of their own data and particular contexts. Given the amount of contextual mining involved in SCLA, and the sensitive nature of many of the constructs involved (emotional regulation, mental health, etc.), it is of critical importance that technologies take privacy and ethical issues into account, if we want to avoid dystopian futures of SCLA usage (Ferguson et al., 2016). Platforms that enable learner ownership of their data, are customizable to doctoral students' own (and potentially very varied) objectives, and provide interpretable results, will be sorely needed. Thus, the extrapolation of recent efforts into explainable LA (De Laet et al., 2018) and value-sensitive design of LA systems (Chen & Zhu, 2019) is a necessary element of future SCLA efforts in this area.
- *Research instruments and frameworks for doctoral SEL.* Although there exists now a broad body of work on the benefits and intervention programs to support SEL in schools, much less work and research tooling is available for adults. Given the increasing heterogeneity of the doctoral student population (often comprising foreign students from different cultures, who may or may not have had any prior SEL education), it is unclear whether research instruments designed to measure whether (and how much) SEL skills have improved in children (e.g., SELS, Thomas et al., 2021) could be easily adapted to this particular adult population. It is not even clear whether the same developmental frameworks and competences for SEL defined for children, should be directly applied to doctoral students (who are themselves of very different ages and stages in life), and will require further exploration by researchers.

- *Lack of a single clear theoretical framing.* As noted by Conley's (2015) review of SEL in higher education, multiple kinds of interventions exist within the umbrella of SEL, which operate under very different theoretical and psychological assumptions (from cognitive-behavioral therapy to mindfulness and many others). The rest of the chapters in this book also suggest a variety of theoretical perspectives that LA could use as a base to support non-cognitive learning aspects. This theoretical heterogeneity can be seen as a challenge impeding accumulation of knowledge in this area, but it could also be leveraged by LA solutions that combine multiple theories for a richer understanding of phenomena (see, e.g., Bauer et al., 2019), perform "theoretical triangulation", and engage in what some authors have called "theoretical ecumenism" (Dillenbourg et al., 2018).
- *Going beyond the individual.* Although a doctoral dissertation is seen as primarily an individual endeavor, many other actors play critical roles in this long learning process: (co-)supervisors are a clear influence in the process, but also peers, and even families and friends. The use of these other actors as alternative data sources and to provide additional or contrasting perspectives, can be essential for the development of doctoral SEL skills, and could prevent perceptual biases that are common at this level of education (see, e.g., the 'impostor syndrome' so prevalent in doctoral students, e.g., Pervez et al., 2021). In this sense, the application of co-regulation or socially-shared regulation of learning theories (Hadwin et al., 2011) to the doctoral learning process can also be a beneficial path for future SCLA research.

This chapter has only scratched the surface of the rich, largely unexplored area of research at the crossroads of doctoral education, SEL and learning analytics. Indeed, a single-case approach to learning analytics (SCLA) could also be beneficial for other educational levels and settings where individual differences and context are of critical importance, such as special education or lifelong workplace/professional development. We focus on doctoral education first because the inquisitive nature and analytical skills of doctoral students make this population the most likely "low-hanging fruit". In other areas with different learner populations, data sources beyond written reflections may need to be used (cf. the use of multimodal learning analytics to understand embodied aspects of learning, see Andrade et al., 2016), and devices other than data visualizations and quantitative models will undoubtedly be needed as outputs. There is still a lot of work to be done to realize this vision of SCLA systems that help each learner understand their socio-emotional experience. We hope you, the reader, will take this challenge up and join us in this path towards a better socio-emotional support of our future researchers, and for every learner out there.

Acknowledgements This research has received funding from the European Union's Horizon 2020 research and innovation programme under grant agreement No. 669074. It has also received funding from the European Union's Erasmus Plus programme, grant agreement 2019-1-NO01-KA203-060280. The Universidad de Valladolid co-authors acknowledge funding of the European Regional Development Fund and the National Research Agency of the Spanish Ministry of Science, Innovation and Universities, under project grant TIN2017-85179-C3-2-R, and PID2020-112584RB-C32, the European Regional Development Fund and the Regional Government of Castile and Leon, under project grant VA257P18.

References

Amabile, T., & Kramer, S. (2011). *The progress principle: Using small wins to ignite joy, engagement, and creativity at work*. Harvard Business Press.

Andrade, A., Delandshere, G., & Danish, J. A. (2016). Using multimodal learning analytics to model student behaviour: A systematic analysis of behavioural framing. *Journal of Learning Analytics, 3*, 282–306.

Bair, C. R., & Haworth, J. G. (2004). Doctoral student attrition and persistence: A meta-synthesis of research. In *Higher education: Handbook of theory and research* (pp. 481–534). Springer.

Bar-On, R., Handley, R., & Fund, S. (2006). The impact of emotional intelligence on performance. In *Linking emotional intelligence and performance at work: Current research evidence with individuals and groups* (pp. 3–19). Psychology Press.

Barry, K. M., Woods, M., Warnecke, E., et al. (2018). Psychological health of doctoral candidates, study-related challenges and perceived performance. *Higher Education Research and Development, 37*, 468–483.

Barry, K. M., Woods, M., Martin, A., et al. (2019). A randomized controlled trial of the effects of mindfulness practice on doctoral candidate psychological status. *Journal of American College Health, 67*, 299–307.

Bauer, E., Sailer, M., Kiesewetter, J., et al. (2019). Using ENA to analyze pre-service teachers' diagnostic argumentations: A conceptual framework and initial applications. In *International Conference on Quantitative Ethnography* (pp. 14–25). Springer.

Beardsley, M., Vujovic, M., Portero-Tresserra, M., & Hernández-Leo, D. (2019). ClassMood app: A classroom orchestration tool for identifying and influencing student moods. In *European conference on technology enhanced learning* (pp. 723–726).

Boud, D., & Lee, A. (2006). *What counts as practice in doctoral education. Quality in postgraduate research: Knowledge creation in testing times CEDAM* (pp. 45–54). The Australian National University, Canberra.

Boyle, T., & Ravenscroft, A. (2012). Context and deep learning design. *Computers & Education, 59*, 1224–1233.

Brohinsky, J., Marquart, C., Wang, J., et al. (2021). Trajectories in epistemic network analysis. In A. R. Ruis & S. B. Lee (Eds.), *Advances in quantitative ethnography* (pp. 106–121). Springer International Publishing.

Bronkhorst, L. H., & Akkerman, S. F. (2016). At the boundary of school: Continuity and discontinuity in learning across contexts. *Educational Research Review, 19*, 18–35.

Cahyani, A. D., Marshall, L., & Forshaw, M. (2019). Students' perception on data sources from outside virtual learning environment for learning analytics. In *Proceedings of the 2019 11th international conference on education technology and computers* (pp. 165–170). ACM.

Cai, Z., Siebert-Evenstone, A., Eagan, B., et al. (2019). nCoder+: A semantic tool for improving recall of nCoder coding. In *International Conference on Quantitative Ethnography* (pp. 41–54). Springer.

CASEL. (2003). *Safe and sound: An educational leader's guide to evidence-based social and emotional learning (SEL) programs*. Author.

CASEL. (2013). *Effective social and emotional learning programs: Preschool and elementary school edition*. Collaborative for Academic, Social, and Emotional Learning (CASEL).

Chen, B., & Zhu, H. (2019). Towards value-sensitive learning analytics design. In *Proceedings of the 9th international conference on Learning Analytics & Knowledge* (pp. 343–352). ACM.

Cirillo, F. (2006). The pomodoro technique (the pomodoro). *Agile Processes in Software Engineering and, 54*, 35.

Collie, R. J., Shapka, J. D., & Perry, N. E. (2012). School climate and social–emotional learning: Predicting teacher stress, job satisfaction, and teaching efficacy. *Journal of Educational Psychology, 104*, 1189.

Conley, C. S. (2015). SEL in higher education. In *Handbook of social and emotional learning: Research and practice* (pp. 197–212).

Creswell, J. W., & Clark, V. L. P. (2007). *Designing and conducting mixed methods research*. Sage Publications.
D'Mello, S., Dieterle, E., & Duckworth, A. (2017). Advanced, analytic, automated (AAA) measurement of engagement during learning. *Educational Psychologist, 52*, 104–123. https://doi.org/10.1080/00461520.2017.1281747
De Clercq, M., Frenay, M., Azzi, A., et al. (2021). All you need is self-determination: Investigation of PhD students' motivation profiles and their impact on the doctoral completion process. *International Journal of Doctoral Studies, 16*, 189–209.
De Laet, T., Broos, T., & Duorado, R. et al. (2018). *Explainable learning analytics: Challenges and opportunities of this emerging research line*.
Devos, C., Boudrenghien, G., Van der Linden, N., et al. (2017). Doctoral students' experiences leading to completion or attrition: A matter of sense, progress and distress. *European Journal of Psychology of Education, 32*, 61–77.
Di Mitri, D., Scheffel, M., Drachsler, H., et al. (2017). Learning pulse: A machine learning approach for predicting performance in self-regulated learning using multimodal data. In *Proceedings of the seventh international learning analytics & knowledge conference* (pp. 188–197). ACM.
Dillenbourg, P., Prieto, L. P., & Olsen, J. K. (2018). Classroom orchestration. In *International Handbook of the Learning Sciences*. Routledge.
Duckworth, A. L., & Yeager, D. S. (2015). Measurement matters: Assessing personal qualities other than cognitive ability for educational purposes. *Educational Researcher, 44*, 237–251. https://doi.org/10.3102/0013189X15584327
Edwards, M. S., King, J. B., & Ashkanasy, N. M. (2019). Social and emotional learning in graduate school to improve student Well-being and performance: A proposed training program. *New Directions for Teaching and Learning, 2019*, 91–105. https://doi.org/10.1002/tl.20367
Eklund, K., Kilpatrick, K. D., Kilgus, S. P., & Aqdas, H. (2018). A systematic review of state-level social–emotional learning standards: Implications for practice and research. *School Psychology Review, 47*, 316–326.
Evans, T. M., Bira, L., Gastelum, J. B., et al. (2018). Evidence for a mental health crisis in graduate education. *Nature Biotechnology, 36*, 282–284.
Ferguson, R., Brasher, A., Clow, D., et al. (2016). Learning analytics: Visions of the future. In *Proceedings of the sixth international learning analytics & knowledge conference*.
Field, J. (2012). Is lifelong learning making a difference? Research-based evidence on the impact of adult learning. In *Second international handbook of lifelong learning* (pp. 887–897). Springer.
González-Betancor, S. M., & Dorta-González, P. (2020). Risk of interruption of doctoral studies and mental health in PhD students. *Mathematics, 8*, 1695.
Hadwin, A. F., Järvelä, S., & Miller, M. (2011). Self-regulated, co-regulated, and socially shared regulation of learning. In *Handbook of self-regulation of learning and performance* (pp. 65–84).
Hakami, E., & Hernandez-Leo, D. (2021). Investigating the well-being impacts of educational technologies supported by learning analytics: An application of the initial phase of IEEE P7010 recommended practice to a set of cases. In *LAK21: 11th international learning analytics and knowledge conference* (pp. 269–279).
Howells, K., Stafford, K., Guijt, R., & Breadmore, M. (2017). The role of gratitude in enhancing the relationship between doctoral research students and their supervisors. *Teaching in Higher Education, 22*, 621–638. https://doi.org/10.1080/13562517.2016.1273212
Hsieh, H.-F., & Shannon, S. E. (2005). Three approaches to qualitative content analysis. *Qualitative Health Research, 15*, 1277–1288.
Hyndman, R. J., & Athanasopoulos, G. (2018). *Forecasting: Principles and practice*. OTexts.
Jonassen, D. H., & Grabowski, B. L. (2012). *Handbook of individual differences, learning, and instruction*. Routledge.
Knight, S., Wise, A. F., & Chen, B. (2017). Time for change: Why learning analytics needs temporal analysis. *Journal of Learning Analytics, 4*, 7–17.

Kovanović, V., Joksimović, S., Mirriahi, N., et al. (2018). Understand students' self-reflections through learning analytics. In *Proceedings of the 8th international conference on learning analytics and knowledge* (pp. 389–398).

Levecque, K., Anseel, F., Beuckelaer, A. D., et al. (2017). Work organization and mental health problems in PhD students. *Research Policy, 46*, 868–879. https://doi.org/10.1016/j.respol.2017.02.008

Liu, M.-C., & Huang, Y.-M. (2017). The use of data science for education: The case of social-emotional learning. *Smart Learn Environ, 4*, 1. https://doi.org/10.1186/s40561-016-0040-4

Lockyer, L., Heathcote, E., & Dawson, S. (2013). Informing pedagogical action: Aligning learning analytics with learning design. *American Behavioral Scientist, 57*, 1439–1459.

Milicev, J., Biello, S., & Gardani, M. (2020). Evaluating mental health and wellbeing of postgraduate researchers: Prevalence and contributing factors. *PsyArXiv.* https://doi.org/10.31234/osf.io/k92hq

Perry, S. M., & Bass, K. M. (2008). *Impact of social-emotional learning software on objective school outcomes among diverse adolescents: A summary analyses of six randomized controlled trials*. Research Associates Rockman et al., Ripple Effects.

Pervez, A., Brady, L. L., Mullane, K., et al. (2021). An empirical investigation of mental illness, impostor syndrome, and social support in management doctoral programs. *Journal of Management Education, 45*, 126–158. https://doi.org/10.1177/1052562920953195

Prieto, L. P., Magnuson, P., Dillenbourg, P., & Saar, M. (2020a). Reflection for action: Designing tools to support teacher reflection on everyday evidence. *Technology, Pedagogy and Education, 29*, 279–295. https://doi.org/10.1080/1475939X.2020.1762721

Prieto, L. P., Rodríguez-Triana, M. J., Ley, T., & Eagan, B. (2020b). The value of epistemic network analysis in single-case learning analytics: A case study in lifelong learning. In *International conference on quantitative ethnography (ICQE20)*. Springer.

Prieto, L. P., Pishtari, G., Rodríguez-Triana, M. J., & Eagan, B. (2021). Comparing natural language processing approaches to scale up the automated coding of diaries in single-case learning analytics.

Rodríguez-Triana, M. J., Martínez-Monés, A., Asensio-Pérez, J. I., & Dimitriadis, Y. (2015). Scripting and monitoring meet each other: Aligning learning analytics and learning design to support teachers in orchestrating CSCL situations. *British Journal of Educational Technology, 46*, 330–343.

Saqr, M., & López-Pernas, S. (2021). Idiographic learning analytics: A single student (N= 1) approach using psychological networks. In *Proceedings of the NetSciLA21 workshop.*

Shaffer, D.W. (2017). *Quantitative ethnography*. Lulu.com

Shaffer, D. W., & Ruis, A. (2017). Epistemic network analysis: A worked example of theory-based learning analytics. In *Handbook of learning analytics*. Society for Learning Analytics Research.

SoLAR. (2011). What is learning analytics? In: Society for Learning Analytics Research (SoLAR). https://www.solaresearch.org/about/what-is-learning-analytics/. Accessed 8 July 2021.

Stern, R., Harding, T., Holzer, A., & Elbertson, N. (2015). Current and potential uses of technology to enhance SEL: what's now and what's next. *Handbook of Social and Emotional Learning: Research and Practice*, 516–531.

Swiecki, Z., & Shaffer, D. W. (2020). iSENS: An integrated approach to combining epistemic and social network analyses. In *Proceedings of the Tenth International Conference on Learning Analytics & Knowledge* (pp. 305–313).

Taylor, S. (2021). *Towards describing the global doctoral landscape*. UK Council for Graduate Education.

Terrell, S. R., Snyder, M. M., Dringus, L. P., & Maddrey, E. (2012). A grounded theory of connectivity and persistence in a limited residency doctoral program. *Qualitative Report, 17*, 62.

Thomas, C. L., Zolkoski, S. M., & Sass, S. M. (2021). Investigating the psychometric properties of the social-emotional learning scale. *Assessment for Effective Intervention, 1534508420984522.*

Walton, K. E., San Pedro, S., Whitmer, J., et al. (2020). *Using feature engineering from online learning environments to observe social and emotional skills and academic performance*. ACT, Inc.

Williamson, B. (2017a). Decoding ClassDojo: Psycho-policy, social-emotional learning and persuasive educational technologies. *Learning, Media and Technology, 42*, 440–453. https://doi.org/10.1080/17439884.2017.1278020

Williamson, B. (2017b). Moulding student emotions through computational psychology: Affective learning technologies and algorithmic governance. *Educational Media International, 54*, 267–288. https://doi.org/10.1080/09523987.2017.1407080

Winn, W. (1996). Instructional design and situated learning: Paradox or partnership? In *Instructional design fundamentals: A reconsideration* (pp. 159–170).

Wollast, R., Boudrenghien, G., Van der Linden, N., et al. (2018). Who are the doctoral students who drop out? Factors associated with the rate of doctoral degree completion in universities. *International Journal of Higher Education, 7*, 143–156.

Yin, R. K. (1994). *Case study research: Design and methods*. Sage.

Zins, J. E., & Elias, M. J. (2007). Social and emotional learning: Promoting the development of all students. *Journal of Educational and Psychological Consultation, 17*, 233–255.

Chapter 13
Investigating the Potential of AI-Based Social Matching Systems to Facilitate Social Interaction Among Online Learners

Qiaosi Wang, Ida Camacho, and Ashok K. Goel

Abstract Online education has been growing in demand over the years across universities and colleges. However, online learners frequently experience social isolation, which negatively impacts their learning experience and outcome. In this chapter, we investigate the design of AI-based social matching systems to help foster social connections among online learners in higher education context. Specifically, we seek to answer three core design questions: (1) What data should be collected to facilitate students' social interaction process? (2) How to design technology to support students' interactions with one another? (3) What are students' concerns about the use of AI-based social matching systems? We begin by exploring the feasibility, design, and concerns of AI-based social matching through existing literature. We then present our ongoing work on the design and use of AI conversational agents as social matching systems in the online learning context. Finally, we outline future directions for research on designing human-centered social matching systems in online learning.

Keywords AI-based social matching systems · Social interaction · Online learners · Learning analytics

13.1 Introduction

With growing demand for online for-degree programs as well as non-degree courses across universities and colleges, online learning has become critical in shaping the landscape of higher education. The success of online learning depends on multiple

---

Q. Wang (✉) · I. Camacho · A. K. Goel
School of Interactive Computing, Georgia Institute of Technology, Atlanta, GA, USA
e-mail: qswang@gatech.du; ashok.goel@cc.gatech.edu

Y. E. Wang et al. (eds.), *Social and Emotional Learning and Complex Skills Assessment*, Advances in Analytics for Learning and Teaching,
https://doi.org/10.1007/978-3-031-06333-6_13

factors, one of which is the degree of social connectedness among online learners (Aldosemani et al., 2016; Arbaugh et al., 2008). Strong social ties among online learners are crucial to raise students' perceived satisfaction (Hostetter & Busch, 2006; Rovai, 2001), reduce dropout rates (Rovai, 2002), and stimulate intellectual exchange by providing a safe atmosphere (Aldosemani et al., 2016; Rovai, 2001). Many learning scientists consider learners' social presence an integral part of student success in online learning (Arbaugh et al., 2008; Garrison & Arbaugh, 2007; Lave et al., 1991).

While the social dimension of online learning has received much more attention amid the increasing focus on social and collaborative learning, few studies investigated the design of technical systems that could facilitate online learners' social interaction process *before* learning takes place (Kreijns et al., 2003; Sun et al., 2019; Wang et al., 2020b). For example, recent research in Computer-Supported Collaborative Learning (CSCL) has focused on building and evaluating systems that can improve learners' social and collaborative learning process (Sparks et al., 2019; Tubman et al., 2019). Other research has offered strategies to improve students' social connectedness during learning processes through improvements on what the instructors could do (e.g., share personal stories, use humor and emoticons), what the students could do (e.g., contribute to discussion boards), and how the course design should be changed (e.g., limit class size, structure collaborative learning activities) (Aragon, 2003; Irwin & Berge, 2006; Johnson, 2001). These studies often hold the assumption that collaborative and social learning are different means to an end of helping online learners feel socially connected (Kreijns et al., 2003). Yet researchers have suggested that strong social bonds between students are in fact the foundation of optimal social and collaborative learning (Garrison et al., 2001; Kreijns et al., 2003; Rourke, 2000) instead of the other way around. For learners to be open to making mistakes and for them to willingly exchange ideas, they need to have a certain level of trust in each other, feel a sense of social belonging in the learning community, and feel close to each other for more risk-taking and adventurous learning attitudes (Aldosemani et al., 2016; Kreijns et al., 2003). *In this chapter, with the goal of building a strong social foundation for future learning processes, we thus explore the design of information technology that can help online learners build social relationships before learning happens.*

One promising way to facilitate online learners' social interaction process *before* learning takes place is to help online learners develop affinity for one another through the discovery of shared identity (Sun et al., 2019; Wang et al., 2020b). This discovery of shared identity can be facilitated through the use of **social matching systems**. A social matching system is a particular type of recommender system that aims at providing recommendations of people that might be of interest for someone to connect with (Mayer et al., 2015; Terveen & McDonald, 2005). Social matching systems, while prominently used in the context of online dating (Zytko et al., 2018), have also been employed to rediscover old friends on social networks (Chen et al., 2009; Motoyama & Varghese, 2009), link job seekers with potential employees (Olsson et al., 2020) and connect academic researchers to local community collaborators (Zytko & DeVreugd, 2019).

Many social matching systems follow a five-stage process: profiling users, computing matches, introduction, interaction, and feedback (Terveen & McDonald, 2005). To help support online learners' social interaction process, a social matching system must first build a *profile* of each learner through collecting relevant data and information that could be useful in finding matches, potentially based on the learner's background, geographical location, interests and hobbies, classes taken, progress in the course and the program, etc. This is the stage where Artificial Intelligence (AI) technology could come in and infer more information about the students through their digital footprints. Using this profile, the system could *compute matches* for the learner based on some criteria either explicitly set by the student (e.g., want to connect with students located in the same city) or implicitly inferred by systems powered by AI technology (e.g., connecting students who are going through the same type of issues with the assignment). After the matches are computed, the system should *introduce the matches* together in some form, for instance, directly putting matches into contact or providing the matches' contact information to one another. Depending on the learners and the learning context, the system could also intervene during learners' *interaction* processes, for example, post ice breaker questions to help them start a conversation. Finally, given that learners' profile might change over time or they might not like the recommended matches, they should be able to provide *feedback* to the system to optimize future matches.

While this basic process offers a general model of how a social matching system could operate in an online learning context, with the potential introduction of AI technology aimed at improving the social matching process among online learners, new design requirements are needed to create AI-based social matching systems that can tailor to learners' preferences and needs while also addressing privacy and other ethical considerations. *Thus the core question that we seek to explore in this book chapter is how to design AI-based social matching systems for enhancing social interactions among online learners from a human-centered perspective.* To investigate this question in the context of the five-stage process described above, we break it down into three sub-questions and map the sub-questions to different stages of the process as illustrated in Fig. 13.1:

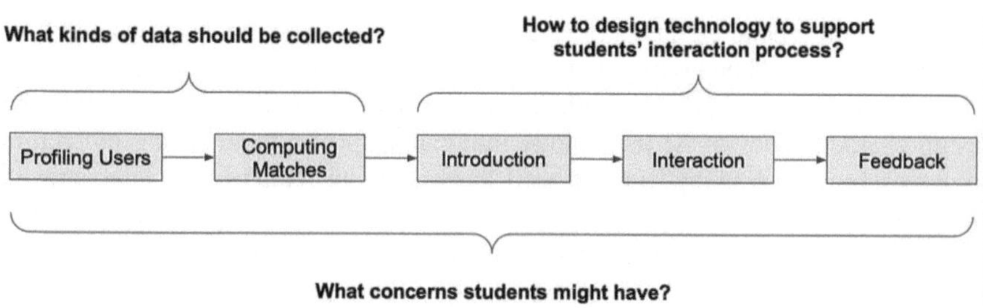

Fig. 13.1 Three design questions to be explored on data, interaction, and concerns at different stages of social matching process in online learning context

- What kinds of data should be collected to help online students make social connections?
- How to design AI-based social matching systems to support the processes of social interaction among online learners?
- What ethical concerns might students have for AI-based social matching systems in online learning environments?

To explore these questions, we first draw from relevant work in a variety of fields related to learning analytics, Computer-Supported Cooperative Work (CSCW), and Computer-Supported Collaborative Learning (CSCL) to identify relevant design implications of social matching systems in online learning environments. Building upon existing literature, we present our ongoing work and latest findings on the design and development of an AI-based social matching system in the context of Georgia Tech's Online Master of Science in Computer Science (OMSCS) program (Galil, 2020; Joyner et al., 2019). While social matching systems can take many forms, we specifically focus on using Conversational Agents (CAs) as social matching systems given CAs' success in providing emotional and social support within online communities (Narain et al., 2020; Nordberg et al., 2019). This work builds on but is different and separate from our earlier work on the AI teaching assistant named Jill Watson (Goel & Polepeddi, 2016, 2019; Wang et al., 2021): while Jill answered learners' questions on discussion forums of online classes and thereby enhanced teacher presence, the present work addresses the issue of promoting social interactions among the online learners. Taking a human-centered design perspective, we investigate the potential of AI-based social matching systems in online learning environments based on empirical evaluation and deployment of CAs as social matching systems. We then highlight future directions of designing AI-based social matching systems in online learning environments.

13.2 Related Work

In this section, we review relevant work to understand the potential of AI-based social matching systems in online learning environments. Following our three questions mapped out to the basic model of social matching systems (Fig. 13.1), we first discuss several crucial elements of establishing interpersonal connections and how these elements could be highlighted and inferred from online students' digital footprints. Next, drawing upon theoretical frameworks from CSCW, we highlight the design characteristics of AI technology that can support remote social interaction processes. Finally, drawing from the well-known ethical challenges and concerns in the use of AI technologies powered by users' online data, we discuss the potential concerns AI-based social matching systems might raise for online learners.

13.2.1 Profiling and Computing Matches for Online Learners

While establishing connections online can be very different from and more challenging than in in-person context, many elements of interpersonal connections are shared across both settings. Social psychologists have been studying interpersonal attraction for decades and identified different types of factors that are crucial to building an amicable interpersonal relationship (Terveen & McDonald, 2005): personal characteristics (e.g., personal preferences, personality), demographics (e.g., gender, profession), and familiarity (e.g., time spent together). Based on prior research, people are more prone to connect with those who share similarities in personal characteristics and demographics (Gilbert & Karahalios, 2009; Granovetter, 1973), as well as high levels of familiarity (Kraut et al., 1988). On top of establishing interpersonal connections, cooperative actions also require that the individuals are likely to meet again in the future, the individuals can identify each other prior to interactions, and the individuals possess adequate amount of information of people's past behaviors (Kollock, 1997; Terveen & McDonald, 2005).

Perhaps one of the biggest challenges for online social interactions is that most people find that inferring this kind of insight is difficult, sometimes impossible, based on solely on another person's online behavior (Kehrwald, 2008). While in in person interactions, most people can efficiently, and often accurately, gauge people's personality, characteristics, and age based on their appearances and behaviors, online environment is often text-based, stripping people's ability to make inferences about another person's demographics and other characteristics.

Yet online environments also present many opportunities for AI systems to make such inferences: it is easier to capture and retain information in online environments than in in-person contexts, which can enable AI systems to make inferences from people's online behaviors. For example, based on people's digital footprints in online environments, researchers were able to use AI techniques to infer people's mental states (e.g., stress) from online forum data (De Choudhury et al., 2013; Saha & De Choudhury, 2017), predicting people's personality from social media cues (Farnadi et al., 2016; Skowron et al., 2016), and inferring about people's interpersonal ties using social media data (Gilbert & Karahalios, 2009). These studies all point to the feasibility of compensating for the lost social cues in online social interactions using people's digital footprint.

Leveraging people's online information to infer behaviors and personal characteristics is hardly an uncharted area in online education—the field of learning analytics and educational data mining have been analyzing online learners' data to make inferences about students for many years (Avella et al., 2016; Du et al., 2021). The overall objectives for learning analytics is to leverage online learners' data to predict learner performance, offer decision support for teachers and learners, detect behavioral patterns and learner models, as well as predict dropout rates (Avella et al., 2016; Du et al., 2021). To accomplish these objectives, researchers have been able to leverage many data sources readily available in online learning—students' educational records, demographics, textual data of online discussions, facial expression, frequency of logins, duration of content accessed—that can shed light on students' learning progress, learning patterns, learning behaviors, etc. (Du et al., 2021). However, these efforts at using learning analytics for enhancing online learning have mostly focused on the cognitive aspects of learning; the potential of using learning analytics approaches to support students' social interaction process requires further exploration.

13.2.2 Designing Technology-Mediated Remote Social Interactions

Decades of CSCW research has produced several well-established theoretical frameworks to guide the design of technologies in supporting remote interactions. Among these theoretical frameworks, both Ackerman's social-technical gap (Ackerman, 2000) and Erickson and Kellogg's social translucence (Erickson & Kellogg, 2000) draw inspiration from in-person social interactions to design technology that can support remote social interactions.

Ackerman defines social-technical gap as *"the great divide between what we know we must support socially and what we can support technically"* (Ackerman, 2000). In his seminal work, Ackerman points out that when technology mediates remote interactions, they are often designed to be rigid, reductionist, and do not allow sufficient ambiguity compared to in-person interactions (Ackerman, 2000). Much research has since adopted this framework and identified the social-technical gap in a variety of contexts such as health tracking (Chung et al., 2017), collaboration among telesurgery teams (Duysburgh et al., 2014), and online collaborative consumption (Gheitasy et al., 2015). In his original piece, Ackerman proposed first order approximations—solutions that partially solve the problem but with known trade-offs—to help bridge the social-technical gap. One optimal approximation is to design augmentative information technology, for example, by offering advice to users (Ackerman, 2000). This potentially can be accomplished through the use of CAs (Lee et al., 2017).

While the idea of social-technical gap typically acts as a general guide and call-to-action in CSCW research, to bridge this gap between social and technical requirement, Erickson and Kellogg go a step further and outline detailed principles on designing towards socially translucent systems to support natural online interactions (Erickson & Kellogg, 2000). Specifically, socially translucent systems have three characteristics: visibility, awareness, and accountability. *Visibility* refers to system's ability of making social information more visible; *Awareness* refers to people's ability to know each others' existence; *Accountability* refers to system's ability to hold people accountable for their behavior by generating and enforcing social rules. Erickson and Kellogg posit that these three characteristics allow people to observe, imitate, aware, and interact with others socially in in-person context, and thus building socially translucent system is a fundamental requirement for people to carry out normal interactions online (Erickson & Kellogg, 2000). Since then, social translucence has been often employed in the design of technology-mediated interactions. For example, prior research has developed methods to support collective awareness through creating common repository to generate mutual understanding for members of globally distributed teams (Bjørn & Ngwenyama, 2009) and conducting synchronous coding sessions for learner engagement (Byun et al., 2020).

In summary, this body of work emphasizes the lack of naturalness in remote interactions compared to in-person interactions—social-technical gap —and how technology can be designed to be socially translucent in order to create the naturalness in online environment. While these two theoretical frameworks have not been widely used in research on online learning, several studies on online learners' social presence and social interactions have offered some support to the generalizability of these issues in the online learning context: the lack of visibility of social cues (Kehrwald, 2008; Sun et al., 2019) and the diminished accountability and motivation in reaching out to others (Kehrwald, 2008) all contribute to students' feeling of social isolation in online learning environments.

13.2.3 Potential Challenges in Social Matching Among Online Learners

Like many other AI systems that leverage big data, social matching systems and learning analytics approaches present several potential ethical challenges and concerns. Well known ethical concerns such as privacy, consent, anonymity, and accuracy of data are shared across the use of social matching systems and interventions based on learning analytics (Avella et al., 2016; Terveen & McDonald, 2005; Wang et al., 2020c).

Both social matching systems and learning analytics interventions produce results based on either user data that are voluntarily offered by the users or user data that are available but not explicitly consented to by the users such as postings on a public forum. Given that humans are social creatures, one ethical dilemma social matching systems face is the fact that many people often are okay with their sensitive personal information being used in specific contexts—and sometimes even voluntarily offer it—but this can lead to oversharing (Terveen & McDonald, 2005). For example, sensitive information such as student grades is commonly collected in learning analytics approaches, often in order to assess students' learning progress. However, it remains questionable whether students are aware of the extent to which their data is being collected and analyzed in online learning environments since usually only instructors and institutions have access to the data and the results (Slade & Prinsloo, 2013). Designing social matching systems in online learning environment thus would require transparency of the processes of data collection and analysis, as well as careful informed consent procedures to address privacy and ethical concerns.

One common pitfall for AI systems that are powered by big data is the fact that sometimes individuals are treated more like data points than humans with identity and agency. One important characteristic to keep in mind of is that people's identity is often transient and temporal (Slade & Prinsloo, 2013; Terveen & McDonald, 2005): students' learning behaviors as well as their preferences can change over time. Feedback from the students regarding their social matches thus can play a crucial role for the system to update and caliberate future recommendations (Terveen & McDonald, 2005). In learning analytics approches, treating students as agents could mean asking for their collaboration throughout the analytical process (Slade & Prinsloo, 2013). This not only means performing data collection, analysis, and usage only with students' explicit and specific consent, but also to ensure that the system can leverage information students voluntarily offer to help them achieve their own learning goals (Slade & Prinsloo, 2013).

The most basic functionality of social matching systems is to recommend and match people with similarities (Terveen & McDonald, 2005), which is based on people's natural similarity-seeking behaviors during in-person interactions (Olsson et al., 2020). In online environment this tendency towards similarity still persists—prior research has found that in online team formation, people tend to team up with those who are similar to them and thus lead to non-diverse and segregated teams (Gómez-Zará et al., 2019). Previous research has pointed out that this fundamental design characteristic of social matching systems can lead to ethically concerning consequences such as the creation of echo chambers and polarization in the community (Olsson et al., 2020). Olsson et al. further argue that recommendation systems should not encourage biased human behaviors and that one potential solution is to enable social serendipity and random encounters in online social matching (Olsson et al., 2020).

13.2.4 Summary

To investigate the potential of AI-based social matching systems in online learning contexts, we first reviewed relevant literature to explore the three core design questions about data, design, and concerns (see Fig. 13.1). Based on the existing literature, we found that understanding and identifying similarities in personal characteristics, demographic, and familiarity is crucial in establishing social connections, both in-person and online (Granovetter, 1973; Terveen & McDonald, 2005). In online contexts, recent development in AI and Natural Language Processing (Avella et al., 2016; Du et al., 2021; Saha & De Choudhury, 2017) allow fairly accurate inference of such social information and thus should be leveraged to collect relevant data in profiling and computing matches when designing social matching systems in online learning. To design social matching systems that can support students' social interactions, principles of social translucence (Erickson & Kellogg, 2000) and social-technical gap (Ackerman, 2000) could be applied to help replicate the naturalness of in-person interactions to online environment. Concerns regarding AI-based social matching systems could arise at any stage of the social matching process, specifically, privacy, consent (Slade & Prinsloo, 2013), oversharing of personal information (Terveen & McDonald, 2005), updating students' transient identities (Slade & Prinsloo, 2013; Terveen & McDonald, 2005), and the unintended creation of echo chambers (Olsson et al., 2020) are all valid concerns and should be taken into account when designing social matching systems for online learners.

Based on these design considerations that we identified through existing literature, we designed and deployed a community-facing AI conversational agent called SAMI (Social Agent Mediated Interaction) (Goel, 2020; Wang et al., 2020a, b) to perform social matching among online students. In the next section, we describe our design and deployment of SAMI in an online learning context.

13.3 SAMI: Conversational Agents as Social Matching Systems

Due to its human-like characteristics and the ability to converse with people, CAs have been widely used to provide social and emotional support in both dyadic interactions and community contexts. Prior research has demonstrated the positive effect of using CAs to facilitate mental health patients' self-disclosure during consultations (Lee et al., 2020), help healthcare professionals manage occupational stress (Yorita et al., 2020), and provide social support to older adults who are socially isolated (Ring et al., 2015; Simpson et al., 2020). Designing social matching systems as CAs is thus a promising way to support online learners who feel socially isolated yet also requires more design explorations.

Inspired by the prior research's usage of textual data in online discussion forum, SAMI leverages students' self-introduction posts on the discussion forum, where online learners usually conduct class-related discussions and posting self-introductions at the beginning of the semester. Specifically, SAMI utilizes Natural Language Processing to extract different entities such as hobby, city, country from students' self-introduction posts in order to build a profile for each online student. Online students can opt-in to receive SAMI response by adding "#ConnectMe" in their introduction post as seen in Fig. 13.2.

The current version of SAMI matches students shared similarities among students such as proximate geographical locations and similar hobbies, etc. (see Fig. 13.2). After identifying students' preferred matching criteria, SAMI creates a private group of all students with commonalities and then invites each student to the private group. To further engage students in building connections, SAMI also posts ice-breaker questions within each group (see Fig. 13.3).

We constructed the first basic version of SAMI in 2017 for Georgia Tech OMSCS class on Knowledge-Based AI (Goel & Joyner, 2016, 2017). Since then we have both incrementally enhanced SAMI's capabilities and deployed it in additional OMSCS classes (Goel, 2020). We have also conducted detailed evaluations (Wang et al., 2020b) and collected extensive student feedback on SAMI for future improvements in its design and delivery. We present our findings in the next section.

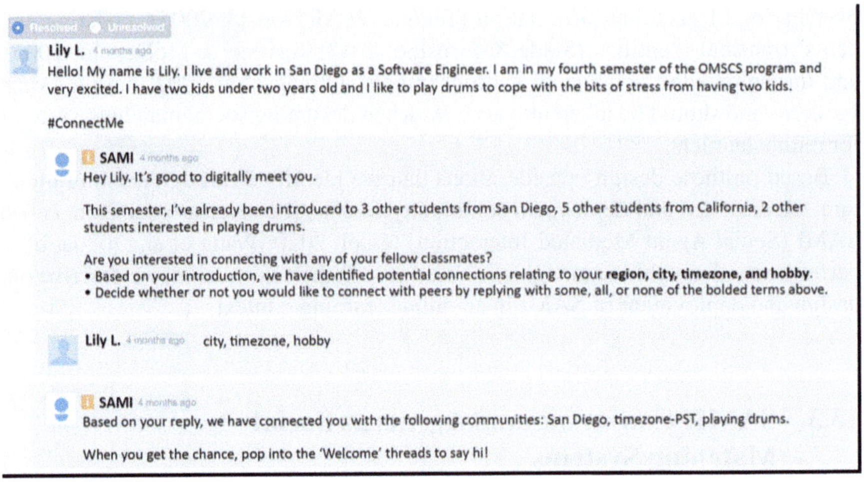

Fig. 13.2 SAMI uses NLP to extract entities from student's post, inquires about student's matching criteria, then puts the student in private groups with other students who share similarities

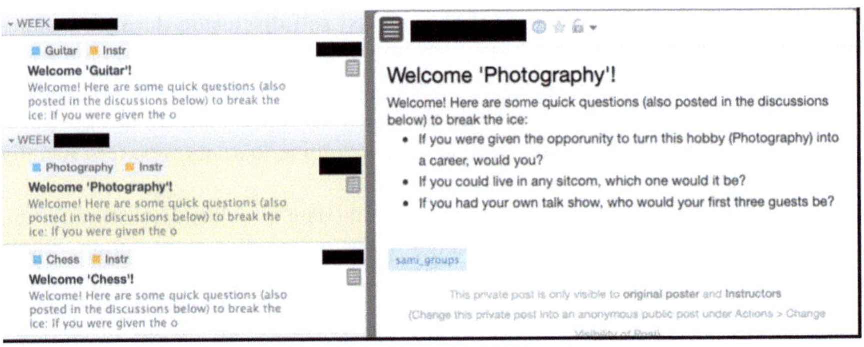

Fig. 13.3 SAMI creates different private groups based on entities identified from students' introduction posts and then put students with similarities in the according groups (e.g., similar hobby). SAMI also posts ice breaker questions in each group to help students start the conversation

13.4 Evaluation of a Social Matching System in Online Learning

In this section, we present and discuss our findings from a series of qualitative surveys with around 400 students (Wang et al., 2020b) and a set of semi-structured interviews with 26 online students (Wang et al., 2020a) in the OMSCS program regarding their existing challenges in building social connections with other students as well as their experience with SAMI. Based on our analysis, we outline design implications for building social matching systems in online learning environments, highlight online learners' concerns regarding the use of social matching systems in online learning, and discuss potential directions for designing AI-based social matching systems for online learners.

13.4.1 Social Matching Among Online Learners: Challenges and Design

Designing AI systems that can support students' interaction with one another is a design problem that could manifest through the introduction, interaction, and feedback stages of the social matching process (see Fig. 13.1). This raises several questions: How much information about a given student should the system share with other students? How should the system introduce the matches to each other? How much should the system intervene during the introduction and interaction stages? In this section, we seek to answer these questions by understanding online learners' current challenges in remote social interactions and interpreting their feedback on SAMI in helping online learners connect with one another. We point out that the design of social matching systems should provide social translucence as well as bringing the randomness and spontaneity commonly seen in in-person interactions into online social interactions.

Making Social Signals Visible Among Online Learners Reaching out and building connections with strangers can be an intimidating process. During in-person interactions, people are able to gain social cues from another person's behaviors or facial expressions. However, most of these social cues become invisible in an online environment (Erickson & Kellogg, 2000). In our study, we found that many online learners are hesitant in reaching out to other people due to the loss of social cues—they do not know whether other students are willing to connect with them or how their messages will be received by other students. After SAMI was deployed in participants' online classes, students were more readily able to identify social signals identified by SAMI. Some students interpreted the "#ConnectMe" in other students' self-introduction posts as a signal of students' willingness to build social connections. When designing social matching systems in online learning environment, making social signals visible is an important starting point and could be implemented easily by adding simple features. For example, one potential feature is adding icons on students' avatars to indicate students' willingness to connect with others to improve visibility of social cues (Szostek et al., 2008).

Raising Awareness of Potential Social Companions One of the main goals and advantages of online learning is to help education scale by giving more students the opportunities to learn (Larreamendy-Joerns & Leinhardt, 2006). One result of this scaling is that online classes usually have hundreds or even thousands of students per class. The downside of having these many students per class is that the class size reduces students' awareness of other students' existence in the program/class, which negatively contributes to online students' social interaction process. This diminished awareness poses several challenges in the learners' social interaction processes. First, the overwhelming number of students and activities within each class made it extremely difficult for online students to identify individual students that they could potentially build social connections with. Second, with hundreds of students communicating via the class discussion forum and chat group, these main communication channels quickly became a wall of text, which makes most interactions there seemed impersonal. With the deployment of SAMI, participants became more conscious of other students who shared similar backgrounds, interests or experiences with them. Even without the personalized recommendations, students said that just scrolling through SAMI's replies to students self-introduction posts made them realize there were other students with different or similar experiences, which made the online learning environment seem more personable and personalized. Future social matching systems thus could raise online student's awareness of other students by highlighting students' shared identities. Offering statistics of the entire class or the program to demonstrate the diverse student population could also help students to increase the personable feelings in online learning environment.

Providing Accountability to the Social Interaction Process Erickson and Kellogg pointed out that while awareness and acccountability often co-occur in physical world, they are not usually coupled in the online spaces (Erickson & Kellogg, 2000). Accountability is often fostered through the creation of social norms in a community that hold people accountable for their social behaviors (Erickson & Kellogg, 2000). In online learning environments, both the existence and the lack of social rules could prohibit online learners' social interactions. While the intention of the online class discussion forum is to replicate the physical classroom where students could have interactions and discussions about and beyond academics, the implicit social rule to use online class discussion forum only for academic discussions makes students feel accountable to only have academic discussions on the forum instead of casual conversations. Although online learners often get the chance to get to know fellow students through group projects, after the semester ends, they don't usually "encounter" one another anymore, which reduces students' feeling of accountability to talk with each other again. One feature of SAMI is to put students with similarities directly into a private group and post ice breaker questions to help students start the conversation. Participants in our study believed that by putting students directly in touch with each other, SAMI not only alleviated students' mental barrier in initiating the conversation, but also made students feel accountable to start building the connections because SAMI already "started" the conversations between students. Providing accountability in the social matching process thus could also take the form of the AI agent initiating the conversation between the matches or offering timely nudges for individuals to start building the connections.

Creating Randomness and Spontaneity in Remote Social Interaction While many challenges online students encounter during their social interaction process originate from the lack of social translucence in online learning environment—visibility, awareness, accountability, we also identified another set of challenge that originated from the social-technical gap (Ackerman, 2000) in online social interactions. According to our participants, the randomness and spontaneity that were crucial and inherent in in-person environment currently could not be supported in the online learning environment. For example, in in-person educational environment, students can often randomly run into each other on campus or having work conversations that organically lead to more social activities. However, in online learning programs, the social and learning aspects typically are separated, especially when compared to traditional in-person educational programs. Instead of forming social connections organically during the process of taking classes or walking around campus that are inherently built into the on-campus educational experience, online learners have to establish social connections in a more intentional way (e.g., driving for an hour to attend a local meet-up with other online students). While highlighting shared identities or explicitly expressing social signals could help create social translucence into the online social interaction process, the nuances and subtlety of in-person interactions should also be preserved when performing remote social matching (Ackerman, 2000; Olsson et al., 2020). A potential direction for future designs of online social matching systems among online learners is to intentionally create seemingly random matches (e.g., students who seem very different at first but have "deep" similarities) or introduce matches in seemingly random encounters (e.g., introduce students who are reviewing the same lecture materials).

Summary of Design Implications The design of social matching systems for online learners should cater to students' existing challenges and needs. We highlight the lack of social translucence (Erickson & Kellogg, 2000) and the existence of social-technical gap (Ackerman, 2000) in existing online learning platforms that hinder online learners' social interaction process. AI-based social matching systems in online learning environment thus could aim at increasing visibility of social signals, raising awareness of potential social companions, providing accountability to the social interaction process, and introducing randomness and spontaneity into online learning environment. Echoing with prior research (Kreijns et al., 2003), our work provides further empirical evidence that social interactions in online learning environment cannot be taken for granted to naturally happen just because the platforms allow it. Future research can explore the design question of how to replicate the randomness and spontaneity of in-person social interactions to online learning environments.

13.4.2 Towards Collaborative Social Matching in Online Learning

Throughout the model of the social matching system process (see Fig. 13.1), the design questions regarding both data and interactions are targeted at specific stages: questions about data could manifest from profiling users and computing matches;

and questions about designing interactions could exhibit in the introduction, interaction, and feedback stages. Concerns that online students have regarding social matching systems in online learning environment come from issues surrounding both data and interaction design. Based on students' feedback on SAMI, we found that students are concerned about losing agency during the social matching process when it is mediated by an AI agent. We also found that students prefer the agent to be more transparent about the matching process and mechanism. A little to our surprise, the students in our prior studies (Wang et al., 2020a, b) did not express many concerns regarding data privacy. Based on our prior findings, we propose the future direction of designing social matching process as a collaborative process between the system and the students to mitigate students' concerns regarding both data and interactions as discussed below.

Preference in System Transparency In our prior studies, many participants suggested that SAMI could be more transparent about its decision-making process and working mechanism for a smoother interaction between students and SAMI. This preference stems from students' belief that if they could better communicate with SAMI using similar vocabularies, the matching results could be more accurate. In human-AI collaborative decision-making processes, transparency and the willingness to collaborate are crucial for a desirable collaborative experience and outcome (Cai et al., 2019). Fortunately, aligned with prior literature (Jhaver et al., 2019; Liao et al., 2020), participants in our prior studies indicated their willingness to understand the AI agent's vocabulary beforehand to adjust their choice of words during communication in order to improve the accuracy of matching results. Echoing with prior research that urges designing collaborative learning analytics interventions (Slade & Prinsloo, 2013), designing collaborative social matching systems with online learners could also be a promising direction to offer more system transparency as well as ensuring the accuracy of matching results.

Concerns About Losing Agency in Building Social Connections In our prior studies, one concern that was raised by online students about the prospect of continuing usage of SAMI in the online learning program is the possibility of losing agency in making decisions on building social connections with other online students. This concern was mostly based on SAMI's feature of directly putting students in private groups. This feature, while created an adequate amount of social pressure for students to start the initial conversation, was also critiqued by students who said they wanted more freedom in choosing which group they could join or whom they should connect with. Building social connections with others is inherently a very personal decision-making process. While social matching systems can provide convenience and efficiency in facilitating online learners' social interaction process by suggesting matches that students otherwise would not be able to find, we found that students are unwilling to cede control of the decision-making process in choosing whom they should connect with (Sundar, 2020). This would require the social matching system to work with the students collaboratively throughout the decision-making process to maintain students' sense of agency. This could be accomplished by having the system communicate with the students about all the progress that has been made in computing matches, asking students to set and revise the matching criteria, and incorporating students feedback in future matching computations. However, it is important to keep in mind that the social matching system should also provide an adequate level of social pressure and accountability, such as putting students directly into groups, for students to initiate the conversation. This design issue of balancing between maintaining student's agency in decision-making and creating social pressure and accountability in social matching process thus requires further exploration.

Concerns About Privacy Even though privacy is often a concern for AI systems that leverage public data (Fiesler et al., 2016), most participants in both of our prior studies did not express privacy concerns regarding SAMI. In fact, many online students indicated that they were willing to offer more information for SAMI to find more connections for them, which also aligns with prior work that users are often willing to provide information to social matching systems to get connected (Terveen & McDonald, 2005). Currently, SAMI only obtains public information presented on the online forum, which might have alleviated online students' privacy concerns—some students in the study believed that the goal of posting on public forum was for others to see it. To achieve high accuracy in social matching systems, accessing latent behavioral data that users don't explicitly consent to would be inevitable and might result in violations of user privacy (Fiesler et al., 2016; Fiesler & Proferes, 2018). This is especially important for social matching systems as users are often more likely to disclose sensitive information for more accurate matching results (Terveen & McDonald, 2005). When CAs perform social matching for students, privacy issue would require more scrutiny as CAs possess human-like characteristics that could encourage people's self-disclosure during conversations (Følstad et al., 2018; Ischen et al., 2019; Lee et al., 2020) which might lead students to unintentionally disclose sensitive information that could be used by the CA to improve matching accuracy. The balance between privacy and accuracy in CAs as social matching systems in online communities thus requires further exploration.

Directions for Future Designs Based on students' preferences and concerns regarding the use of AI-based social matching systems in online learning, we highlight the design direction of designing towards a collaborative social matching process between the students and the social matching system. Collaborative social matching could not only offer transparency about the social matching process, potentially increase matching accuracy, but also could help mitigate students' concerns about losing agency in building social connections. While privacy concerns were not commonly expressed by students in both of our prior studies, designers should be cautious about the use of both public and private information, especially users' tendency to overshare sensitive personal information to ensure matching accuracy. We also want to raise several design issues that future research should explore such as user vs. system control over decision-making in human-AI collaboration and balancing between maintaining user agency and creating social pressure in the social matching process.

13.5 Conclusions

As online learning is adopted by increasingly large number of educational institutions, the social dimension of online learning requires more attention from researchers in a variety of fields. AI-based social matching systems as an information technology that can support online learners' social interaction process is a promising first step to help reduce online learners' feelings of social isolation and acts as a social foundation for future learning. However, as with many data-driven AI approaches, the potential of AI-based social matching systems in online learning requires further examination to cater to online learners' challenges and needs in remote social interactions and mitigating potential privacy and ethical concerns. In this chapter, we explored the potential of AI-based social matching systems through three core design questions about data, design, and concerns. Drawing upon relevant literature, we established the feasibility of inferring social information from online learners' digital footprints, discussed related theoretical frameworks in designing technology to support remote social interactions, and presented existing concerns and challenges in AI-based approaches that leverage student data. We further elaborated on this initial design implication drawn from existing literature through a discussion of our ongoing work on the design and evaluation of an AI conversational agent as a social matching system in an online learning context. Based on findings from our prior studies, we outlined the design implications of designing AI-based social matching systems to provide social translucence as well as to create randomness and spontaneity in remote social interactions. We then pointed out directions for future work on building collaborative social matching systems to mitigate students' concerns and potential challenges.

References

Ackerman, M. S. (2000). The intellectual challenge of CSCW: The gap between social requirements and technical feasibility. *Human–Computer Interaction, 15*(2–3), 179–203.

Aldosemani, T. I., Shepherd, C. E., Gashim, I., & Dousay, T. (2016). Developing third places to foster sense of community in online instruction. *British Journal of Educational Technology, 47*(6), 1020–1031.

Aragon, S. R. (2003). Creating social presence in online environments. *New Directions for Adult and Continuing Education, 2003*(100), 57–68.

Arbaugh, J. B., Cleveland-Innes, M., Diaz, S. R., Garrison, D. R., Ice, P., Richardson, J. C., & Swan, K. P. (2008). Developing a community of inquiry instrument: Testing a measure of the community of inquiry framework using a multi-institutional sample. *The Internet and Higher Education, 11*(3–4), 133–136.

Avella, J. T., Kebritchi, M., Nunn, S. G., & Kanai, T. (2016). Learning analytics methods, benefits, and challenges in higher education: A systematic literature review. *Online Learning, 20*(2), 13–29.

Bjørn, P., & Ngwenyama, O. (2009). Virtual team collaboration: Building shared meaning, resolving breakdowns and creating translucence. *Information Systems Journal, 19*(3), 227–253.

Byun, J., Park, J., & Oh, A. (2020). Cocode: Co-learner screen sharing for social translucence in online programming courses. In *Extended abstracts of the 2020 CHI conference on human factors in computing systems* (pp. 1–4).

Cai, C. J., Winter, S., Steiner, D., Wilcox, L., & Terry, M. (2019). "hello ai": Uncovering the onboarding needs of medical practitioners for human-ai collaborative decision-making. *Proceedings of the ACM on Human-computer Interaction, 3*(CSCW), 1–24.

Chen, J., Geyer, W., Dugan, C., Muller, M., & Guy, I. (2009). Make new friends, but keep the old: Recommending people on social networking sites. In *Proceedings of the SIGCHI conference on human factors in computing systems* (pp. 201–210).

Chung, C. F., Gorm, N., Shklovski, I. A., & Munson, S. (2017). Finding the right fit: Understanding health tracking in workplace wellness programs. In *Proceedings of the 2017 CHI conference on human factors in computing systems* (pp. 4875–4886).

De Choudhury, M., Gamon, M., Counts, S., & Horvitz, E. (2013). Predicting depression via social media. In *Proceedings of the international AAAI conference on web and social media* (Vol. 7).

Du, X., Yang, J., Shelton, B. E., Hung, J. L., & Zhang, M. (2021). A systematic meta-review and analysis of learning analytics research. *Behaviour & Information Technology, 40*(1), 49–62.

Duysburgh, P., Elprama, S. A., & Jacobs, A. (2014). Exploring the social-technological gap in telesurgery: Collaboration within distributed or teams. In *Proceedings of the 17th ACM conference on computer supported cooperative work & social computing* (pp. 1537–1548).

Erickson, T., & Kellogg, W. A. (2000). Social translucence: An approach to designing systems that support social processes. *ACM transactions on computer-human interaction (TOCHI), 7*(1), 59–83.

Farnadi, G., Sitaraman, G., Sushmita, S., Celli, F., Kosinski, M., Stillwell, D., Davalos, S., Moens, M. F., & De Cock, M. (2016). Computational personality recognition in social media. *User Modeling and User-Adapted Interaction, 26*(2), 109–142.

Fiesler, C., & Proferes, N. (2018). "Participant" perceptions of twitter research ethics. *Social Media+ Society, 4*(1), 2056305118763366.

Fiesler, C., Lampe, C., & Bruckman, A. S. (2016). Reality and perception of copyright terms of service for online content creation. In *Proceedings of the 19th ACM conference on computer-supported cooperative work & social computing* (pp. 1450–1461).

Følstad, A., Nordheim, C. B., & Bjørkli, C. A. (2018). What makes users trust a Chatbot for customer service? An exploratory interview study. In *International conference on internet science* (pp. 194–208). Springer.

Galil, Z. (2020). OMSCS: The revolution will be digitized. *Communications of the ACM, 63*(8), 27–29.

Garrison, D. R., & Arbaugh, J. B. (2007). Researching the community of inquiry framework: Review, issues, and future directions. *The Internet and Higher Education, 10*(3), 157–172.

Garrison, D. R., Anderson, T., & Archer, W. (2001). Critical thinking and computer conferencing: A model and tool to assess cognitive presence.

Gheitasy, A., Abdelnour-Nocera, J., & Nardi, B. (2015). Socio-technical gaps in online collaborative consumption (OCC) an example of the etsy community. In *Proceedings of the 33rd annual international conference on the Design of Communication* (pp. 1–9).

Gilbert, E., & Karahalios, K. (2009). Predicting tie strength with social media. In *Proceedings of the SIGCHI conference on human factors in computing systems* (pp. 211–220).

Goel, A. (2020). Ai-powered learning: Making education accessible, affordable, and achievable. *arXiv preprint arXiv*, 200601908.

Goel, A., & Joyner, D. (2016). An experiment in teaching artificial intelligence online. *International Journal for the Scholarship of Technology-Enhanced Learning, 1*(1), 1–27.

Goel, A. K., & Joyner, D. A. (2017). Using AI to teach AI: Lessons from an online AI class. *AI Magazine, 38*(2), 48–59.

Goel, A. K., & Polepeddi, L. (2016). *Jill Watson: A virtual teaching assistant for online education* (Technical report). Georgia Institute of Technology.

Goel, A. K., & Polepeddi, L. (2019). Chapter 7: Jill Watson: A virtual teaching assistant for online education. In C. Dede, J. Richards, & B. Saxberg (Eds.), *Learning engineering for online education: Theoretical contexts and design-based examples*. Routledge.

Gómez-Zará, D., Paras, M., Twyman, M., Lane, J. N., DeChurch, L. A., & Contractor, N. S. (2019). Who would you like to work with? In *Proceedings of the 2019 CHI conference on human factors in computing systems* (pp. 1–15).

Granovetter, M. S. (1973). The strength of weak ties. *American Journal of Sociology, 78*(6), 1360–1380.

Hostetter, C., & Busch, M. (2006). Measuring up online: The relationship between social presence and student learning satisfaction. *Journal of the Scholarship of Teaching and Learning, 6*, 1–12.

Irwin, C., & Berge, Z. (2006). Socialization in the online classroom. *E-Journal of Instructional Science and Technology, 9*(1), n1.

Ischen, C., Araujo, T., Voorveld, H., van Noort, G., & Smit, E. (2019). Privacy concerns in chatbot interactions. In *International workshop on Chatbot research and design* (pp. 34–48). Springer.

Jhaver, S., Birman, I., Gilbert, E., & Bruckman, A. (2019). Human-machine collaboration for content regulation: The case of reddit automoderator. *ACM Transactions on Computer-Human Interaction (TOCHI), 26*(5), 1–35.

Johnson, C. M. (2001). A survey of current research on online communities of practice. *The Internet and Higher Education, 4*(1), 45–60.

Joyner, D. A., Isbell, C., Starner, T., & Goel, A. (2019). Five years of graduate cs education online and at scale. In *Proceedings of the ACM conference on global computing education* (pp. 16–22).

Kehrwald, B. (2008). Understanding social presence in text-based online learning environments. *Distance Education, 29*(1), 89–106.

Kollock, P. (1997). *Design principles for online communities*. Peter Kollock.

Kraut, R., Egido, C., & Galegher, J. (1988). Patterns of contact and communication in scientific research collaboration. In *Proceedings of the 1988 ACM conference on computer-supported cooperative work* (pp. 1–12).

Kreijns, K., Kirschner, P. A., & Jochems, W. (2003). Identifying the pitfalls for social interaction in computer-supported collaborative learning environments: A review of the research. *Computers in Human Behavior, 19*(3), 335–353.

Larreamendy-Joerns, J., & Leinhardt, G. (2006). Going the distance with online education. *Review of Educational Research, 76*(4), 567–605.

Lave, J., Wenger, E., et al. (1991). *Situated learning: Legitimate peripheral participation*. Cambridge University Press.

Lee, M., Frank, L., Beute, F., De Kort, Y., & IJsselsteijn, W. (2017). Bots mind the social-technical gap. In *Proceedings of 15th European conference on computer-supported cooperative work-exploratory papers*. European Society for Socially Embedded Technologies (EUSSET).

Lee, Y. C., Yamashita, N., & Huang, Y. (2020). Designing a chatbot as a mediator for promoting deep self-disclosure to a real mental health professional. *Proceedings of the ACM on Human-Computer Interaction, 4*(CSCW1), 1–27.

Liao, Q. V., Gruen, D., & Miller, S. (2020). Questioning the AI: Informing design practices for explainable AI user experiences. In *Proceedings of the 2020 CHI Conference on Human Factors in Computing Systems* (pp. 1–15).

Mayer, J. M., Jones, Q., & Hiltz, S. R. (2015). Identifying opportunities for valuable encounters: Toward context-aware social matching systems. *ACM Transactions on Information Systems (TOIS), 34*(1), 1–32.

Motoyama, M., & Varghese, G. (2009). I seek you: Searching and matching individuals in social networks. In *Proceedings of the eleventh international workshop on Web information and data management* (pp. 67–75).

Narain, J., Quach, T., Davey, M., Park, H. W., Breazeal, C., & Picard, R. (2020). Promoting well-being with sunny, a Chatbot that facilitates positive messages within social groups. In *Extended abstracts of the 2020 CHI conference on human factors in computing systems* (pp. 1–8).

Nordberg, O. E., Wake, J. D., Nordby, E. S., Flobak, E., Nordgreen, T., Mukhiya, S. K., & Guribye, F. (2019). Designing Chatbots for guiding online peer support conversations for adults with ADHD. In *International workshop on Chatbot research and design* (pp. 113–126). Springer.

Olsson, T., Huhtamäki, J., & Kärkkäinen, H. (2020). Directions for professional social matching systems. *Communications of the ACM, 63*(2), 60–69.

Ring, L., Shi, L., Totzke, K., & Bickmore, T. (2015). Social support agents for older adults: Longitudinal affective computing in the home. *Journal on Multimodal User Interfaces, 9*(1), 79–88.

Rourke, L. (2000). Operationalizing social interaction in computer conferencing. In *Proceedings of the 16th annual conference of the Canadian Association for Distance Education* (pp. 335–353).

Rovai, A. P. (2001). Building classroom community at a distance: A case study. *Educational Technology Research and Development, 49*(4), 33.

Rovai, A. P. (2002). Development of an instrument to measure classroom community. *The Internet and Higher Education, 5*(3), 197–211.

Saha, K., & De Choudhury, M. (2017). Modeling stress with social media around incidents of gun violence on college campuses. *Proceedings of the ACM on Human Computer Interaction, 1*(CSCW), 1–27.

Simpson, J., Gaiser, F., Macík, M., & Breßgott, T. (2020). Daisy: A friendly conversational agent for older adults. In *Proceedings of the 2nd conference on conversational user interfaces* (pp. 1–3).

Skowron, M., Tkalčič, M., Ferwerda, B., & Schedl, M. (2016). Fusing social media cues: Personality prediction from twitter and Instagram. In *Proceedings of the 25th international conference companion on world wide web* (pp. 107–108).

Slade, S., & Prinsloo, P. (2013). Learning analytics: Ethical issues and dilemmas. *American Behavioral Scientist, 57*(10), 1510–1529.

Sparks, J. R., Coiro, J., Castek, J., Kiili, C., Holland, B., & Lee, C. (2019). Assessing collaborative online inquiry and social deliberation in digital environments. In Lund, K., Niccolai, G. P., Lavoué, E., Hmelo-Silver, C., Gweon, G., & Baker, M. (Eds.), *A Wide Lens: Combining Embodied, Enactive, Extended, and Embedded Learning in Collaborative Settings, 13th International Conference on Computer Supported Collaborative Learning (CSCL) 2019*, Volume 2 (pp. 939–940). Lyon, France: International Society of the Learning Sciences.

Sun, N., Wang, X., & Rosson, M. B. (2019). How do distance learners connect? In *Proceedings of the 2019 CHI conference on human factors in computing systems* (pp. 1–12).

Sundar, S. S. (2020). Rise of machine agency: A framework for studying the psychology of human–AI interaction (HAII). *Journal of Computer-Mediated Communication, 25*(1), 74–88.

Szostek, A. M., Karapanos, E., Eggen, B., & Holenderski, M. (2008). Understanding the implications of social translucence for systems supporting communication at work. In *Proceedings of the 2008 ACM conference on computer supported cooperative work* (pp. 649–658).

Terveen, L., & McDonald, D. W. (2005). Social matching: A framework and research agenda. *ACM Transactions on Computer-Human Interaction (TOCHI), 12*(3), 401–434.

Tubman, P., Benachour, P., & Oztok, M. (2019). *Interactive visualizations to enhance social learning practices in MOOC platforms*. International Society of the Learning Sciences.

Wang, Q., Camacho, I., & Goel, A, (2020a). *Designing for agent-mediated online social connections: Lessons learned and potential challenges*. http://speechinteraction.org/CSCW2020/papers/3-Wang.pdf. Presented at the Conversational User Interface Workshop at Computer Supported Cooperative Work (CSCW).

Wang, Q., Jing, S., Camacho, I., Joyner, D., & Goel, A. (2020b). Jill Watson SA: Design and evaluation of a virtual agent to build communities among online learners. In *Extended abstracts of the 2020 CHI conference on human factors in computing systems* (pp. 1–8).

Wang, Q., Jing, S., Joyner, D., Wilcox, L., Li, H., Plötz, T., & Disalvo, B. (2020c). Sensing affect to empower students: Learner perspectives on affect-sensitive technology in large educational contexts. In *Proceedings of the seventh ACM conference on learning at scale* (pp. 63–76).

Wang, Q., Saha, K., Gregori, E., Joyner, D., & Goel, A. (2021). Towards mutual theory of mind in human-AI interaction: How language reflects what students perceive about a virtual teaching assistant. In *Proceedings of the 2021 CHI conference on human factors in computing systems* (pp. 1–14).

Yorita, A., Egerton, S., Chan, C., & Kubota, N. (2020). Chatbot for peer support realization based on mutual care. In *2020 IEEE symposium series on computational intelligence (SSCI)* (pp. 1601–1606). IEEE.

Zytko, D., & DeVreugd, L. (2019). Designing a social matching system to connect academic researchers with local community collaborators. *Proceedings of the ACM on Human-Computer Interaction, 3*(GROUP), 1–15.

Zytko, D., Regalado, V., Grandhi, S. A., & Jones, Q. (2018). Supporting online dating decisions with a prompted discussion interface. In *Companion of the 2018 ACM conference on computer supported cooperative work and social computing* (pp. 353–356).

Chapter 14
Developing Social Interaction Metrics for an Active, Social, and Case-Based Online Learning Platform

Brent Benson and Mohamed Houtti

Abstract There is strong evidence that social and active learning techniques improve outcomes for students taking classes in traditional classroom settings and reduce the achievement gap for underrepresented students in STEM courses. However, bringing social and active learning to an online context has presented many challenges. The Harvard Business School Online Course Platform was developed to bring an active, social, and case-based pedagogy to online business learning and to carry HBS's case-based, inductive teaching approach from the physical classroom to a mostly-asynchronous online audience. Initial results for this active-learning approach to an online learning context are positive, but deeper evaluation is required. The study of social learning analytics is in its early stages and much of the focus has been on networks of social interaction. A set of metrics for measuring social interaction and activity patterns have been developed and implemented in this online learning platform and are being used to understand how active and social learning features relate to desired outcomes like course completion and assessment scores.

Keywords Social learning analytics · Active learning · Social learning · Social behavior

B. Benson (✉)
Harvard Business School Online, Boston, MA, USA
e-mail: bbenson@hbs.edu

M. Houtti
University of Minnesota, Minneapolis, MN, USA
e-mail: houtt001@umn.edu

Y. E. Wang et al. (eds.), *Social and Emotional Learning and Complex Skills Assessment*, Advances in Analytics for Learning and Teaching,
https://doi.org/10.1007/978-3-031-06333-6_14

14.1 Active Learning

Active and social approaches to teaching in the classroom have been shown to produce better outcomes than the more popular lecture-based or *sage on the stage* approach (Freeman et al., 2014). Adding even a small component of active learning or social interaction improves outcomes like test scores and completion rates by significant, measurable amounts.

A more recent study (Theobald et al., 2020) shows that active learning techniques not only improve student performance overall, but actually decreases the achievement gap in underrepresented groups like African Americans, Hispanics, first generation college students, and low-income students. These results should be pushing all education in the direction of active learning approaches.

The consensus definition of active learning used in the Freeman and Theobald studies is:

> Active learning engages students in the process of learning through activities and/or discussions in class, as opposed to passively listening to an expert. It emphasizes higher-order thinking and often involves group work.

This practice of engaging the learner in activity that reinforces the learning is key, and often involves a social or group aspect. In this work we focus on the social aspects of active learning in an adult learning context (college age and up), while acknowledging that non-social active learning is also important and is used in tandem in the HBS Online Course Platform discussed in this work.

14.2 Social Learning Analytics

The area of *Social Learning Analytics* is a part of the *Learning Analytics* field that deals with measuring and evaluating social interactions between learners. At its most effective, education is not an individual effort, but exists in a space of social interaction and collaboration (Shum and Ferguson, 2012).

There are some initial findings that point to amounts of social interaction having a positive effect on outcomes in both in-person (Okita, 2012; Hurst et al., 2013) and online (Hernández-García et al., 2015) learning contexts. However the types of social interaction in many studies are ad-hoc in nature, including message boards and forums, chat applications, and other out-of-platform mechanisms for interaction. Much of the work has been focused on Social Network Analysis, understanding the networks and relationships around the interactions between students.

Studying a learning platform that is designed with social interaction as a key part of its pedagogy offers a way to study social interaction specifically around the platform affordances like shared reflections, polls, and associated comments that encourage focused social interaction around course content.

14.3 Active, Social, and Case-Based Online Learning

The HBS Online Course Platform was initially developed by a team at Harvard Business School in 2013–2014 in an attempt to bring the HBS case method (Ellet, 2007) to a mostly-asynchronous online setting for adult learners in a professional, tertiary, or continuing education context. The goal of the project has been to increase the reach of the school to people who may find it difficult or impossible to attend in person. A parallel team developed a synchronous online platform originally called *HBX Live*, built around audio/visual technology to support remote case-based teaching with synchronous participants dialing in from around the world.

The creators of the Course Platform faced the question of how to implement an asynchronous version of case learning, when the primary tool of the case method is a guided, Socratic discussion led by a professor. In order to allow scaling to thousands of participants, the decision was made to use mostly-asynchronous social interactions between the students—guided by the platform content, rather than by an in-person instructor—to provide the necessary interaction, assumption questioning, and sharing of personal experience.

None of the existing online learning management systems (LMSs) supported in-platform social interaction (other than a separate forums sections) or the immersive types of interactive teaching elements that were required. A new platform was built, using some of the technical architecture of other LMSs (Python/Django front end, MySQL/MongoDB back end), but otherwise created in a bespoke fashion for the specific pedagogy being created by the team.

The key features designed to promote social engagement and collaborative learning in the Course Platform are presented in the following sections.

14.3.1 Cohort: The Primary Unit of Social Interaction

The intent of the HBS Online Course Platform is to provide social interaction with a set of peer participants. This requires identifying a set of participants that progress through the course at approximately the same pace with the same course start and end dates. The idea of allowing students to start at any time on a rolling basis was rejected as not providing the ability to develop academic relationships within the learning group.

How many students should be in the peer group? Because the platform is mostly-asynchronous, participants aren't necessarily logged in and working at the same time. In addition, students may have different interests and levels of experience.

Observation of early cohorts and how many students in the cohort were logged in at a particular time led to a target number of 300–400 participants in each social cohort in order to ensure a critical mass of learners on the platform for interaction purposes. If an offering of a course has more than 600 or so enrollees, they are divided into two social cohorts.

A cohort size of several hundred students is robust in the face of time zone and work habit differences, and minimizes the chance that a participant will not have people engaging with their ideas and offerings on the platform.

14.3.2 Some Synchrony: Common Deadlines and Locked Content

While the HBS Online Course Platform courses are asynchronous—students log in whenever they want—there is a certain level of synchrony imposed by common deadlines and locked content.

Each course module is unlocked at a fixed time, determining when participants can begin work on that module. Within a module, content is locked/not visible until the student completes the previous content, meaning that students can look back at previous work, but cannot skip ahead. There is also a shared module deadline by which all work in the module must be completed. A typical module is available for 2 weeks.

The combination of start/end times for modules and locked content may seem restrictive, but it encourages a flow that reinforces the case-based narrative which is being revealed one piece at a time. The storytelling flow of the course material is also characterized by smaller pieces of content where each piece of content, video, html, interactive graph, etc. takes at most a couple of minutes. There are no 30 min videos followed by a 20 min quiz.

Figure 14.1 shows the activities of a student in the HBS Online course *Economics for Managers* compared to a traditional Microeconomics MOOC. Notice the variety and number of activities in the HBS Online course.

14.3.3 Shared Reflections/Polls/Cold Calls

The narrative flow created by the deadlines, incremental unlocking, and fast pace is punctuated by moments of reflection where the student is asked to consider what has been discussed so far, to process and synthesize that information, and provide an opinion in the form of written text that can be seen and considered by others in the cohort (see Fig. 14.2 example).

Once the student has provided a written response, this response can be commented on and liked/starred by others in the cohort. This is the key element of social interaction in the HBS Online Course Platform.

There are variations on the simple shared reflection response. The reflection can be preceded by a poll which codifies a specific stance or answer across the cohort. The individual then expounds on their choice in the reflection text.

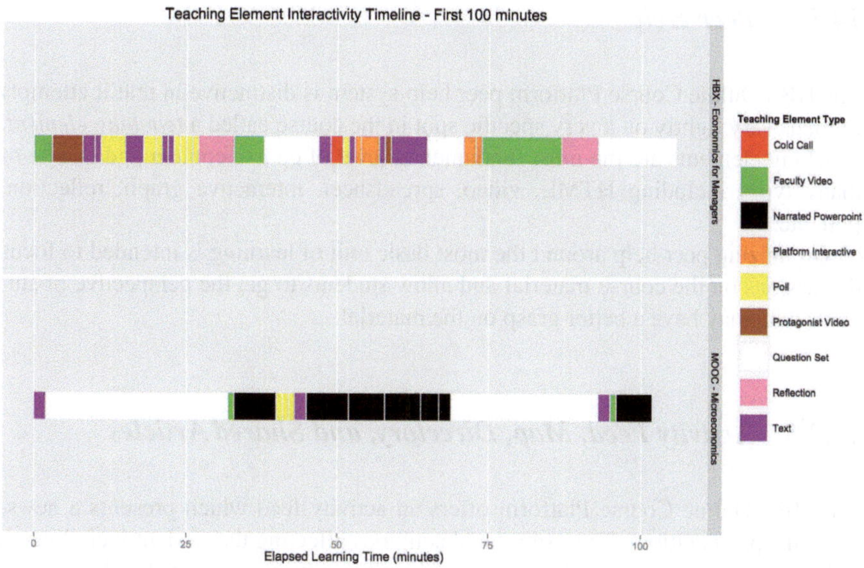

Fig. 14.1 Teaching element interactivity

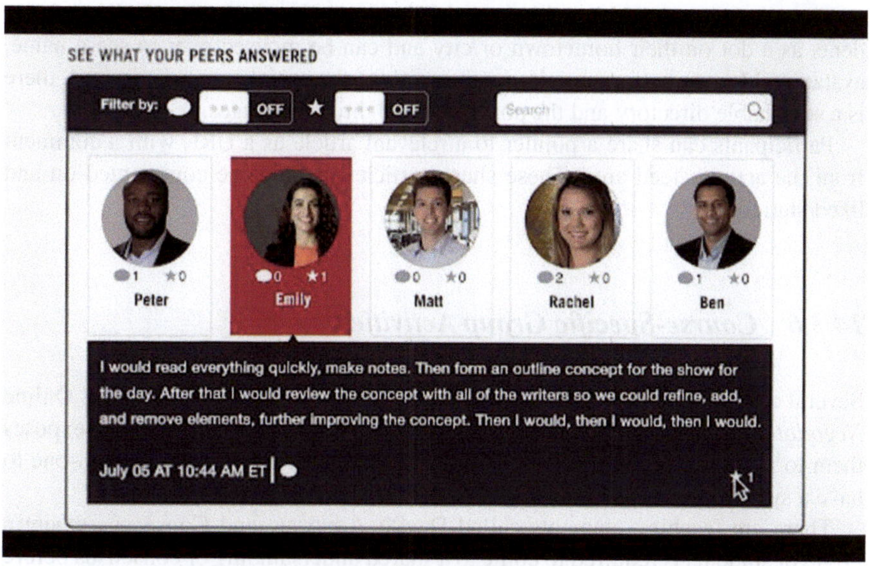

Fig. 14.2 Example of shared reflection social interaction

There is also a shared reflection variant called a *cold call* which is modeled after the practice of calling on students in class without prior notice. A cold call adds a timed element to the shared reflection, requiring the response to be written within several minutes with a count-down timer.

14.3.4 Peer Help

The HBS Online Course Platform peer help system is distinctive in that it attempts to focus very tightly on a very specific spot in the course called a *teaching element*. Teaching elements are the most fundamental level of course content and can be of many types including HTML, video, spreadsheet, interactive graph, reflection, poll, etc.

Organizing peer help around the most basic unit of learning is intended to focus discussions on the course material and allow students to get the perspective of students who may have a better grasp on the material.

14.3.5 Activity Feed, Map, Directory, and Shared Articles

The HBS Online Course Platform offers an activity feed which presents a newsfeed-like presentation of course-related activities affecting the student including an indication with a link if a shared reflection written by the student has been commented on or liked, a pointer to a peer help response given by the user that has received a response, platform announcements, etc.

There is an interactive world map which shows all of the currently logged students as a dot on their hometown or city and can be moused over to see a name, avatar, and location. If a more in-depth search of the social cohort is desired, there is a searchable directory and the ability to send direct messages.

Participants can share a pointer to a relevant article as a URL with a comment from the activity feed area. These shared articles can also be commented on and liked/starred.

14.3.6 Course-Specific Group Activities

Several courses offer synchronous interactions in smaller groups. The HBS Online *Negotiation Mastery* course groups students into buyer and seller groups, exposes them to asymmetric information, and then matches up the students one-on-one to have a synchronous text-based negotiation to try and consummate a deal.

There are teaching elements called Devil's Advocate and Consensus where a group of students is required to come to a shared understanding or consensus before finishing the exercise. The *Leadership Principles* course uses a Video Assessment Teaching Element which allows participants to upload a video to be assessed in a configurable way by others in the social cohort. The feedback is provided anonymously. There are plans to develop additional Course Platform features around small group formation and facilitation of small group activities.

14.3.7 Grading, Completion, and Credential vs. Certificate

The courses offered by HBS Online are grouped into two categories: credential and certificate.

The flagship credential program is called CORe (Credential Of Readiness) and consists of three courses: *Business Analytics*, *Economics for Managers*, and *Financial Accounting*. This program provides a comprehensive overview of business topics and also has graded quizzes and a high stakes final examination. Completing students are given a grade of *Fail*, *Pass*, *Honors*, and *High Honors*. The rigor and extra assessment of this program makes it fertile ground for studying the effects of social interaction and outcomes, and also puts it in a separate class in terms of completion rates.

The certificate programs fall into more of an executive education model where the certificate is given if the student meets the deadlines of the course and completes the required materials. These certificate programs do not have a high stakes final exam.

Course completion is measured as a desired outcome across both credential and certificate programs for the purposes of this research, while quiz grades and final grade are also used for the CORe credential program.

14.4 Social Interaction Metric Definitions

The primary goal of this research was to identify on-platform student social behaviors that can be measured, compared and used as quantitative metrics for social interaction. The goal when deciding on these metrics was to capture a wide range of student social interaction in a set of *easily interpretable* numbers that (1) are useful across both credential courses and certificate offerings and (2) can be used to examine/compare students' social interaction levels at any point in a course.

Investigation began with interaction data from the peer help system. However, it was observed that peer help data is quite sparse—i.e., many students do not interact with the system at all—making it difficult to use peer help participation as a consistent predictor of student success. As mentioned previously, shared reflections, polls, and cold calls represent the key element of social interaction in the HBS Online Course Platform. These elements are also distributed consistently throughout each course, providing a consistent stream of social interaction data that can be examined at multiple time points within any given course. Therefore, shared reflections, polls, and cold calls became the focus of measurement around students' relative levels of on-platform social activity.

Responses to reflections, polls, and cold calls can be commented on and liked/starred by others in the cohort. These two behaviors—commenting and liking—form the basis of our first two metrics: *number of comments posted* and *number of likes given*. Both commenting and liking are entirely optional; a student could hypothetically complete any of our certificate or credential courses without submitting a single comment or liking a single response. Despite this, the overwhelming majority of HBS Online students do engage with their peers in these ways. Next was an examination of the hypothesis that the level of visible on-platform engagement correlates with completion and performance outcomes within cohorts.

There is prior work showing that the vast majority of interactions with online platforms do not result in the creation of any visible product, such as a comment or like (Schneider et al., 2009). More recent work finds that students' identities can also affect how they interact with each other in online learning platforms (Morales-Martinez et al., 2020). Women, for example, might be more reluctant to participate publicly in a forum that is majority male. This led to the importance of measuring *latent* social interaction—i.e., on-platform social interaction that is not visible to the students' peers or to the course administrators. The third and final metric, therefore, is the *number of views* on peers' responses to reflections, polls, and cold calls. Viewing peers' responses is also entirely optional; it represents an invisible type of active engagement that goes beyond what is strictly required for completing the course and for which students get no credit, even from their peers. The hypothesis held again that the level of invisible on-platform engagement correlates with course completion and performance within cohorts.

14.5 Social Interaction Metric Measurements

It was important to address the challenge of how to scale these social interaction metrics across different course offerings, time periods, and social cohort sizes. The number of social teaching elements varies between different course offerings (and even within different modules of the same course offering), so comparing the raw number of comments, likes, and views between offerings or modules would not be indicative of actual differences in social interaction as much as mere differences in course content. Furthermore, some students drop out part-way through a course. It is important to include these students in the analyses, but it doesn't make sense to compare the raw amount of social engagement between students who participated in a course for 2 weeks to those who participated for 6 weeks.

The solution is to divide the raw counts by the total number of social teaching elements the student has encountered so far. For example, the *number of likes given* metric becomes the *number of likes given per social teaching element*. One can also think about this as corresponding to the ratio of non-mandatory-to-mandatory social interaction on-platform, where a higher ratio indicates greater social interaction.

There was another question as to whether the social metrics should also be normalized by cohort size. After all, a larger cohort means a greater availability of peer responses the student can comment on, like, or view. For example, if there are only two peer responses available for a student to comment on, they may submit only two comments because there just isn't any more interesting material to comment on. That same student might have submitted 10 comments if there were 10 peer responses available on that social teaching element. However, further examination made it clear that there was no correlation between cohort size and the proposed metrics. It seems that all of the cohorts were "big enough." In other words, while the difference between 10 peer responses and 2 peer responses is substantive, the difference between 800 and 300 is not; in both of the latter cases, the student already has more than enough content to view, like, and comment on, and the presence of additional peer content does not cause them to spend more time interacting with social teaching elements. Therefore, there is no need or benefit to normalizing by cohort size.

The metrics were examined across 232 cohorts, corresponding to course offerings from 2017 to 2021. Fifty-three of these cohorts were for credential course offerings and the other 173 were for certificate course offerings. Within each of these cohorts, there were statistically significant differences in all of the social metrics between students with different outcomes. For example, in a recent credential cohort picked out for closer inspection, the median views per social teaching element for students who passed was slightly below the *bottom quartile* for students who passed "with high honors"— 3.16 and 3.30, respectively (see Fig. 14.3). This pattern holds consistently between completers and non-completers in certificate courses as well for each of the outlined metrics.

Next was a comparison of the levels of social interaction between completion-based certificate courses and the more rigorous credential courses to establish baselines for each. While certificate courses had 13% and 19% greater views and comments, respectively, per social teaching element, students in credential courses gave 50% more likes per social teaching element. Credential course students are told that their levels of social interaction within the course can affect their grades. Though students are not given any specific knowledge about how social interaction is measured for grading purposes, there are anecdotal reports that some students attempt to artificially increase their levels of social interaction, with the hope that doing so will affect their final grades. It remains to be investigated whether grading guidance causes credential students to click "like" on social teaching elements much more often, as doing so is both a *low-effort* and *visible* way to interact socially on the platform. This further highlights the importance of using latent social interaction metrics alongside visible ones.

Next were measurements as to whether these social metrics are useful as predictors of student success. Was there enough data at the end of the first module to gauge likely success or failure?

Starting with a recent credential course and looking at just the data available at the end of the first module, a regression model was used to predict performance in subsequent modules, as measured by the average quiz score in those modules. (The first quiz score was excluded from the final average so we could use it as a predictor.) The first predictor was a logistic regression model based on social interaction metrics, which achieved a root-mean-squared error (RMSE) of 12.94 percentage points (see Fig. 14.4). Using the student's score on the first-module quiz instead gave a better RMSE of 10.52. Using both first quiz scores and social interaction metrics as predictors reduced the RMSE slightly to 10.36. This shows that, while frequent assessments are an overall more powerful tool for gauging a student's

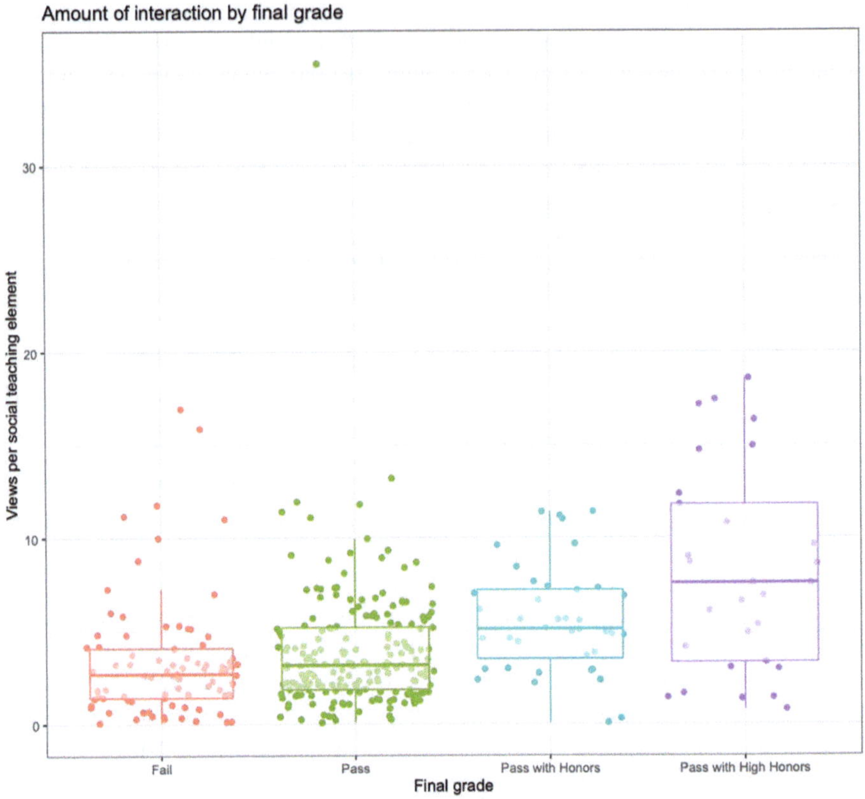

Fig. 14.3 Amount of student interaction by final grade

future performance, social interaction data can be useful as supplemental predictors. Interestingly, a model with only two predictors—the first quiz score and number of views per social teaching element—achieved a RMSE of 10.35, indicating that most of the increase in predictive power provided by social interaction data, at least in this use case, comes from latent data about number of views rather than the other two, more visible, social metrics.

14.6 Conclusion

A key contribution of this work is a set of scaled metrics around in-platform social behaviors that are useful for comparing interaction levels between students, modules, courses, and offerings. These metrics are the number of likes/stars given, number comments given, and number of responses viewed, which are scaled and normalized by dividing by the number of socially shared responses that are available for interaction.

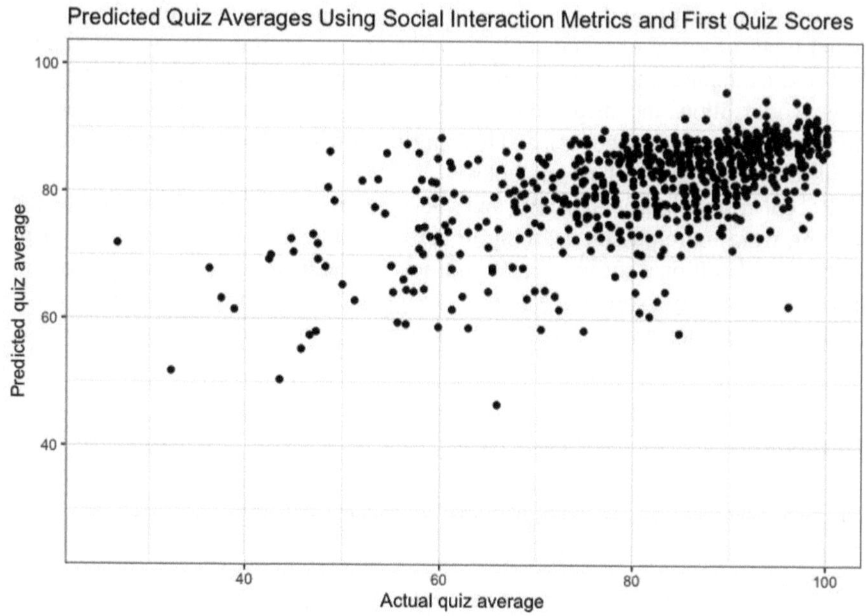

Fig. 14.4 Predicted vs. actual quiz average

These interaction metrics were tested for relationships to desired student outcomes like course completion and course grade. The primary finding is that the degree of social activity—especially when measured by a latent metric like how many of your social cohorts responses you have read—has a strong relationship with desired outcomes like course grade and course completion in the studied platform which is designed to encourage social learning and discussion.

Other social metrics, like peer help activity, don't seem to have a similar relationship to the measured outcomes, primarily because there isn't a consistent availability across all students.

14.7 Future Work

The identified social engagement metrics, combined with non-social indicators like procrastination and early quiz performance, could be used to build models to help identify and help students at risk of not completing, or failing a course.

Work has begun on classifying what types of student responses drive the most social interaction and engagement. Additional work is being done to use Natural Language Processing and Machine Learning classifiers to understand how characteristics like personal identification or emotion are related to interaction and interest.

Course and platform designers would like to create more opportunities for additional types of social engagement by introducing in-platform contexts for students to work, learn, and interact together in small group divisions of their social cohort. This would allow the measurement of whether these small group interactions improve outcomes for students, and to experiment with cohort assignment and composition to facilitate the best outcomes.

Recent research around active and social learning supports the inference that moving from more passive learning to active and social contexts should improve outcomes. Additional measurement and development of social learning analytics has the opportunity to provide a more thorough understanding of how to most effectively match students with effective active learning opportunities.

References

Ellet, W. (2007). *The case study handbook: How to read, discuss, and write persuasively about cases*. Harvard Business School Press.

Freeman, S., et al. (2014). Active learning increases student performance in science, engineering, and mathematics. *Proceedings of the National Academy of Sciences, 111*(23), 8410–8415. https://doi.org/10.1073/pnas.1319030111

Hernández-García, Á., et al. (2015). Applying social learning analytics to message boards in online distance learning: A case study. *Computers in Human Behavior, 47*, 68–80. https://doi.org/10.1016/j.chb.2014.10.038

Hurst, B., Wallace, R., & Nixon, S. B. (2013). The impact of social interaction on student learning. *Reading horizons, 52*(4), 375.

Morales-Martinez, G., Latreille, P., & Denny, P. (2020). Nationality and gender biases in multicultural online learning environments: The effects of anonymity. In *Proceedings of the 2020 CHI Conference on Human Factors in Computing Systems* (pp. 1–14). Association for Computing Machinery. https://doi.org/10.1145/3313831.3376283

Okita, S. Y. (2012). Social interactions and learning. In N. M. Seel (Ed.), *Encyclopedia of the sciences of learning* (pp. 3104–3107). Boston, MA. https://doi.org/10.1007/978-1-4419-1428-6_1770

Schneider, F., et al. (2009). Understanding online social network usage from a network perspective. In *Proceedings of the 9th ACM SIGCOMM conference on Internet measurement* (Vol. IMC '09, pp. 35–48). Association for Computing Machinery. https://doi.org/10.1145/1644893.1644899

Shum, S. B., & Ferguson, R. (2012). Social learning analytics. *Educational Technology & Society, 15*(3), 3–26.

Theobald, E. J., et al. (2020). Active learning narrows achievement gaps for underrepresented students in undergraduate science, technology, engineering, and math. *Proceedings of the National Academy of Sciences, 117*(12), 6476–6483. https://doi.org/10.1073/pnas.1916903117

Chapter 15
Network Climate Action Through MOOCs

Yue Li and Marianne E. Krasny

Abstract In the context of global environmental change and climate change in particular, engaging the public in climate actions is important especially if actions can be scaled up through social networks. The current study examined MOOC participants' climate actions and how they influenced their networks to take the same actions in countries across the globe. The research is based on a Cornell University MOOC entitled *Network Climate Action: Scaling Up Your Impact.* We conducted a post course survey including checkbox and open-ended questions to explore participants' climate actions and their networks as MOOC learning outcomes. Participants commonly chose reducing food waste and plant-rich diets as their climate actions, and applied social influence research to persuade their family and friends to take the same actions. The study helps us understand how a MOOC can foster participants taking climate actions and helping spread those actions through their social networks.

Keywords Online learning · Climate actions · Social influence · Social networks

15.1 Introduction

Online learning plays an increasing role in environmental, sustainability, and climate education for both the educators and the general public. Online learning provides accessible professional development opportunities for environmental educators, encourages the general public to take environmental actions, and creates social learning opportunities among all participants as they interact with and influence each other. In response to the COVID-19 pandemic, schools around the world are moving online, which forces researchers and educators to explore new ways to

Y. Li (✉) · M. E. Krasny
Cornell University, Ithaca, NY, USA
e-mail: yueli@cornell.edu; mek2@cornell.edu

Y. E. Wang et al. (eds.), *Social and Emotional Learning and Complex Skills Assessment*, Advances in Analytics for Learning and Teaching,
https://doi.org/10.1007/978-3-031-06333-6_15

engage audiences through the online learning environment (Crawford et al., 2020). Instead of simply transferring teaching materials online, we need sound pedagogies for motivating participants to engage in online learning, improve practices, and take environmental actions beyond the online courses.

The definition of success for course participants in a course like the *Network Climate Action: Scaling Up Your Impact* MOOC offered by Cornell University goes beyond mastery of the course materials; success encompasses thoughtful reflection on how to connect learning materials to real-life climate actions and committing to or actually conducting those actions. Such a model for engaging people at scale in climate actions outside of the course is sometimes referred to as "courses for a cause" (Krasny et al., 2020). The goal of the *Network Climate Action: Scaling Up Your Impact* MOOC is to engage participants in taking acclimate actions and influencing their networks to take the same actions. Unlike many MOOCs that depend on self-paced learning and automatically graded quizzes (Admiraal et al., 2015), this course takes a project-based learning approach that guides course participants to develop an action plan for themselves and their networks. The action plan gives participants an opportunity to reflect on the concepts and principles that they have learned in the course and how they can apply them outside of the course. How an online course can engage people at scale in sharing ideas and making actionable plans for conducting climate actions beyond the course warrants further investigation to advance network climate actions through MOOCs worldwide.

The proposed research is to examine MOOC participants' climate actions and how they influence people around them to take the same actions. The MOOC project is based on the hypothesis that social influence through tight networks plays a key role in forming individuals' behavior (Centola, 2018, 2021; Christakis & Fowler, 2013; Frank, 2020; Friedkin & Johnsen, 1990). The specific research questions are: (1) What climate actions do MOOC participants choose to take? (2) Which networks do MOOC participants choose to influence? (3) How do MOOC participants apply research-based strategies learned from the course to influence their networks to take the same actions? The results of the study will not only provide strategies to engage the public in climate actions through participating in MOOCs, but also help us understand how participants influence others' behaviors through the online environment.

15.2 Literature Review

Online courses not only provide an opportunity for participants to access content and instructors, but also can facilitate participant-participant interactions (Moore, 1989). For example, courses use learning management system discussion boards (Skrypnyk et al., 2015; Zhang et al., 2016), social media (DuBois et al., 2019), and local and interest study groups (Brinton et al., 2014; Krasny et al., 2018) to facilitate social learning. Such online informal interactions can enhance the learning experience and provide a space to post personal feelings or reflections (Liu et al., 2016a), as well as lead to formation of online communication networks (Kellogg et al., 2014) and offline collaboration networks (Li & Krasny, 2020) among participants. Through these interactions, communities of practice (Wenger, 1998) may form in which students exchange information about their practices. For example, environmental education professionals in online courses emphasizing participant-participant interactions co-authored an eBook (Russ, 2015), adapted or intended to adapt learned ideas into their environmental education practice (Li et al., 2016), and changed their practices to a limited extent by the end of the course (Li & Krasny, 2019). Because the influence of these interactions may vary depending on cultural (Liu et al., 2016b) and language background (Colas et al., 2016), MOOCs provide a context for exploring how students with varying learning preferences related to cultural backgrounds (Hofstede, 2011; Parrish & Linder-VanBerschot, 2010) participate in social learning.

Peer-peer communications, knowledge co-creation, and engaging in sustainable behaviors as a result of online courses suggest that such courses can promote social learning and social emotional learning goals. Originally conceived as learning approaches that go beyond instructor-driven content to incorporate relationship building and developing the capacity to make choices and engage in behaviors that promote individual, community, and environmental health (Weissberg et al., 2015), social and social emotional learning have long been a part of online learning as evidenced by the early cMOOCs (Cormier & Siemens, 2010). However, as the more content-driven xMOOCs gained popularity, ideas about social learning and community-oriented outcomes gave way to an emphasis on individual learning that would promote students' career prospects (Stewart, 2013). More recently, online learning scholars and providers have sought to leverage MOOCs' ability to reach large numbers of participants globally to help address social and environmental problems, in part through social learning approaches (Ferguson & Shum, 2012; Joksimovic et al., 2020; Krasny et al., 2018).

Evaluating learning outcomes that go beyond the course to encompass network formation and behavior change presents its own set of challenges. Much research has focused on students' learning process and short-term outcomes. Course completion is one of the most commonly used outcome measures in online education research, but reflects different types and levels of engagement depending on the particular course. Kahan et al. (2017) called for attention to various learning behaviors that capture different types of online learning engagement. Project-based MOOCs, in particular, focus on participants' knowledge construction (Barak & Watted, 2017) and collaborative learning (Spoelstra et al., 2014). Longer-term outcomes beyond the course remain challenging to measure. Only a few studies have measured MOOC participants' post-course development such as knowledge transfer (Chen et al., 2016), knowledge co-creation (Krasny et al., 2018), teaching practice (Napier et al., 2020) and career advancement (Wang et al., 2014). Krasny et al. (2020) used seven examples from active MOOC participants to show how MOOCs could spark innovative sustainability actions in local communities. Further research is needed to explore participants' actions beyond the course, and how these actions could scale up to make a larger impact globally.

The *Network Climate Action: Scaling Up Your Impact* MOOC that is the focus of this study was developed to foster the exchange of practical ideas for climate solutions among course participants from diverse cultural backgrounds, and prepare and encourage them to take climate actions as well as influence their networks to take the actions. Thus, we used structured weekly discussions to reflect on and share ideas about course materials and project-based learning to apply learned ideas into actions. At the end of the course, we conducted a post survey to evaluate participants' climate actions and applications of learned ideas in their actions. The goal of the study is to investigate the action outcomes of a MOOC in motivating participants' actions outside of the course. Moreover, this study takes a social influence approach to examine how MOOC participants' individual actions scale up through their networks.

15.3 Methods

We conducted a post course survey including both checkbox and open-ended questions to examine MOOC participants' climate actions and their networks.

15.3.1 Online Course and Participants

The Cornell University online course *Network Climate Action: Scaling Up Your Impact* was offered for the first time from April 7 to May 12, 2020, using the edX Edge course platform for participants outside of China and Xiaoe-tech course platform for participants in China. The goal of the course was to motivate participants to choose a climate action they can take themselves and apply social influence research to persuade their family, friends, social media followers, or other social network to also take that action. The course materials include video lectures, readings, discussions and webinars. The weekly video lectures and readings cover topics including social networks and spread of complex behaviors (Centola, 2018), social mobilization (Rogers et al., 2018), social norms (Nolan et al., 2008; Sparkman & Walton, 2017), persuasion (Cialdini, 2001), social marketing (McKenzie-Mohr & Smith, 2009; Weinreich, 2015), and social media (Young et al., 2017). We also briefly introduced motivations for pro-environmental behaviors (Sussman et al., 2016) and choice architecture (Garnett et al., 2019). The webinar series focused on plant-rich diet as one of the most popular climate actions. We also used social media including Facebook, WhatsApp and WeChat as optional social learning and interaction platforms for participants. Participants could choose to pay the standard registration fee ($60), any amount they could afford, or nothing; in this way, we earn income to pay instructors' salaries while making the course accessible to anyone globally.

Among 624 registrants, 365 participants enrolled in edX Edge and 259 participants enrolled in Xiaoe-tech. The average age of the participants was 32 years old. A total of 66.35% were women and 76.92% had a bachelor's or higher degree. Among 50 countries, most participants were from China (39.42%), the US (18.59%), Iran (16.03%) and Nigeria (6.57%). The most frequently reported occupations were university students (31.57%), business employees (12.02%), self-employed (9.00%) and non-formal educators (8.81%). The most frequently reported organization types were universities (31.73%) and NGOs (25.64%).

15.3.2 Data Collection and Analysis

To guide the course participants take climate actions and influence their networks, we asked them to fill out a climate action plan in week 1, and keep updating their action plan during weeks 2–4 with guiding questions according to the weekly topics. At the end of the course, we conducted a post survey with the course participants to report their actions and the network, such as family, friends, or colleagues, they chose to influence (Appendix). First, we used checkbox questions to assess participants' climate actions, demographics of the networks they influenced or were planning to influence, and means of communication. The question about climate actions provided a list of 13 types of most common actions (https://drawdown.org/) individuals could do or support (e.g., through donations) in their daily life including plant-rich diets, tree planting, mass transit, walking or biking, household recycling, ridesharing, reducing food waste, LED lighting, water saving, composting, solar energy, health and education (especially for women), advocacy, and an option for other actions. We also asked the demographics of their networks, for example, age range (multiple choices), gender (multiple choices), occupation (single choice), and urban or rural (single choice). Further, we used open-ended questions to ask participants the strategies they applied from the course materials to influence their networks (e.g., norm messages).

We conducted descriptive analysis of all checkbox questions to provide the percent of participants who chose each option. Then the first author coded all open-ended questions using predetermined codes (Saldaña, 2013) based on the course lecture topics (Table 15.1) in Google Spreadsheets. The major themes include spread of complex behaviors (Centola, 2018), social mobilization (Rogers et al., 2018), social norms (Nolan et al., 2008; Sparkman & Walton, 2017), persuasion (Cialdini, 2001), social marketing (McKenzie-Mohr & Smith, 2009; Weinreich, 2015), social media (Young et al., 2017), pro-environmental behaviors (Sussman et al., 2016), and choice architecture (Garnett et al., 2019).

Table 15.1 Predetermined codes from the course lectures

Theme	Sub-themes
Spread of complex behaviors	Strategic complementarity, Credibility, Legitimacy, Emotional contagion, Strong and weak ties
Social mobilization	Personal, Accountable, Normative, Identity Relevant, Connected
Social norms	Descriptive norms, Injunctive norms, Dynamic/trending norms
Persuasion	Liking, Reciprocity, Social proof, Consistency, Authority, Scarcity
Social marketing	Focus on behavior change, Understand your audience, Keep the social in social marketing, Commitment, Prompts, Incentives
Other	Social media, Motivations, Choice architecture

15.4 Results

Among 320 participants who submitted their final report (response rate 51.28%), 103 participants implemented their climate actions personally and 198 participants implemented their climate actions both personally and with their networks. We report results on climate actions, the networks participants chose to influence, and strategies used to influence networks.

15.4.1 Climate Actions

The number of actions participants chose to take ranged between 1 and 14 (mean = 5). The two most popular actions participants chose are reducing food waste (79.69%), and plant-rich diets (73.12%), followed by water saving (51.25%), household recycling (46.25%), public transit (44.38%), walking or biking (43.75%), advocacy (33.75%), LED lighting (30.63%), health and education (especially for women) (29.69%), tree planting (25.62%), composting (21.88%), ridesharing (19.69%) and solar energy (17.50%) (Table 15.2). Participants also mentioned other types of actions (10.00%) including biodiversity/pollinators, conservation agriculture, purchasing electric vehicles, smart technologies, conscious consumption, and organic food.

15.4.2 Networks

The number of networks participants chose to influence ranged between 1 and 6 (mean = 3). The most popular networks participants chose were family (86.25%) and friends (85.00%) followed by social media followers (51.88%), fellow students (43.44%) and colleagues (36.56%) (Fig. 15.1). Participants also mentioned other types of networks (11.56%) including children, city managers, employers, volunteers, local clubs, neighbors, NGO members, restaurants, students, community members, school teachers/parents, and youth camp members.

Table 15.2 The percentage of participants who chose each climate action

Climate action	Percentage of participants (%, N = 320)
Reducing food waste	79.69
Plant-rich diets	73.12
Water saving	51.25
Household recycling	46.25
Public transit	44.38
Walking or biking	43.75
Advocacy	33.75
LED lighting	30.63
Health and education (especially for women)	29.69
Tree planting	25.62
Composting	21.88
Ridesharing	19.69
Solar energy	17.50
Other	10.00

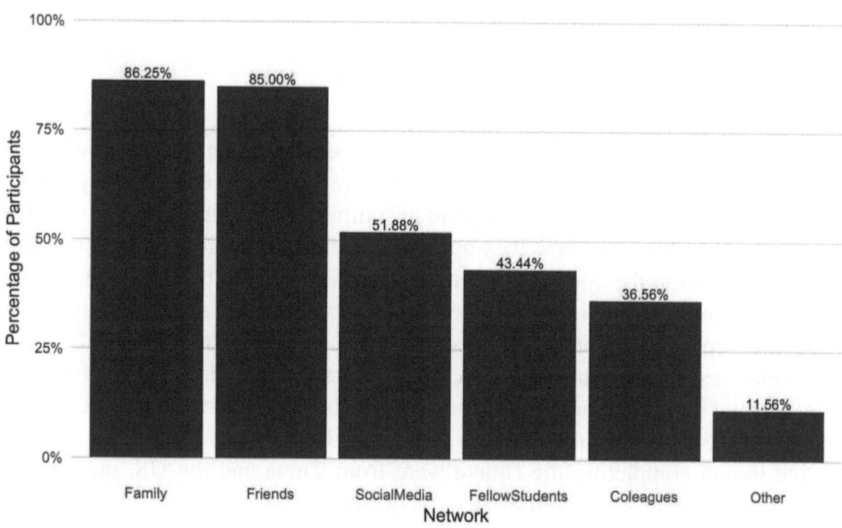

Fig. 15.1 The percentage of participants who chose each network to influence

In terms of demographics of MOOC participants' influence networks, the most popular age range participants chose to influence was 18–30 years (78.68%), followed by 30–50 years (58.62%), under 18 years (34.17%) and over 50 years (25.71%). In terms of occupation of MOOC participants' influence networks, students (73.67%) and professionals (73.67%) were most popular, followed by volunteers (33.23%) and the retired (27.59%). Most participants chose mixed men and women (67.78%) and those in mostly urban areas (83.07%) as their networks to influence for their climate actions (Fig. 15.2).

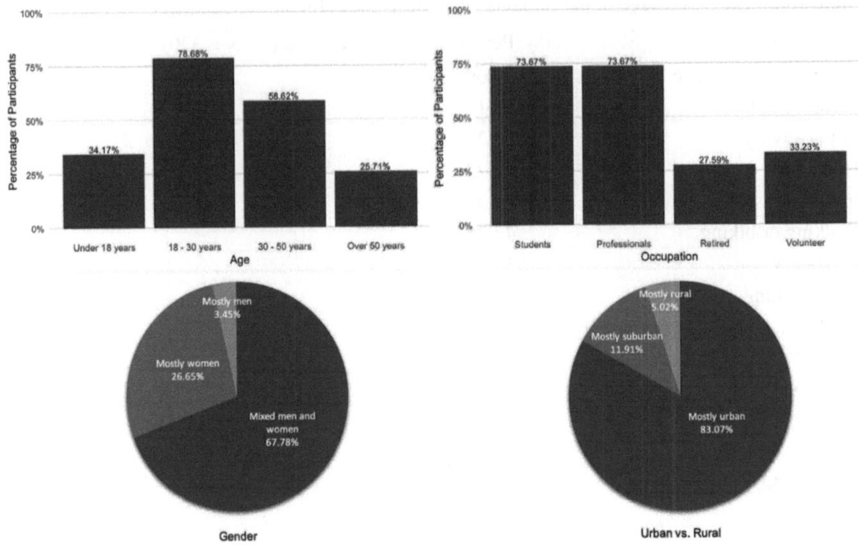

Fig. 15.2 Demographics of the networks participants chose to influence (N = 319)

15.4.3 Strategies to Influence Networks

Most respondents reported they used a mix of in-person and online communication (62.27%) to influence their networks. Other respondents used mostly online (24.21%) or mostly face-to-face (13.52%) to communicate with their networks. The respondents also described how they applied specific communication or persuasion strategies learned in the course to influence their networks to take climate actions. The most popular strategies mentioned were social norms, social mobilization, spread of complex behaviors, social marketing and social media. A few participants also mentioned other strategies including persuasion, motivations for pro-environmental behaviors, and choice architecture, but they did not elaborate on how they used these strategies in their network actions. Because most of the participants enrolled in and completing the course were from China and the US, most of the quotes we show below are from these two countries.

15.4.3.1 Social Norms

Specifically, 32.81% respondents mentioned they applied or were planning to apply social norms including descriptive (11.56%), trending/dynamic (10.31%), or injunctive (2.50%) norms to influence their networks. For example, a participant who chose reducing food waste and plant-rich diets as her action described how she used both descriptive and trending norms in influencing her friends and social media followers:

Social norms: the descriptive norms such as Beijing's garbage sorting, Empty Plate Action policies, and public account articles...through trending social norms...More and more people join the plant-rich diets and choose to support organic food". (female, 34 yrs, China)

A participant who chose solar energy as his action mentioned how trending norms are more effective than descriptive norms in influencing his family, friends and faith community:

Talking about actions as joining a "growing number" of people making that environmental related decision is more powerful than saying "please join the 1% of home owners of Oregon who have installed solar panels." (male, 65 yrs, US)

A participant who chose plant-rich diets as her action mentioned how she applied injunctive norms in influencing her family and friends:

I made a video call with them last week and explained to them why they should do it, to reduce climate change effects, impact the lowest in the environment, and also because it is healthier for them. In that part, I applied the injunctive norm. (female, 30 yrs, US)

Compared to descriptive and trending norms, fewer participants selected injunctive norms perhaps because telling people what they should do could be inappropriate in some cultural contexts. For example, a participant who chose plant-rich diets and reducing food waste as his action mentioned:

I used the descriptive norms to tell my family what others do, why we need to protect the environment, etc. The reason why I don't use injunctive norms is because I'm a junior in my family. Also, everyone in my family is used to the eating behavior in the past, so I need to persuade and adapt slowly. (male, 18 yrs, China)

15.4.3.2 Social Mobilization

For social mobilization, respondents mentioned using personal (21.25%), accountable (9.69%), normative (25.00%), identity relevant (11.56%) and connected (10.00%) principles to influence their networks. For example, a participant who chose reducing food waste applied personal and accountability principles to influence her family:

Through our personal conversations, I have attempted to convey my sincerity and commitment to my own journey of reducing food waste which I hope will mean they are more likely to consider the ideas I put forward. I've also used Accountability by checking in weekly with my family via Zoom to ask how their food waste journey is going. (female, 51 yrs, New Zealand)

A participant who chose reducing food waste as her action mentioned how she used several social mobilization principles to influence her friends:

Many of my persuasion strategies came from the PANIC [Personal, Accountable, Normative, Identity Relevant, Connected] principles. I made the behavior change very personal by having the food swaps occur face to face at one time with everyone present. I also used the concept of accountability to make sure that all of my roommates showed up to the food swaps and participated. It was obvious if one person was missing from the group of six so the social pressure to participate worked in my favor. I also made the behavior identity rel-

> evant. Many of my roommates identify as environmentalists so I framed the action as something an environmentalist would do, and something they should do if they identify themselves with that social group. I enacted the food swap every week after our group dinners, so it made the behavior normative and integrated into my roommates' routines. (female, 25 yrs, US)

A participant who chose reducing food waste, plant-rich diets and composting as her actions described how she planned to use social mobilization principles to influence her friends and colleagues:

> I plan to use the PANIC principles to address my target audience. First to personalize the approach (family v. friends v. coworkers) Second to encourage them to share what they are doing (online, with me or with each other) … as a way to be observable and hopefully accountable. To have a sort of social marketing campaign that looks at how these behaviors are becoming more normal and how they contribute to a certain identity (family: good cook, friends: active participants in community and coworkers: avid environmentalist and educators). (female, 35 yrs, US)

15.4.3.3 Spread of Complex Behavior

Participants applied the literature on spread of complex behaviors in influencing their networks. Specifically, 14.06% respondents mentioned using theories of strong and weak ties. For example, a participant who chose reducing food waste and plant-rich diets applied strong ties to influence her roommates:

> Strong ties: I choose my housemates because we spend the most time together (strong ties), which provides the opportunity for multiple exposure, trust, and the social aspects. I didn't choose my parents even though they are also strong ties due to our limited daily interactions. (female, 27 yrs, China)

Respondents also mentioned four mechanisms of spread of complex behavior. For example, a participant who chose plant-rich diets, tree planting, public transit and walking as her actions described how she applied each of the mechanisms to influence her family and friends.

> First, strategic complementarity: There is a small program of Ant Forest on Alipay platform. By walking, paying online through Alipay, or other behaviors can be converted into energy … transformed into offline tree planting. Many people conduct tree planting activities remotely through online behaviors. At the same time, coupled with the participation and recommendation of many celebrities, it further expands influence and action in the fan community. There are more and more people adopting this method and putting it into practice, and the value of this program also increases. Second, trust. When I took some environmental protection activities in my family, such as choosing public transportation or walking, the family members gradually turned to be consistent with me under my influence. Because the trust of family members is much higher than the call on the Internet… Third and fourth: Legitimacy and emotional contagion. For some people, when they see some environmental protection calls on social platforms, they will always wonder if it is a pre-arranged or sponsored event, which will doubt its authenticity. Therefore, they will not really think about the meaning of environmental protection and consider environmental protection behaviors. It is different if it is in a dormitory. For example, I will practice vegetarian behavior with other roommates in the dorm. It is difficult for one person to persist, but it is relatively easy to be

with a roommate. If two people participate in vegetarian activities together, it will be relatively easy for the influence to spread to the entire dormitory. Everyone will be emotionally infected, and the saying that eating less meat and dairy products is good for physical and mental health will be more acceptable to roommates. (female, 20 yrs, China)

15.4.3.4 Social Marketing

Participants applied social marketing strategies. For example, a participant who chose reducing food waste, plant-rich diets and public transit as her actions applied the social marketing principle of understanding audience to influence her family and social media followers:

I will understand the needs of the target audience. There are two types of my audience, one is my family, and the second is my WeChat friends. For example, family members need to lose weight. For them, a vegetarian diet is a better way to lose weight. At the same time, family members, especially parents, are not too concerned about climate change, so I will combine a vegetarian diet with weight loss and talk less about the relationship with climate change. (female, 26 yrs, China)

A participant who chose reducing food waste, plant-rich diets and recycling as her actions was planning to apply commitment and incentives to influence family friends and fellow students:

Social marketing strategy: prepare to create a group for friends who are interested in trying vegetarian food and reduce trash and establish a few guidelines. Everyone must make a commitment in the group, and regularly share their results in the form of pictures…use the incentives, those who complete the commitment will finally get a small prize. (female, 21 yrs, China)

A participant from China who chose reducing food waste as her action applied prompts to influence family friends:

Prompts strategy: often remind friends to insist on actions to reduce food waste and give them some stickers to remind them. (female, 22, China)

Most respondents (68.03%) chose to provide recognition, rewards or incentives to members of their networks, including offering praise for contributions, sending hand-made cards, giving reusable bags or water bottles, sharing recipes, providing plant seeds or seedlings, or inviting for meals. For example, a participant mentioned:

Give home-garden seed pack, Trees for tree planting, and home emissions checking tool, arrange training and workshop to knowledge sharing, publish book and deliver to all members of the group. (male, 27 yrs, Sri Lanka)

15.4.3.5 Social Media

Participants also mentioned using social media to connect with their networks, including WeChat (57.55%), WhatsApp (28.25%), Instagram (28.30%), Facebook (27.04%), QQ (21.70%), ZOOM/Skype (18.56%), Telegram (13.21%), Twitter

(10.69%), Tik Tok (6.92%), and other tools such as LinkedIn, Weibo, podcasts, and YouTube. For example, a participant who chose reducing food waste and plant-rich diets as her actions used WeChat to influence her family and friends.

> Through the establishment of a WeChat group to contact event participants, and in the WeChat group, you can also see the pictures and frequency of vegetarian meals that other participants daka [check in]. Based on the pictures of everyone daka, we will further explain the true meaning and content of the vegetarian meal, and praise and reward members who daka frequently…summarize the daka of everyone every week, and finally display the results in the WeChat moments.(female, 26 yrs, China)

A participant who chose plant-rich diets as his action used WhatsApp and Instagram to influence his students:

> We have a group in WhatsApp and Instagram. I define a project to my student to investigate the influence of changing diet on the environment. I asked them to present it in class and with planning every day … 2 students come and present their lecture. After the end of class all of students must contribute to the discussion that holds in WhatsApp and Instagram groups. In these groups, I present slides and much information about this issue, and students must express their idea about this issue. These discussions in WhatsApp and Instagram have a great impact on students to change their diet to plant-rich. (male, 31 yrs, Iran)

A participant who chose reducing food waste and plant-rich diets as her actions used FaceTime and phone calls with her family, and Facebook and Instagram with her friends and social media followers:

> Because of COVID, I am unable to do anything in person. So I decided to utilize the PANIC principles to determine how I communicate with each of my targets. My in-laws, I use FaceTime and phone calls because they don't engage in Social Media. I also surveyed folks through Facebook and Instagram and then based on their answers, met them where they are in terms of how they prefer to receive communications. (female, 26 yrs, US)

A participant who chose plant-rich diets as her action mentioned how she used videocall, google drive and WhatsApp in influencing her family and friends:

> After I convinced them with the videocall, I used two other social media to continue with my action plan: First, I included 10 recipes and, a table of equivalences between animal and vegetal protein in a google drive document. They can access easily these documents to find easy and quick recipes to cook so they do not have to lose time searching for them, and they can also edit the document and add more recipes. The second social media I am using is WhatsApp, I created a WhatsApp group to share photos, articles, jokes, or anything related to a plant-rich diet. I think all of the 3 social media that I am using in my action plan are efficient because my friends have started adding more recipes and they have shared photos of their meals and cheers other people to try new things. (female, 30, US)

15.5 Discussion

The current study shows that the *Network Climate Action: Scaling Up Your Impact* MOOC participants took climate actions and applied strategies learned from the course to influence their family, friend and other networks. The research adds to current online learning literature by showing that MOOC participants can transform learning into actions, thus going beyond content learning outcomes to incorporate social learning (Wals, 2007). The results also suggest that actions can be scaled up by spreading of behaviors through networks of MOOC participants. Below we discuss these contributions, explain why MOOC participants chose certain climate actions and networks, as well as provide implications for future MOOC design.

MOOCs can foster participant action outcomes in addition to knowledge and skills. The fact that the MOOC participants took climate actions beyond the course reflects the idea of courses for a cause (Krasny et al., 2020). In the *Network Climate Action: Scaling Up Your Impact* MOOC, providing a list of practical actions through video lectures and readings, and asking students to develop a network action plan and revise it each week, likely helped participants to think about different actions they could take in their own context. Further, we provided guided questions each week to help participants update their action plans based on the social influence lectures and readings of that week, and to identify their influence networks and ways to measure their impact. Finally, students interacted with other course participants through discussion boards and social media groups, which fostered participants' exchange of ideas and provided support in implementing their actions. The MOOC in the current study provides a unique model for online courses that aim to foster actions and scale up action impact through the course participants' networks.

We provide explanations for the most common climate actions and networks participants chose. First, the most common climate actions participants chose were reducing food waste and plant-rich diets. These two actions are relatively easy to implement at the individual level compared to other types of actions such as solar energy, which requires infrastructure or systemic change. Also, the weekly webinar series focusing on plant-rich diets during the course may have influenced participants' choices of actions. Second, the two most common types of networks participants chose to influence were family and friends, which may reflect the course emphasis on close networks with strong ties as suggested by course materials (Centola, 2018). Social media as used by most of the participants in this study was a key means for participants to communicate with their networks, but participants did not choose large, loose social media networks to scale up their actions.

This study has shown that MOOCs can promote participants taking a sustainability action and influencing others to take that action, thus adding to previous literature on MOOC influence outside of a course (Wang et al., 2014). In so doing, we have demonstrated that MOOCs can indeed become "courses for a cause" although the cause may be different from the original vision of providing access to high quality education for anyone around the world (Koller, 2012), a claim that has been contested due to limited access to high speed internet, lack of familiarity with technology, and differences in learning styles among students in poor and non-Western countries (Jung & Gunawardena, 2014; Krasny et al., 2020; Liyanagunawardena et al., 2014) In short, we believe that MOOCs are a social good, but that the "good" provided is evolving over time.

15.6 Limitations

The study only focused on one course and used a post-survey to measure participants' climate action and their application of the course ideas into practice, which does not show a causal relationship between the course and the outcomes. Further, we did not explore deeply the cultural differences in terms of actions and networks participants chose. Finally, we did not follow up with MOOC participants to examine the long-term impact of the course on their actions several months after the course ended.

15.7 Conclusion

This study demonstrated that online course participants can not only acquire knowledge within the course, but also apply learned ideas to taking sustainability actions themselves and influencing their tight social networks to take such actions. By empowering MOOC participants to choose their action and network, our "course for a cause" is adaptable to different cultural contexts globally. Although the course reflects several principles of social and social emotional learning, future teaching and research could apply and assess the impacts of different aspects of such learning more systematically.

In an effort to move beyond correlational research and understand in more depth how MOOCs can change sustainability behaviors and practices, we are working with our Information Science colleagues to conduct controlled experiments across different online courses with action outcomes ("courses for a cause"). In this way we can examine the causal relationships of aspects of social learning, for example norms and social accountability, with learning within the course as well as sustainability practices beyond the course. We are also examining differences in cultural learning preferences among students in China and the US, which will enable us to understand how network climate actions vary in different cultural contexts. MOOCs, due to the large sample sizes and diversity of participants, afford opportunities for this and related research, which will shed light on the societal goals of social and emotional learning.

Appendix

Post Survey

1. Have you already implemented your network climate action?
   (a) No, I haven't implemented it yet.

(b) Yes, I have begun to implement it personally.
(c) Yes, I have begun to implement it both personally and with my network(s).

2. Which climate action(s) is/are included in your plan?

   (a) Plant-rich diets
   (b) Reducing food waste
   (c) Composting
   (d) Tree planting
   (e) Public transit
   (f) Walking or biking to work/school
   (g) Ridesharing or carpooling
   (h) Recycling
   (i) LED lighting (household)
   (j) Water saving (household)
   (k) Solar energy
   (l) Health and education (especially for women)
   (m) Advocacy
   (n) Other

3. Personal implementation: How did you or will you implement the solution personally?
4. Network implementation: Please select group or groups you tried or will try to persuade to take the action.

   (a) Family members
   (b) Friends
   (c) Work colleagues
   (d) Fellow students
   (e) Social media followers
   (f) Other, please describe

5. Do members of your implementation network trust your opinions about climate change and the need for action?
6. What are the demographics of the group you have influenced or are hoping to influence?

   6.1 Implementation group – Age

      (a) Under 18 years
      (b) 18–30 years
      (c) 30–50 years
      (d) Over 50 years

   6.2 Implementation group – Gender

      (a) Mixed men and women
      (b) Mostly men
      (c) Mostly women

6.3 Implementation group – Occupation

(a) Students
(b) Professionals
(c) Retired
(d) Volunteer

6.4 Implementation group – Urban vs. Rural

(a) Mostly urban
(b) Mostly rural
(c) Mostly suburban

7. How did you or will you communicate with your network?

(a) Mostly face-to-face (in-person)
(b) Mostly online
(c) Mixture of in-person and online

8. If you communicate with your network online, which tool(s) have you used or will you use?

(a) Not applicable
(b) Facebook
(c) WhatsApp
(d) Instagram
(e) Twitter
(f) Telegram
(g) WeChat
(h) QQ
(i) ZOOM/Skype
(j) Tik Tok
(k) Other, please specify

9. What communication/persuasion strategies that you learned about in the course have you used or will you use with your network?
10. Have you or will you provide recognition, rewards, or incentives to members of your network for their climate actions?

(a) Yes, please briefly describe
(b) No

References

Admiraal, W., Huisman, B., & Pilli, O. (2015). Assessment in Massive Open Online Courses. *Electronic Journal of E-learning, 13*(4), 207–216.

Barak, M., & Watted, A. (2017). *Project-Based MOOC: Enhancing Knowledge Construction and Motivation to Learn. In Digital tools and solutions for inquiry-based STEM learning* (pp. 282–307). IGI Global.

Brinton, C. G., Chiang, M., Jain, S., Lam, H., Liu, Z., & Wong, F. M. F. (2014). Learning about Social Learning in MOOCs: From Statistical Analysis to Generative Model. *IEEE Transactions on Learning Technologies, 7*(4), 346–359. https://doi.org/10.1109/TLT.2014.2337900

Centola, D. (2018). *The truth about behavioral change*. MIT Sloan Management Review.

Centola, D. (2021). *Change: How to make big things happen*. Little, Brown Spark.

Chen, G., Davis, D., Hauff, C., & Houben, G.-J. (2016). *Learning transfer: Does it take place in MOOCs? An investigation into the uptake of functional programming in practice.* Paper presented at the Proceedings of the third ACM Conference on Learning@ Scale.

Christakis, N. A., & Fowler, J. H. (2013). Social contagion theory: Examining dynamic social networks and human behavior. *Statistics in Medicine*. https://doi.org/10.1002/sim.5408

Cialdini, R. B. (2001). The science of persuasion. *Scientific American, 284*(2), 76–81.

Colas, J.-F., Sloep, P., & Garreta-Domingo, M. (2016). The effect of multilingual facilitation on active participation in MOOCs. *The International Review of Research in Open and Distributed Learning, 17*(4), 280–313.

Cormier, D., & Siemens, G. (2010). The open course: Through the open door–open courses as research, learning, and engagement. *EDUCAUSE Review, 45*(4), 30–32.

Crawford, J., Butler-Henderson, K., Rudolph, J., Malkawi, B., Glowatz, M., Burton, R., … Lam, S. (2020). COVID-19: 20 countries' higher education intra-period digital pedagogy responses. *Journal of Applied Learning & Teaching, 3*(1), 1–20. https://doi.org/10.37074/jalt.2020.3.1.7

DuBois, B., Krasny, M. E., & Russ, A. (2019). Online professional development for environmental educators: Strategies to foster critical thinking and social interactions. *Environmental Education Research, 25*, 1–15. https://doi.org/10.1080/13504622.2018.1564247

Ferguson, R., & Shum, S. B. (2012). *Social learning analytics: Five approaches*. Paper presented at the Proceedings of the 2nd international conference on learning analytics and knowledge, Vancouver, British Columbia, Canada. https://doi.org/10.1145/2330601.2330616

Frank, R. H. (2020). *Under the influence: Putting peer pressure to work*. Princeton University Press.

Friedkin, N. E., & Johnsen, E. C. (1990). Social influence and opinions. *Journal of Mathematical Sociology, 15*(3-4), 193–206.

Garnett, E. E., Balmford, A., Sandbrook, C., Pilling, M. A., & Marteau, T. M. (2019). Impact of increasing vegetarian availability on meal selection and sales in cafeterias. *Proceedings of the National Academy of Sciences, 116*(42), 20923–20929. https://doi.org/10.1073/pnas.1907207116

Hofstede, G. (2011). Dimensionalizing cultures: The Hofstede model in context. *Online Readings in Psychology and Culture, 2*(1)., 2307-0919.1014.

Joksimovic, S., Siemens, G., Wang, Y. E., San Pedro, M. O. Z., & Way, J. (2020). Editorial: Beyond cognitive ability. *Journal of Learning Analytics, 7*(1), 1–4. https://doi.org/10.18608/jla.2020.71.1

Jung, I., & Gunawardena, C. N. (2014). *Culture and online learning: Global perspectives and research*. Stylus Publishing, LLC.

Kahan, T., Soffer, T., & Nachmias, R. (2017). Types of participant behavior in a massive open online course. *International Review of Research in Open and Distributed Learning: IRRODL, 18*(6), 1–18. https://doi.org/10.19173/irrodl.v18i6.3087

Kellogg, S. B., Booth, S., & Oliver, K. (2014). A social network perspective on peer supported learning in MOOCs for educators. *The International Review of Research in Open and Distributed Learning, 15*(5), 263–289. https://doi.org/10.19173/irrodl.v15i5.1852

Koller, D. (2012). *What we're learning from online education*. Retrieved from https://www.ted.com/talks/daphne_koller_what_we_re_learning_from_online_education/transcript?language=en
Krasny, M. E., DuBois, B., Adameit, M., Atiogbe, R., Baih, L., Bold-erdene, T., … Yao, Y. (2018). Small Groups in a Social Learning MOOC (slMOOC): Strategies for fostering learning and knowledge creation. *Online Learning Journal, 22*(2). https://doi.org/10.24059/olj.v22i2.1339
Krasny, M. E., Golshani, Z., Hampton, B. L., Coleman, J. F. R. M., Schrenk, M., Warakula, M. R., … Yu, Y. (2020). Courses for a cause: MOOC contributions to a "better place for all". In K. Zhang, C. J. Bonk, T. C. Reeves, & T. H. Reynolds (Eds.), *MOOCs and open education in the global South: Challenges, successes, and opportunities*. Routledge.
Li, Y., & Krasny, M. E. (2019). Practice change in environmental education: Lessons from professional development. *Environmental Education Research, 25*(7), 1119–1136. https://doi.org/10.1080/13504622.2018.1540033
Li, Y., & Krasny, M. E. (2020). Development of professional networks among environmental educators. *Professional Development in Education, 46*(2), 337–353. https://doi.org/10.1080/19415257.2018.1562957
Li, Y., Krasny, M. E., & Russ, A. (2016). Interactive learning in an urban environmental education online course. *Environmental Education Research, 22*(1), 111–128. https://doi.org/10.1080/13504622.2014.989961
Liu, M., McKelroy, E., Kang, J., Harron, J., & Liu, S. (2016a). Examining the use of Facebook and Twitter as an additional social space in a MOOC. *American Journal of Distance Education, 30*(1), 14–26. https://doi.org/10.1080/08923647.2016.1120584
Liu, Z., Brown, R., Lynch, C. F., Barnes, T., Baker, R., Bergner, Y., & McNamara, D. (2016b). *MOOC learner behaviors by country and culture; an exploratory analysis*. Paper presented at the 9th international conference on Educational Data Mining, Raleigh, NC.
Liyanagunawardena, T. R., Williams, S., & Adams, A. A. (2014). The impact and reach of MOOCs: A developing countries' perspective. *eLearning Papers*, 38–46. Retrieved from http://centaur.reading.ac.uk/38250/
McKenzie-Mohr, D., & Smith, W. (2009). *Fostering sustainable behavior: Community-based social marketing*. Retrieved from http://www.cbsm.com/book/preface
Moore, M. G. (1989). Editorial: Three types of interaction. *The American journal of distance education, 3*(2), 1–7. https://doi.org/10.1080/08923648909526659
Napier, A., Huttner-Loan, E., & Reich, J. (2020). Evaluating learning transfer from MOOCs to workplaces: A case study from teacher education and launching innovation in schools. *RIED. Revista Iberoamericana de Educación a Distancia, 23*(2), 45–64. https://doi.org/10.5944/ried.23.2.26377
Nolan, J. M., Schultz, P. W., Cialdini, R. B., Goldstein, N. J., & Griskevicius, V. (2008). Normative social influence is underdetected. *Personality and Social Psychology Bulletin, 34*(7), 913–923.
Parrish, P., & Linder-VanBerschot, J. (2010). Cultural dimensions of learning: Addressing the challenges of multicultural instruction. *The International Review of Research in Open and Distributed Learning, 11*(2), 1–19. https://doi.org/10.19173/irrodl.v11i2.809
Rogers, T., Goldstein, N. J., & Fox, C. R. (2018). Social mobilization. *Annual Review of Psychology, 69*, 357–381.
Russ, A. (2015). *Urban environmental education*. Cornell University Civic Ecology Lab/NAAEE.
Saldaña, J. (2013). *The coding manual for qualitative researchers* (2nd ed.). Sage.
Skrypnyk, O., Joksimović, S. K., Kovanović, V., Gašević, D., & Dawson, S. (2015). Roles of course facilitators, learners, and technology in the flow of information of a cMOOC. *The International Review of Research in Open and Distributed Learning, 16*(3). https://doi.org/10.19173/irrodl.v16i3.2170
Sparkman, G., & Walton, G. M. (2017). Dynamic norms promote sustainable behavior, even if it is counternormative. *Psychological Science, 28*(11), 1663–1674. https://doi.org/10.1177/0956797617719950
Spoelstra, H., Van Rosmalen, P., & Sloep, P. (2014). Toward project-based learning and team formation in open learning environments. *Journal of Universal Computer Science, 20*(1), 57–76.

Stewart, B. (2013). Massiveness+ openness= new literacies of participation. *Journal of Online Learning and Teaching, 9*(2), 228–238.

Sussman, R., Gifford, R., & Abrahamse, W. (2016). *Social mobilization: How to encourage action on climate change*. Pacific Institute for Climate Solutions.

Wals, A. E. (2007). *Social learning towards a sustainable world: Principles, perspectives, and praxis*. Wageningen Academic Publishers.

Wang, Y., Paquette, L., & Baker, R. (2014). A longitudinal study on learner career advancement in MOOCs. *Journal of Learning Analytics, 1*(3), 203–206.

Weinreich, N. (2015). *Supercharge your social impact: Using social marketing for behavior change (eBook).*

Weissberg, R. P., Durlak, J. A., Domitrovich, C. E., & Gullotta, T. P. (2015). Social and emotional learning: Past, present, and future. In J. A. Durlak, C. E. Domitrovich, R. P. Weissberg, & T. P. Gullotta (Eds.), *Handbook of social and emotional learning: Research and practice* (pp. 3–19). The Guilford Press.

Wenger, E. (1998). *Communities of practice: Learning, meaning, and identity* (In. ed.). Cambridge University Press.

Young, W., Russell, S. V., Robinson, C. A., & Barkemeyer, R. (2017). Can social media be a tool for reducing consumers' food waste? A behaviour change experiment by a UK retailer. *Resources, Conservation and Recycling, 117*, 195–203.

Zhang, J., Skryabin, M., & Song, X. (2016). Understanding the dynamics of MOOC discussion forums with simulation investigation for empirical network analysis (SIENA). *Distance Edcuation, 37*(3), 270–286. https://doi.org/10.1080/01587919.2016.1226230

Index

Y. E. Wang et al. (eds.), *Social and Emotional Learning and Complex Skills Assessment*, Advances in Analytics for Learning and Teaching,
https://doi.org/10.1007/978-3-031-06333-6

FSC
www.fsc.org
MIX
Papier aus verantwortungsvollen Quellen
Paper from responsible sources
FSC® C105338

If you have any concerns about our products,
you can contact us on
ProductSafety@springernature.com

In case Publisher is established outside the EU,
the EU authorized representative is:
Springer Nature Customer Service Center GmbH
Europaplatz 3, 69115 Heidelberg, Germany

Printed by Libri Plureos GmbH
in Hamburg, Germany